Biological Activity of Metal Complexes

Biological Activity of Metal Complexes

Guest Editor

Vinay K. Sharma

MDPI

Basel • Beijing • Wuhan • Barcelona • Belgrade • Novi Sad • Cluj • Manchester

Guest Editor
Vinay K. Sharma
Department of Biological
Engineering
Massachusetts Institute
of Technology
Cambridge
USA

Editorial Office
MDPI AG
Grosspeteranlage 5
4052 Basel, Switzerland

This is a reprint of the Special Issue, published open access by the journal *Inorganics* (ISSN 2304-6740), freely accessible at: https://www.mdpi.com/journal/inorganics/special_issues/J4F64NO16B.

For citation purposes, cite each article independently as indicated on the article page online and as indicated below:

Lastname, A.A.; Lastname, B.B. Article Title. *Journal Name* **Year**, *Volume Number*, Page Range.

ISBN 978-3-7258-7120-9 (Hbk)
ISBN 978-3-7258-7121-6 (PDF)
https://doi.org/10.3390/books978-3-7258-7121-6

Contents

About the Editor

Vinay K. Sharma

Vinay K. Sharma is a Human Frontier Science Program Cross-Disciplinary Postdoctoral Fellow in the Department of Biological Engineering at the Massachusetts Institute of Technology. His research integrates synthetic chemistry, nanotechnology, and molecular imaging to develop chemically precise tools for studying disease. He designs nanoscale sensors for noninvasive imaging of calcium and key neurotransmitters such as dopamine and serotonin to better understand how neurochemical signaling regulates neural function, behavior, and cognition. He received his PhD in Chemistry from the Technion Israel Institute of Technology in 2023, where he developed tetrapyrrole-based corroles for cancer diagnosis and therapy, including the first pH-responsive corrole analog for theranostics and non-transition metal corrole sonodynamic agents. He conducted research in the laboratory of Harry B Gray at the California Institute of Technology on the photophysical properties of water-soluble corroles for biomedical use and later developed protein corrole nanoassemblies for targeted cancer therapy. His long-term goal is to establish an interdisciplinary research program focused on nanobiochemical tools for probing and manipulating pathological signaling in cancer and neurological disorders.

Preface

The development of biologically active metal complexes requires more than synthetic innovation; it demands a mechanistic understanding of how coordination structure translates into biological function. Metal centers introduce unique geometric, electronic, and redox features that cannot be replicated by purely organic scaffolds, enabling distinctive modes of biomolecular interaction and therapeutic action.

The contributions gathered in this Reprint illustrate how rational ligand design, controlled metal coordination, and detailed physicochemical analysis can be integrated with biological studies to elucidate mechanisms of action. From DNA-binding and protein interactions to photodynamic applications and imaging platforms, these studies demonstrate that predictive structure–activity relationships remain central to progress in medicinal inorganic chemistry.

By bringing together these multidisciplinary efforts, this Reprint highlights the continued expansion of bioinorganic chemistry as a foundational discipline at the interface of coordination chemistry and biomedical science.

Vinay K. Sharma
Guest Editor

inorganics

MDPI

Editorial

Biological Activity of Metal Complexes

Vinay K. Sharma

Department of Biological Engineering, Massachusetts Institute of Technology, 77 Massachusetts Avenue, Cambridge, MA 02139, USA; vsharma1@mit.edu

Metal complexes play a fundamental role in biological systems and continue to attract sustained interest due to their remarkable potential in therapeutic, diagnostic, and biotechnological applications [1–8]. In recent years, the field of bioinorganic chemistry has advanced rapidly, driven by progress in coordination chemistry, spectroscopy, nanotechnology, and molecular biology [9–22]. These developments have enabled a deeper understanding of how metal ions and complexes interact with biomolecular targets and have opened new avenues for the rational design of metal-based agents for cancer therapy, antimicrobial treatment, imaging, and the study of metal-mediated biochemical processes [23–30].

This Special Issue, "Biological Activity of Metal Complexes", was conceived to bring together recent developments in the synthesis, characterization, and biological evaluation of metal complexes, with particular emphasis on elucidating their mechanisms of action and biomedical relevance. The contributions collected reflect the inherently multidisciplinary nature of this research area, spanning coordination chemistry, nanomaterials, spectroscopy, computational analysis, and cellular studies.

A notable theme is the development of rhenium(I) tricarbonyl complexes as versatile biologically active platforms. Paparidis et al. report anthrapyrazole-functionalized *fac*-tricarbonylrhenium complexes that exhibit strong DNA-binding ability, pronounced cytotoxicity toward tumor cells, and promising pharmacokinetic properties, demonstrated through studies with a ^{99m}Tc-analogue, highlighting their dual therapeutic and imaging potential [24].

In a complementary contribution, Carreño et al. present a comprehensive structural, electronic, and biological investigation of a phenanthroline-based rhenium complex [30]. Through the integration of crystallography, electrochemistry, computational studies, and cell viability assays, the authors provide valuable insight into structure–activity relationships and the luminescent behavior of these complexes in biological environments [30].

The interaction of metal complexes with biological macromolecules, particularly DNA and serum proteins, represents another central focus of this Special Issue. The manganese(II) complexes described by Theodoulou et al. demonstrate strong intercalative binding to calf-thymus DNA and reversible association with serum albumins, offering insight into how ligand design influences biomolecular affinity and potential pharmacological behavior [27].

Similarly, Alhashmialameer reports the synthesis of quinolinyl–imine complexes of Cr(III), Mn(II), and Pd(II) via a sonochemical route, followed by biological evaluation and molecular docking studies. These results correlate the coordination structure with antimicrobial, antioxidant, and potential anticancer activities, further emphasizing the importance of ligand architecture in modulating the biological response [28].

Photodynamic therapy and targeted therapeutic strategies are represented by the work of Mantareva et al., who investigate Zn(II)- and Ga(III)-phthalocyanines in combination with the proteolytic enzyme α-chymotrypsin [23]. Their findings demonstrate that

https://doi.org/10.3390/inorganics14020061

enzyme–photosensitizer conjugation can enhance phototherapeutic efficiency while reducing toxicity, illustrating an innovative approach to improving the selectivity of metal-based photosensitizers [23].

The Special Issue also highlights the growing role of inorganic nanomaterials in biological detection and imaging. Phuong et al. describe optimized $NaYF_4:Er^{3+}/Yb^{3+}$ upconversion nanocomplexes with controlled morphology and surface functionalization, followed by antibody conjugation for the effective labeling of cancer cells [26]. This work demonstrates how the precise control of nanoparticle structure and surface chemistry translates into improved luminescent performance and bio-recognition capabilities [26].

Beyond therapeutic and imaging applications, understanding the chemical fate of metal-containing compounds in biological environments remains critically important. Degorge et al. investigate the degradation of the organomercurial compound thimerosal by glutathione and cysteine at physiological pH, providing valuable insight into the bioinorganic reactivity of metal species with endogenous thiols and their potential toxicological implications [29].

Collectively, the contributions in this Special Issue demonstrate that the development of biologically active metal complexes requires a balanced integration of synthetic design, physicochemical characterization, and detailed biological evaluation. From DNA binding and protein interactions to photodynamic activity, imaging applications, and toxicological studies, these works highlight the breadth and relevance of research in this rapidly evolving field.

The Guest Editor hopes that this Special Issue will serve as a valuable reference for researchers in bioinorganic and medicinal inorganic chemistry and stimulate further interdisciplinary efforts toward the development of next-generation metal-based diagnostic and therapeutic agents.

Data Availability Statement: This article is an editorial and does not report original research data. No new data were created or analyzed in this study. Data sharing is not applicable.

Acknowledgments: The Guest Editor sincerely thanks all authors for their valuable contributions, the reviewers for their careful and constructive evaluations, and the editorial team of *Inorganics* for their professional support throughout the preparation of this special issue.

Conflicts of Interest: The author declares no conflicts of interest.

References

1. Marker, S.C.; King, A.P.; Granja, S.; Vaughn, B.; Woods, J.J.; Boros, E.; Wilson, J.J. Exploring the in Vivo and in Vitro Anticancer Activity of Rhenium Isonitrile Complexes. *Inorg. Chem.* **2020**, *59*, 10285–10303. [CrossRef]
2. Alderden, R.A.; Hall, M.D.; Hambley, T.W. The Discovery and Development of Cisplatin. *J. Chem. Educ.* **2006**, *83*, 728–734. [CrossRef]
3. Johnstone, T.C.; Suntharalingam, K.; Lippard, S.J. The Next Generation of Platinum Drugs: Targeted Pt(II) Agents, Nanoparticle Delivery, and Pt(IV) Prodrugs. *Chem. Rev.* **2016**, *116*, 3436–3486. [CrossRef] [PubMed]
4. Wilson, J.J.; Lippard, S.J. Synthetic Methods for the Preparation of Platinum Anticancer Complexes. *Chem. Rev.* **2013**, *114*, 4470–4495. [CrossRef]
5. King, A.P.; Gellineau, H.A.; Ahn, J.E.; MacMillan, S.N.; Wilson, J.J. Bis(Thiosemicarbazone) Complexes of Cobalt(III). Synthesis, Characterization, and Anticancer Potential. *Inorg. Chem.* **2017**, *56*, 6609–6623. [CrossRef]
6. Palanimuthu, D.; Shinde, S.V.; Somasundaram, K.; Samuelson, A.G. In Vitro and in Vivo Anticancer Activity of Copper Bis(Thiosemicarbazone) Complexes. *J. Med. Chem.* **2013**, *56*, 722–734. [CrossRef]
7. Anjum, R.; Palanimuthu, D.; Kalinowski, D.S.; Lewis, W.; Park, K.C.; Kovacevic, Z.; Khan, I.U.; Richardson, D.R. Synthesis, Characterization, and in Vitro Anticancer Activity of Copper and Zinc Bis(Thiosemicarbazone) Complexes. *Inorg. Chem.* **2019**, *58*, 13709–13723. [CrossRef]

8. Andres, S.A.; Bajaj, K.; Vishnosky, N.S.; Peterson, M.A.; Mashuta, M.S.; Buchanan, R.M.; Bates, P.J.; Grapperhaus, C.A. Synthesis, Characterization, and Biological Activity of Hybrid Thiosemicarbazone–Alkylthiocarbamate Metal Complexes. *Inorg. Chem.* **2020**, *59*, 4924–4935. [CrossRef]
9. Li, M.; Park, J.Y.; Sheikh, M.; Kayamba, V.; Rumgay, H.; Jenab, M.; Narh, C.T.; Abedi-Ardekani, B.; Morgan, E.; De Martel, C.; et al. Population-Based Investigation of Common and Deviating Patterns of Gastric Cancer and Oesophageal Cancer Incidence across Populations and Time. *Gut* **2023**, *72*, 846–854. [CrossRef] [PubMed]
10. Sen, S.; Won, M.; Levine, M.S.; Noh, Y.; Sedgwick, A.C.; Kim, J.S.; Sessler, J.L.; Arambula, J.F. Metal-Based Anticancer Agents as Immunogenic Cell Death Inducers: The Past, Present, and Future. *Chem. Soc. Rev.* **2022**, *51*, 1212–1233. [CrossRef] [PubMed]
11. Lee, L.C.C.; Lo, K.K.W. Shining New Light on Biological Systems: Luminescent Transition Metal Complexes for Bioimaging and Biosensing Applications. *Chem. Rev.* **2024**, *124*, 8825–9014. [CrossRef]
12. Sharma, V.K.; Assaraf, Y.G.; Gross, Z. Hallmarks of Anticancer and Antimicrobial Activities of Corroles. *Drug Resist. Updates* **2023**, *67*, 100931. [CrossRef]
13. Sharma, V.K.; Gonzalez-Almeyda, N.; Mikhael, S.; Cho, R.H.; Aceves, J.; Ishaya, K.; Kim, S.W.; Wiesenthal, A.; Babajani, A.; Abrol, R.; et al. Next Generation Protein-Corrole Bio-Assemblies Provide Effective Tumoricidal Treatment in a Metastatic Triple-Negative Breast Cancer Model. *bioRxiv* **2026**, 2026.02.03.703292. [CrossRef]
14. Waldron, K.J.; Rutherford, J.C.; Ford, D.; Robinson, N.J. Metalloproteins and Metal Sensing. *Nature* **2009**, *460*, 823–830. [CrossRef]
15. Meggers, E. Targeting Proteins with Metal Complexes. *Chem. Commun.* **2009**, *45*, 1001–1010. [CrossRef]
16. Ferraro, G.; Merlino, A. Metallodrugs: Mechanisms of Action, Molecular Targets and Biological Activity. *Int. J. Mol. Sci.* **2022**, *23*, 3504. [CrossRef]
17. Sharma, V.K.; Mahammed, A.; Soll, M.; Tumanskii, B.; Gross, Z. Corroles and Corrole/Transferrin Nanoconjugates as Candidates for Sonodynamic Therapy. *Chem. Commun.* **2019**, *55*, 12789–12792. [CrossRef]
18. Yadav, P.; Khoury, S.; Fridman, N.; Sharma, V.K.; Kumar, A.; Majdoub, M.; Kumar, A.; Diskin-Posner, Y.; Mahammed, A.; Gross, Z. Trifluoromethyl Hydrolysis En Route to Corroles with Increased Druglikeness. *Angew. Chem.* **2021**, *133*, 12939–12944. [CrossRef]
19. Sharma, V.K.; Stark, M.; Fridman, N.; Assaraf, Y.G.; Gross, Z. Doubly Stimulated Corrole for Organelle-Selective Antitumor Cytotoxicity. *J. Med. Chem.* **2022**, *65*, 6100–6115. [CrossRef]
20. Ghanbari, H.; Derakhshankhah, H.; Bahrami, K.; Keshavarzi, S.; Mohammadi, K.; Hayati, P.; Centore, R.; Parisi, E. Synthesis, Characterization, and Biological Activity of a Fresh Class of Sonochemically Synthesized Cu2+ Complexes. *Sci. Rep.* **2024**, *14*, 21325. [CrossRef]
21. Ruwizhi, N.; Singh, T.; Omondi, B.O.; Bala, M.D. Biological Activity of Late Transition Metal-Based Compounds: From Computational and Theoretical Studies to Laboratory Exploration and Beyond. *Bioinorg. Chem. Appl.* **2024**, *2024*, 2829283. [CrossRef]
22. Towett, E.K.; Tembu, V.J.; Kemboi, D.; Langat, M.K.; Manicum, A.L.E. Review of Recent Medicinal Applications of Rhenium(I) Tricarbonyl Complexes. *Int. J. Mol. Sci.* **2025**, *26*, 7005. [CrossRef]
23. Mantareva, V.; Braikova, D.; Vilhelmova-Ilieva, N.; Angelov, I.; Iliev, I. Accomplishment of α-Chymotrypsin on Photodynamic Effect of Octa-Substituted Zn(II)- and Ga(III)-Phthalocyanines against Melanoma Cells. *Inorganics* **2024**, *12*, 204. [CrossRef]
24. Paparidis, G.; Akrivou, M.; Psomas, G.; Vizirianakis, I.S.; Hatzidimitriou, A.; Gabriel, C.; Sarigiannis, D.; Papagiannopoulou, D. Novel Tricarbonylrhenium-Anthrapyrazole Complexes with DNA-Binding and Antitumor Properties: In Vitro and In Vivo Pharmacokinetic Studies with 99mTc-Analogue. *Inorganics* **2024**, *12*, 254. [CrossRef]
25. Stefanache, A.; Miftode, A.M.; Constantin, M.; Bogdan Goroftei, R.E.; Olaru, I.; Gutu, C.; Vornicu, A.; Lungu, I.I. Noble Metal Complexes in Cancer Therapy: Unlocking Redox Potential for Next-Gen Treatments. *Inorganics* **2025**, *13*, 64. [CrossRef]
26. Phuong, H.T.; Vinh, L.T.; Cong, T.Q.; Tien, T.Q.; Van, N.D.; Hong Ha, V.T.; Phan, V.N.; Hoi, L.T.; Thang, P.D.; Thao, D.T.; et al. Optimized NaYF4: Er3+/Yb3+ Upconversion Nanocomplexes via Oleic Acid for Biomedical Applications. *Inorganics* **2025**, *13*, 140. [CrossRef]
27. Theodoulou, V.; Zianna, A.; Hatzidimitriou, A.G.; Psomas, G. Manganese(II) Complexes with 3,5–Dibromosalicylaldehyde: Characterization and Interaction Studies with DNA and Albumins. *Inorganics* **2025**, *13*, 263. [CrossRef]
28. Alhashmialameer, D. Synthesis of Some Novel Cr(III), Mn(II), and Pd(II) Complexes via the Sono-Chemical Route with a Chlorinated Quinolinyl-Imine Ligand: Structural Elucidation, Bioactivity Analysis, and Docking Simulations. *Inorganics* **2025**, *13*, 271. [CrossRef]

https://doi.org/10.3390/inorganics14020061

29. Degorge, M.F.; Mertz, S.; Gailer, J. Degradation of the Vaccine Additive Thimerosal by L-Glutathione and L-Cysteine at Physiological PH. *Inorganics* **2025**, *13*, 280. [CrossRef]
30. Carreño, A.; Artigas, V.; Ancede-Gallardo, E.; Morales-Guevara, R.; Arce, R.; Leyva-Parra, L.; Martí, A.A.; Videla, C.; Otero, M.C.; Gacitúa, M. Comprehensive Structural, Electronic, and Biological Characterization of Fac-[Re(CO)3(5,6-Epoxy-5,6-Dihydro-1,10-Phenanthroline)Br]: X-Ray, Aromaticity, Electrochemistry, and HeLa Cell Viability. *Inorganics* **2025**, *14*, 3. [CrossRef]

https://doi.org/10.3390/inorganics14020061

inorganics

MDPI

Article

Comprehensive Structural, Electronic, and Biological Characterization of *fac*-[Re(CO)$_3$(5,6-epoxy-5,6-dihydro-1,10-phenanthroline)Br]: X-Ray, Aromaticity, Electrochemistry, and HeLa Cell Viability

Alexander Carreño [1,2,*], Vania Artigas [1], Evys Ancede-Gallardo [1], Rosaly Morales-Guevara [3,4], Roxana Arce [2,5], Luis Leyva-Parra [6], Angel A. Martí [7], Camila Videla [1,8], María Carolina Otero [8] and Manuel Gacitúa [9,10,*]

[1] Laboratory of Organometallic Synthesis (CANS), Departamento de Ciencias Químicas, Facultad de Ciencias Exactas, Universidad Andres Bello, República 330, Santiago 8370186, Chile; v.artigassalinas@uandresbello.edu (V.A.); eancedeg@gmail.com (E.A.-G.); camila.videla.e@gmail.com (C.V.)
[2] Departamento de Ciencias Químicas, Facultad de Ciencias Exactas, Universidad Andres Bello, Av. República 275, Santiago 7550000, Chile; roxana.arce@unab.cl
[3] Departamento de Química de los Materiales, Facultad de Química y Biología, Universidad de Santiago de Chile, Av. Libertador Bernardo O'Higgins 3363, Santiago 9170022, Chile; rmguevara1994@gmail.com
[4] Facultad de Ingeniería, Universidad Finis Terrae, Av. Pedro de Valdivia 1509, Santiago 7501015, Chile
[5] Millenium Institute on Green Ammonia as Energy Vector—MIGA (ICN2021_023), Avenida Vicuña Mackenna 4860, Santiago 7820436, Chile
[6] Centro de Investigación Para el Diseño de Materiales (CEDEM), Facultad de Ciencias Exactas, Departamento de Ciencias Químicas, Universidad Andrés Bello, Avenida República 275, Santiago 837014, Chile; luis.leyva@unab.cl
[7] Department of Chemistry, Rice University, 6100 S. Main St, Houston, TX 77005, USA; aam4@rice.edu
[8] Escuela de Química y Farmacia, Facultad de Medicina, Universidad Andres Bello, República 252, Santiago 8370186, Chile; maria.otero@unab.cl
[9] Laboratorio de Contaminación y Nanoseguridad, Facultad de Ingeniería & Ciencias, Universidad Diego Portales, Ejército 441, Santiago 8370191, Chile
[10] Center for Nanoscience and Nanotechnology (CEDENNA), Manuel Rodríguez Sur 415, Santiago 8320000, Chile
* Correspondence: alexander.carreno@unab.cl (A.C.); manuel.gacitua@mail.udp.cl (M.G.)

Abstract

The rhenium(I) tricarbonyl complex *fac*-[Re(CO)$_3$(5,6-epoxy-5,6-dihydro-1,10-phenanthroline)Br] (ReL) has previously demonstrated promising luminescent properties, enabling its direct application as a probe for walled cells such as *Candida albicans* and *Salmonella enterica*. In this new study, we present a significantly expanded and comprehensive characterization of ReL, incorporating a wide range of experimental and computational techniques not previously reported. These include variable-temperature ^{1}H and ^{13}C NMR spectroscopy, CH-COSY, single-crystal X-ray diffraction, Hirshfeld surface analysis, DFT calculations, Fukui functions, non-covalent interaction (NCI) indices, and electrochemical profiling. Structural analysis confirmed a pseudo-octahedral geometry with the bromide ligand positioned cis to the epoxy group. NMR data revealed the coexistence of cis and trans isomers in solution, with the trans form being slightly more stable. DFT calculations and aromaticity descriptors indicated minimal electronic differences between isomers, supporting their unified treatment in subsequent analyses. Electrochemical studies revealed two oxidation and two reduction events, consistent with ECE and EEC mechanisms, including a Re(I) $\rightarrow$ Re(0) transition at -1.50 V vs. SCE. Theoretical redox potentials showed strong agreement with experimental data. Biological assays revealed a dose-dependent cytotoxic effect on HeLa cells, contrasting with previously reported low toxicity in microbial systems. These findings, combined with ReL's luminescent and antimicrobial

https://doi.org/10.3390/inorganics14010003

properties, underscore its multifunctional nature and highlight its potential as a bioactive and imaging agent for advanced therapeutic and microbiological applications.

Keywords: rhenium tricarbonyl; Fukui; NCI; aromaticity; cyclic voltammetry; X-ray; hirshfeld surface; HeLa

1. Introduction

Extensive research has been dedicated to rhenium(I) tricarbonyl complexes, resulting in various categories of complexes with adjustable structural characteristics [1,2]. The geometrical and optical properties of these complexes have also been explored using computational relativistic methods [3,4]. The first report on the photophysical properties of neutral *fac*-[Re(CO)$_3$(N,N)X] complexes (where N,N is 1,10-phenanthroline and X is Cl) was described by Wrighton, M. & Morse, D. L. in 1974 [5,6]. Since then, compounds of this kind have been extensively studied, resulting in the discovery of numerous luminescent rhenium(I) tricarbonyl complexes with the following architecture: *fac*-[Re(CO)$_3$(N,N)X]^n. Here, N,N signifies a bidentate diimine ligand in an equatorial orientation, X designates a halide ligand serving as an ancillary component, and n represents the charge that can assume values of either 0 or 1+ [7]. These complexes have been found to possess remarkable properties for use in catalysis, as photoluminescent probes, and even in biological applications [8–14]. The most red-shifted emission wavelength of these rhenium(I) tricarbonyl complexes exhibits no discernible correlation with either the solvent polarity or excitation wavelength, revealing a substantial Stokes shift exceeding 200 nm. These characteristics underscore the noteworthy alteration in the dipole moment between the ground and excited states [15–21].

The neutral *fac*-[Re(CO)$_3$(N,N)L] isomer is the most studied complex in this family due to its tunable photophysical properties, which are closely related to its structural variations [22–24]. Recent studies have also explored the potential of rhenium(I) tricarbonyl complexes in biological applications [25,26]. For example, rhenium(I) tricarbonyl complexes with neutral N,N-bidentate formazans have demonstrated potential applications in photodynamic therapy [27,28]. Another study reported unexpectedly high cytotoxicity of rhenium(I) tricarbonyl complexes bearing 1, 10-phenanthroline derivatives [29]. Some studies have shown that these complexes can selectively target cancer cells and induce cell death, as exemplified by the photosensitization of reactive oxygen species and DNA intercalation [29,30]. Additionally, neutral rhenium(I) tricarbonyl complexes have been used as catalysts in organic synthesis, where they have shown high activity and selectivity for various reactions [31,32]. Overall, rhenium(I) tricarbonyl complexes are promising compounds with diverse applications in various fields. Recently, we reported a novel neutral *fac*-[Re(CO)$_3$(5,6-epoxy-5,6-dihydro-1,10-phenanthroline)Br] complex, referred to as ReL [33], Figure 1.

This complex comprises two isomeric forms, *cis* and *trans*, distinguished by the position of the hydrogen atoms in the equatorial ligand with respect to the bromide. At 25 °C, the *trans*-isomeric form predominates [33]. In our previous investigation, we characterized ReL using physicochemical techniques and conducted UV–Vis and photoluminescence experiments in organic solvents of varying polarities. Remarkably, ReL displayed favorable properties as a luminescent probe when examining Gram-negative bacteria (*Salmonella enterica* sv. Typhimurium) and yeasts (*Candida albicans*) using confocal microscopy [26]. Interestingly, although it has been proposed that rhenium(I) complexes must be cationic to stain walled cells (e.g., bacteria and fungi), ReL is neutral [33]. In this study, we present

https://doi.org/10.3390/inorganics14010003

a comprehensive spectroscopic characterization of ReL using 1D and 2D NMR and FTIR spectroscopy. Furthermore, the molecular structure of ReL was determined using single-crystal X-ray diffraction (XRD). To further characterize its molecular properties, Hirshfeld surface analysis and Density Functional Theory (DFT) calculations, including geometry optimization, Fukui function analysis [34], and Noncovalent Interaction (NCI) index calculations [35] were conducted. Additionally, Cyclic Voltammetry (CV) was performed to assess the contribution of the equatorial ligand L (5,6-epoxy-5,6-dihydro-1,10-phenanthroline) to the redox processes of ReL complex. To distinguish the rhenium core-associated redox processes, we compared the CV of ReL with L ligand profiles. Moreover, computational studies were employed to estimate the redox potentials, following a thermodynamic cycle approach to calculate the free energies and electron density plots. This characterization provides valuable insights into the electronic properties of such complexes. Finally, other authors have reported examples of [Re(CO)$_3$(N,N)X] complexes with strong potential for development as chemotherapeutic agents. Based on this, ReL was evaluated for its cytotoxic potential in HeLa cervical cancer cells through the MTT assay.

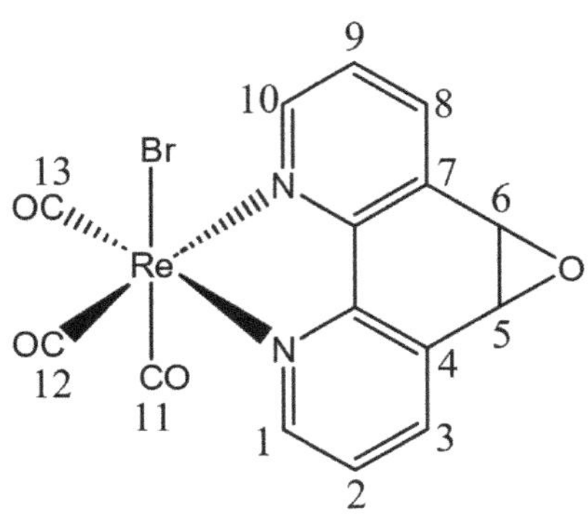

Figure 1. Structure of ReL. Numbers correspond to carbon atoms labels.

2. Results and Discussion

2.1. Characterization of fac-[Re(CO)$_3$(5,6-epoxy-5,6-dihydro-1,10-phenanthroline)Br] (ReL)

The rhenium(I) complex ReL was obtained in good yield by reacting [Re(CO)$_3$Br(THF)]$_2$ with ligand L in a 1:2 ratio. Its structure was confirmed by ^{1}H NMR and FTIR using the KBr pellet method, which revealed three distinct ν(CO) bands, indicating a loss of local symmetry around the Re(CO)$_3$ core. This complements previously published ATR-FTIR data from our group [36,37]. Characteristic features of the fac-isomers were observed, including a narrow symmetric band and two broader antisymmetric bands of comparable intensity (Figure S1). These spectral features align with the fac geometry confirmed by single-crystal X-ray diffraction, discussed below. Importantly, the absence of IR signatures typical of mer-isomers, such as a significantly weaker high-frequency band, supports the exclusive formation of the fac-isomer in this system. The ^{1}H NMR spectrum of ReL in deuterated DMSO at 25 °C (Figure S2a) showed well-resolved signals for the 1,10-phenanthroline ring, consistent with previous reports (Figure S2b). The epoxy group displayed a split signal at approximately 5.09 ppm (5.15 and 5.05 ppm), suggesting the presence of an isomeric mixture. This splitting is attributed to the different spatial orientations of the epoxide relative to the bromide ligand, which supports the coexistence of cis and trans-fac-isomers. Similar behavior has been reported for related rhenium(I) complexes by Martí et al. [38]. In the complex fac-[Re(CO)$_3$(5,6-epoxy-5,6-dihydro-1,10-phenanthroline)Cl], only the trans isomer, where the epoxide group is oriented opposite to the chloride ligand was success-

https://doi.org/10.3390/inorganics14010003

fully isolated and characterized by single-crystal X-ray diffraction, despite the synthesis producing both fac isomers. To confirm the coexistence of these species in solution, ^{1}H NMR spectra were recorded at 25 °C and 60 °C in deuterated DMF. At 25 °C, the protons adjacent to the epoxide appeared as a single resonance near 5.10 ppm, attributed to the cancellation of opposing electronic effects in the cis and trans forms. Upon heating to 60 °C, this signal resolved into two distinct peaks, indicating the presence of both isomers. These species differed in the orientation of the epoxide relative to the chloride ligand (cis/trans). At room temperature, chemical shift differences were effectively compensated, resulting in coincident resonances; increasing the temperature altered solvent properties, reducing this compensation and enabling signal separation. This behavior confirmed that the complex existed as a mixture of two stable fac isomers rather than a fac/mer equilibrium, consistent with IR and ^{13}C NMR data. In our case, the complex ReL crystallized as the cis isomer, as verified by X-ray diffraction. In its ^{1}H NMR spectrum at 25 °C in deuterated DMSO, the vicinal epoxide protons exhibited a narrow splitting at 5.15 ppm and 5.05 ppm, reflecting two distinct environments depending on the epoxide's position relative to the bromide ligand (cis or trans). Variable-temperature ^{1}H NMR experiments at 40 °C, 50 °C, and 60 °C (Figures S3–S5) revealed consistent splitting patterns, further supporting the presence of two stable fac isomers rather than a fac/mer equilibrium. Additionally, the ^{13}C-NMR spectra at 25 °C, 40 °C, 50 °C, and 60 °C (Figures S6–S9) showed a stable resonance at 197.4 ppm assigned to the carbonyl ligands, along with split signals for the phenanthroline moiety. A CH-COSY experiment at 25 °C (Figure S10) confirmed these assignments. Unlike the chloride analog reported by Martí et al., which only crystallized in the trans form, ReL exhibited clear evidence of isomeric coexistence in solution. Notably, no significant spectral changes were observed when lowering the temperature from 60 °C to 25 °C, suggesting that the bromide ligand influences the rhenium core differently than chloride. This hypothesis was further supported by the DFT calculations discussed in the following section. Importantly, the ^{1}H-NMR spectra of ReL showed no detectable impurities (Figure S2a,b), confirming the sample purity. For more details, see the ESI.

2.2. X-Ray Structure of ReL

A single crystal suitable for X-ray diffraction analysis of *fac*-[Re(CO)$_3$(5,6-epoxy-5,6-dihydro-1,10-phenanthroline)Br] (ReL) was successfully grown, exhibiting the *cis* isomer with respect to the bromide and epoxy group positions. ORTEP drawings of the ReL compound and the atom numbering scheme are shown in Figure 2. ReL crystallized in the orthorhombic P212121 space group, with one molecular entity in the asymmetric unit (Table S1). The Re(I) metal center exhibited a slightly distorted octahedral coordination geometry. This geometry is characterized by three facially oriented carbonyl ligands. The diamine L coordinates the complex in the equatorial plane. An axial bromide ancillary ligand is also positioned *cis* to the epoxy group, completing the octahedral geometry. The Re−CO bond distances observed in this study are consistent with those reported for similar Rhenium(I) complexes [39]. Notably, the Re−CO bond length *trans* to the pyridyl group is slightly shorter (1.905 (7) Å) than the Re−CO bonds *trans* to the bromide ligand (1.914 (7) Å). This difference highlights the influence of the epoxy ring, which is positioned *cis* to the bromide ancillary ligand, on the bond length of the entire Re−Br (2.6158 (7) Å) moiety. The bond distances and angles are listed in Table S2. The intermolecular interactions constituting the crystalline packing of the ReL structure reveal that the crystals are stabilized by intermolecular hydrogen bonds C(2)-H(2)···O(1) between O(1) of the epoxide fragment and the C(2) of the phenanthroline fragment, as well as C(6)-H(6)···O(2) among the carboxyl groups in the phenanthroline fragment (Table S3). These interactions are related through a twofold screw axis parallel to the b axis. Additionally, the crystal packing is further

https://doi.org/10.3390/inorganics14010003

stabilized by the C(3)-H(3)···Br(1) intermolecular interactions between the Br(1) ligand and the C(3) of the phenanthroline fragment, which are related through a twofold screw axis parallel to the b axis (Figure 3). Finally, $\pi \cdots \pi$ interactions were observed between the phenanthroline fragments, characterized by an angle of 2.080°, a centroid-centroid distance of 3.857 Å, and a shift distance of 1.639 Å, mediated by a two-fold screw axis parallel to the b-axis.

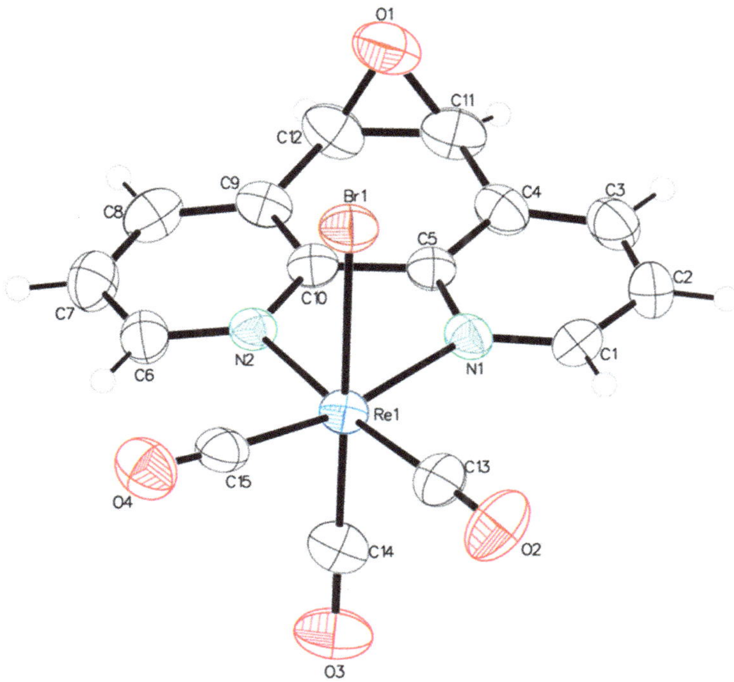

Figure 2. ORTEP view of the molecular structure of ReL. Thermal ellipsoids are drawn at 50% probability.

The crystal packing arrangements of ReL and the analogous complex *fac*-[Re(CO)$_3$(5,6-epoxy-5,6-dihydro-1,10-phenanthroline)Cl] (Figure S11) exhibited notable differences in their $\pi \cdots \pi$ stacking interactions. In ReL, these interactions were more prominent, as indicated by a shorter centroid-to-centroid distance (3.857 Å), a smaller interplanar angle (2.089°), and reduced slippage (1.638 Å). In contrast, the chloride-containing complex displayed slightly weaker $\pi \cdots \pi$ stacking, with corresponding values of 3.942 Å, 4.986°, and 1.733 Å. These structural parameters suggest a more efficient π-stacking arrangement in ReL, which may contribute to enhanced crystal stability and packing density.

A comparative analysis of the dominant intermolecular contacts in both structures highlights the differing contributions of hydrogen bonding and π-stacking interactions to the overall supramolecular architecture. While both types of interactions are present in ReL and its chloride analog, the packing of ReL is primarily governed by $\pi \cdots \pi$ stacking, whereas the chloride complex exhibits a more balanced interplay between hydrogen bonding and π-interactions. These observations underscore the significant influence of halide substitution on modulating non-covalent interactions and crystal packing behavior in rhenium(I) tricarbonyl complexes, suggesting that the bromide ligand influences the rhenium core differently from chloride. On the other hand, the crystal structure of ReL was analyzed using Hirshfeld surface mapping and fingerprint plots, revealing the nature and

https://doi.org/10.3390/inorganics14010003

relative contributions of intermolecular interactions (Figure S12). The dominant contacts were O···H (38.9%), associated with hydrogen bonding, followed by Br···H (10.0%) and C···C (6.5%), linked to dipole–dipole and π–π interactions, respectively. This analysis provides insights into the non-covalent forces stabilizing crystal packing. For more details, see the ESI.

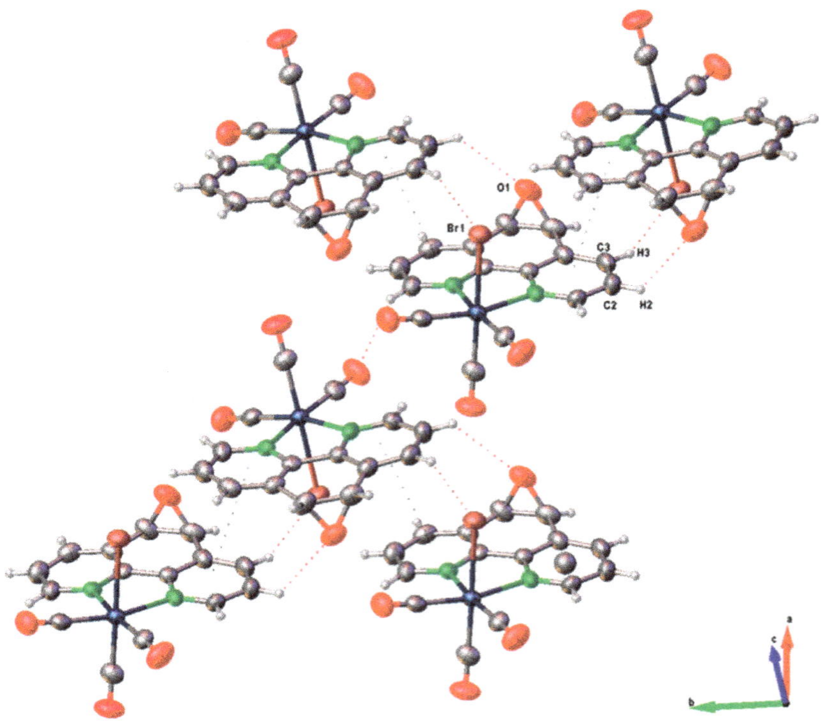

Figure 3. Crystal packing pattern of ReL.

2.3. DFT Studies: Geometries Optimizations, Fukui Index, NCI Analysis and Aromaticity

The structural and electronic features of *fac*-[Re(CO)$_3$(5,6-epoxy-5,6-dihydro-1,10-phenanthroline)Br] (ReL) were examined to clarify the subtle differences between its *cis* and *trans* isomers. These isomers arise from the relative orientation of the bromide ligand in the rhenium tricarbonyl core with respect to the epoxy group of the phenanthroline ligand. As described in Section 2.1, both forms coexist in solution and resist conventional separation, which explains why the electrochemical measurements reflect the properties of their mixture rather than those of the isolated species.

DFT optimization and vibrational analyses revealed no significant electronic or structural differences between the *cis* and *trans* isomers (Figure 4). The computed FTIR spectra reproduce the three characteristic CO stretching bands: the symmetric νCO vibration at 2197.0 cm^{-1} matches the experimental value (Figure S1), whereas the antisymmetric modes are calculated in the 2069.0–2088.0 cm^{-1} range. The third band, absent from the experimental spectrum, can be attributed to the *trans* influence of bromide, which reduces the splitting between the antisymmetric modes to only 19.0 cm^{-1}. A detailed comparison of both forms is provided in the SI (Figures S13 and S14), confirming their near-identical behavior at the molecular level. This effect stems from the reduction in symmetry induced by the epoxy substituent. The enthalpy difference between *cis* and *trans*, only 2.74 kJ·mol^{-1},

https://doi.org/10.3390/inorganics14010003

indicates a slightly greater stability of the *trans* isomer, which is consistent with the minor chemical shift splitting observed in the experimental ^{1}H-NMR spectrum (5.10 and 5.07 ppm for *cis* and *trans* isomers, respectively, see Figure S2a).

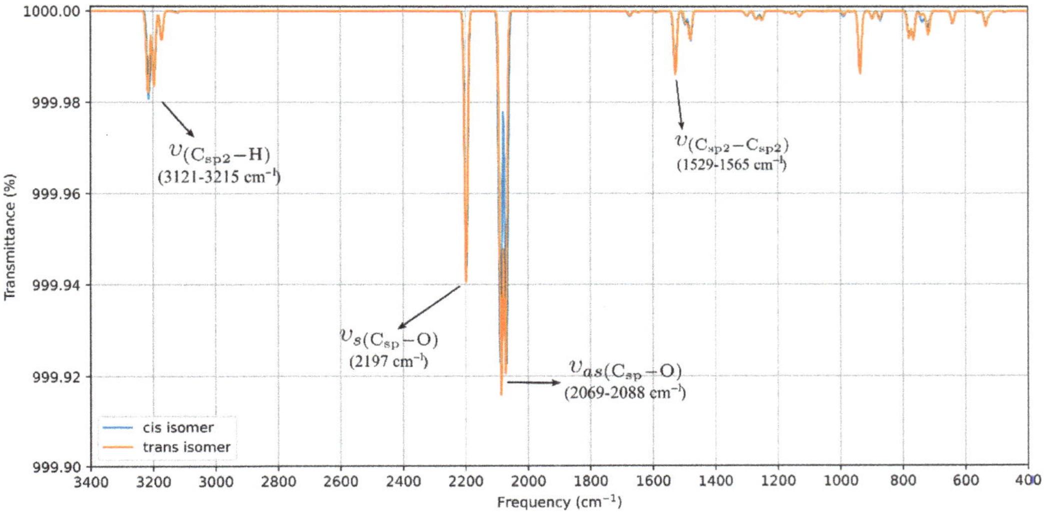

Figure 4. Computed FTIR for ReL of cis and trans isomeric form (with respect to epoxy-1,10-phenanthroline moiety and bromide ancillary ligand).

The electronic reactivity was analyzed using Fukui functions (FF). The FF characterizes the variation in electron density upon the addition or removal of electrons, providing insight into the most electrophilic and nucleophilic regions of a molecule [40–42]. According to Sebthilkumar et al. [43,44], these functions can be computed using specific equations [45,46]. In this study, the FF values for the *cis* and *trans* isomeric forms of ReL were calculated in the gas phase using the BP86 functional and the def2-TZVPP basis set. The SARC-ZORA-TZVPP basis set was employed to account for the relativistic effects on the rhenium atom. For both isomers, the nucleophilic sites (f^{+}) were concentrated on the phenanthroline nitrogen atoms and adjacent carbons, while the electrophilic regions (f^{-}) were located on the bromide ligand and the rhenium center (Figures S15 and S16). The close similarity between the *cis* and *trans* distributions indicates that the orientation of the bromide relative to the epoxy group does not substantially affect the electron-density response.

Noncovalent Interaction (NCI) analyses [47,48] offer a complementary perspective. The NCI index identifies interactions within a chemical system based solely on electron density. A two-dimensional graphical representation provides a detailed characterization of these interactions. Both forms exhibited π–π stacking and halogen bonding consistent with the crystal packing shown in Figure 3, while the reduced density-gradient plots confirmed weak interactions around bromine and essentially no differences in the phenanthroline or epoxy regions (Figure 5). A broader view of the non-covalent interactions can be observed on Figures S17 and S18. A faint Van der Waals contact between bromide and the aromatic framework was observed in the *cis* isomer, but overall, the interaction landscape was nearly identical. Notably, the epoxy oxygen consistently withdraws electron density from the phenanthroline ring, an effect that is insensitive to the position of the neighboring hydrogens.

https://doi.org/10.3390/inorganics14010003

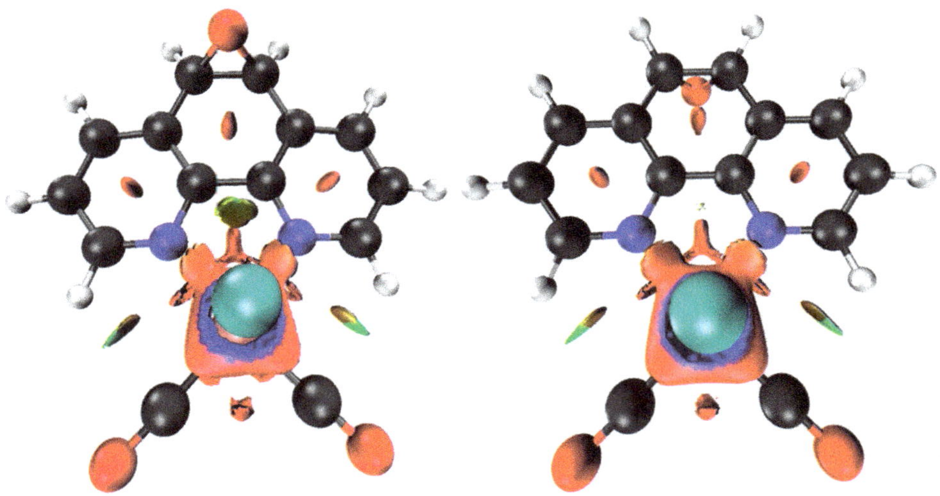

Figure 5. Upper view of the NCI plot for the cis (**left**) and trans (**right**) isomeric forms, revealing different van der Waals interactions. These calculations were performed at the BP86 and def2-TZVPP levels.

Finally, the aromaticity descriptors highlighted the subtle electronic distinction between the isomers. Aromaticity remains a fundamental concept in molecular chemistry as it provides essential insights into the stability, geometry, and magnetic behavior of cyclic systems [49]. Despite the absence of a strictly formalized definition, aromaticity is conventionally assessed through a variety of criteria, including energetic, structural, magnetic, and electronic factors. Each of these criteria reflects distinct aspects of electron delocalization [49,50]. The evaluation of aromaticity was carried out using magnetically induced current density (MICD) calculations at the BP86 level. For Re, a fully relativistic small-core effective potential (ECP60MDF) with its associated segmented valence basis set was employed, whereas the def2-TZVP basis set was used for the remaining atoms. The diatropic (aromatic) and paratropic (antiaromatic) ring currents were assessed using two integration planes. Positive net ring-current strength (RCS_{net}) values are indicative of aromatic character, whereas negative values are associated with antiaromaticity. MICD calculations revealed diatropic currents along the phenanthroline perimeter and paratropic contributions within its internal rings (Figure 6a). In the *cis* isomer, these currents were weak, with RCSnet values of 2.1 $nA \cdot T^{-1}$ for the central ring and nearly negligible contributions from the lateral rings. In contrast, the *trans* isomer sustained stronger local aromaticity in the lateral rings (RCS_{net} = 5.4 $nA \cdot T^{-1}$), while the central ring remained weakly aromatic (1.2 $nA \cdot T^{-1}$). The EDDB values support this view, confirming greater delocalization in the lateral rings of the *trans* isomer, in agreement with Hückel's rule and Clar's description of the sextet localization in phenanthroline (Figure 6b). The slightly higher delocalization in the *trans* isomer correlates with its marginally greater thermodynamic stability.

Taken together, these results indicate that *cis* and *trans* ReL share nearly identical structural and electronic frameworks, with the only appreciable difference being the enhanced local aromaticity of the lateral rings in the *trans* isomer. This distinction, although modest, explains its slight thermodynamic preference and supports treating the isomeric mixture as a unified species in electrochemical and biological studies.

https://doi.org/10.3390/inorganics14010003

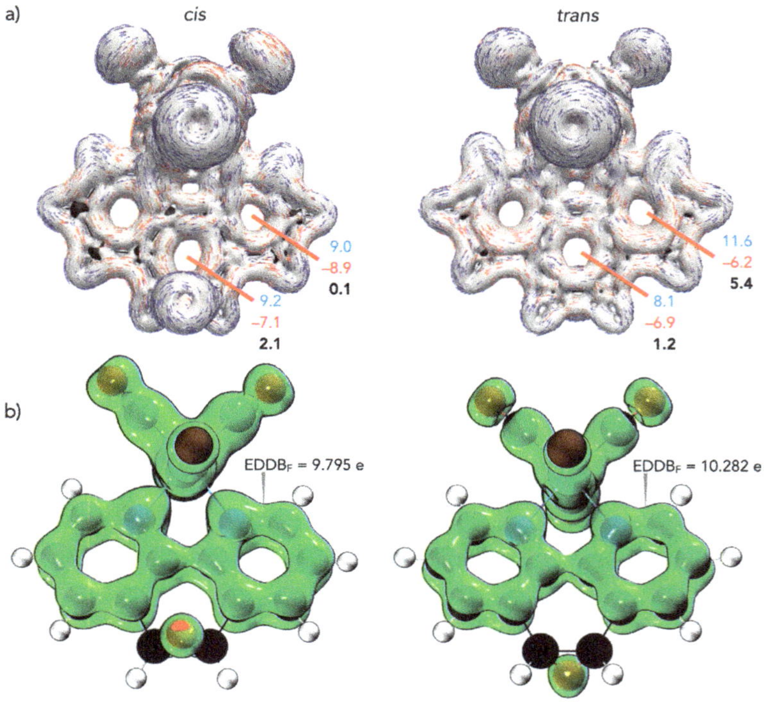

Figure 6. (**a**) Modulus of current density and vector flux with an isovalue of ±0.05 of cis and trans forms of ReL. Integration planes (red) used to determine the intensity of the current densities in nA·T^{-1}. Diatropic values are in blue, paratropic values are in red and net values are in bold. (**b**) $EDDB_F$ isosurfaces with the corresponding electron populations for the cis and trans forms of ReL.

2.4. Cyclic Voltammetry

Electrochemical measurements of ReL were performed on its isomeric mixture, as individual contributions from the *cis* and *trans* forms could not be experimentally resolved. Computational studies revealed negligible differences in energy and non-covalent interactions between the isomers, indicating that the relative orientation of the epoxide and bromide ligands did not significantly influence the redox behavior (See Section 2.3). Therefore, the observed metal-centered oxidation and ligand-centered reduction processes are considered representative of the entire mixture. The redox properties of transition-metal complexes are known to depend on several factors, including the oxidation state of the metal center, the electronic characteristics of the ligands, and the solvent and electrolyte environment [51,52]. The cyclic voltammograms of ReL and its uncoordinated equatorial ligand (L) are shown in Figure 7, enabling a direct comparison of their redox profiles and the identification of metal- and ligand-centered processes.

Here, one can notice the differences between the profiles for a blank solution and the respective compounds. The L displayed oxidation and reduction signals at 1.64 and −0.86 V vs. SCE, respectively. For ReL, two irreversible oxidations were found at 1.56 and 1.93 V, and two quasi-reversible reductions were found at approximately −1.21 and −1.50 V.

A working window study was performed to identify the electrodic processes for the L and ReL, the relation between them, and reversibility, Figure 8.

https://doi.org/10.3390/inorganics14010003

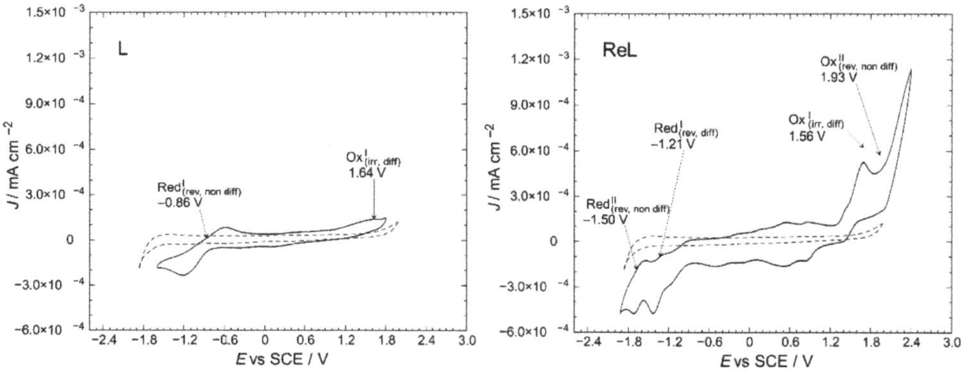

Figure 7. Cyclic voltammetry profiles of the L and ReL compared to blank (dashed line). Interface: Pt | 10^{-3} mol/L of the respective compound +10^{-1} mol/L of $TBAPF_6$ in anhydrous CH_3CN under an argon atmosphere. Scan rate: 200 mV/s.

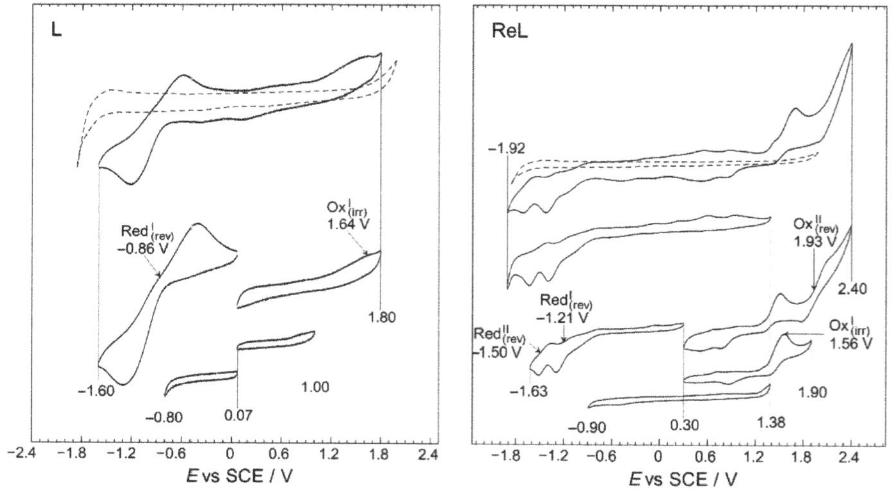

Figure 8. Working-window study CV profiles of the L and ReL compared to blank (dashed line). Interface: Pt | 10^{-3} mol/L of respective compound+10^{-1} mol/L of $TBAPF_6$ in anhydrous CH_3CN under an argon atmosphere. Scan rate: 200 mV/s.

The working-window study (Figure 8) shows that the L and ReL display typical behaviors observed for analogous compounds studied before. The working window study for L exhibited an intense quasi-reversible reduction, Red^I, process at a half-wave potential of −0.86 V and a low-intensity irreversible oxidation at 1.64 V. These signals have been previously identified for similar dinitrogenated (N,N) ligands [16,53–56]. For the ReL complex, various oxidation and reduction processes were observed in the full potential window (−1.92 to 2.40 V vs. SCE) in the cyclic voltammogram shown in Figure 8. However, if the CV profile from the bottom of the figure is inspected (−0.90 to 1.38 V vs. SCE), none of these signals are observed in the wider potential range CV profile and do not correspond to impurities since few were detected by ^1H-NMR (Figure S2a). Precisely, when the anodic limit reaches 1.90 V a reduction is observed 0.85 V after potential-sweep inversion. Thus, this reduction could belong to oxidated products formed after the 1.56 V process, possibly indicating that Ox^I may also be considered a quasi-reversible process. On the other hand, when anodic limit reaches 2.40 V, another reduction signal appears at 0.50 V. Due to the

https://doi.org/10.3390/inorganics14010003

separation between 0.50 and 1.93, it is unlike that the reduction at 0.50 is related to the oxidation's products from 1.93 V; more likely corresponds to a reduction from solvent oxidation by-products. All in all, only two oxidations were observed for ReL: an irreversible process occurring at 1.56 V, followed by another irreversible process at 1.93 V. In addition, two quasi-reversible reductions are observed at −1.21 and −1.50 V, respectively. This behavior is consistent with past outcomes on similar molecules [57,58].

The first oxidation at 1.56 V for ReL corresponds to a (1) single electron irreversible oxidation of rhenium center $Re^{I\rightarrow II}$, followed by an (2) intramolecular fast reduction $Re^{II\rightarrow I}$, with an immediate substitution of bromide radical by a solvent molecule. We then observed a second (3) irreversible oxidation at 1.93 V, producing a dicationic species. Thus, the complete oxidation process followed an electrochemical–chemical–electrochemical steps mechanism (ECE) (Scheme 1), which was previously reported for similar complexes containing different diimine (N,N) ligands [59–61].

$$[Re^{I}(N,N)(CO)_3Br] \rightarrow [Re^{II}(N,N)(CO)_3Br]^{+} + e^{-} \quad (1)$$
$$[Re^{II}(N,N)(CO)_3Br]^{+} + ACN \rightarrow [Re^{I}(N,N)(CO)_3(ACN)]^{+} + Br^{\bullet} \quad (2)$$
$$[Re^{I}(N,N)(CO)_3(ACN)]^{+} \rightarrow [Re^{II}(N,N)(CO)_3(ACN)]^{2+} + e^{-} \quad (3)$$

N,N = 5,6-epoxy-5,6-dihydro-1,10-phenanthroline; ACN= Acetonitrile

Scheme 1. Reaction mechanism for ReL oxidations at 25 °C.

Regarding the reductions in the ReL complex, two quasi-reversible processes are observed in Figures 7 and 8. These reductions have been extensively reported as two single-electron processes: the first one at −1.21 V potential (4) belongs to the quasi-reversible reduction of the L, followed by (5) reversible bromide elimination.

The second at −1.50 V potential (6) corresponds to the reversible reduction from the rhenium center $Re^{I\rightarrow 0}$, generating a dianionic species that may also go through a (7) bromide elimination step. Thus, the reduction follows the electrochemical-electrochemical–chemical mechanism (EEC) with bromide dissociation as the chemical step (Scheme 2). Similar mechanisms have been described for the reduction of other rhenium(I) complexes. Bromide removal is considered reversible because it is coupled to a redox process that generates a coordinatively unsaturated species, which has the capacity to recapture the bromide [57,61].

$$[Re^{I}(N,N)(CO)_3Br] + e^{-} \rightarrow [Re^{I}(N,N)(CO)_3Br]^{-} \quad (4)$$
$$[Re^{I}(N,N)(CO)_3Br]^{-} \rightarrow [Re^{I}(N,N)(CO)_3] + Br^{-} \quad (5)$$
$$[Re^{I}(N,N)(CO)_3Br]^{-} + e^{-} \rightarrow [Re^{0}(N,N)(CO)_3Br]^{2-} \quad (6)$$
$$[Re^{0}(N,N)(CO)_3Br]^{2-} \rightarrow [Re^{0}(N,N)(CO)_3]^{-} + Br^{-} \quad (7)$$

N,N = 5,6-epoxy-5,6-dihydro-1,10-phenanthroline

Scheme 2. Reaction mechanism for ReL reductions at 25 °C.

The general reaction scheme for oxidations and reductions is presented in Figure 9.

The observed reductions of ReL closely align with the position and nature of their individual constituents. Furthermore, computational studies on comparable systems emphasize the evenly distributed LUMO above the (N,N) ligand and the rhenium core [53,57]. Scan rate studies were performed to determine whether the current of the electrodic processes depended on the species diffusion. As shown in Figure S19, the current increased with the scan rate, as expected. As can be seen from the summary presented in Table 1, mass transfer control is present in some of the electrodic processes displayed by the molecules under study.

https://doi.org/10.3390/inorganics14010003

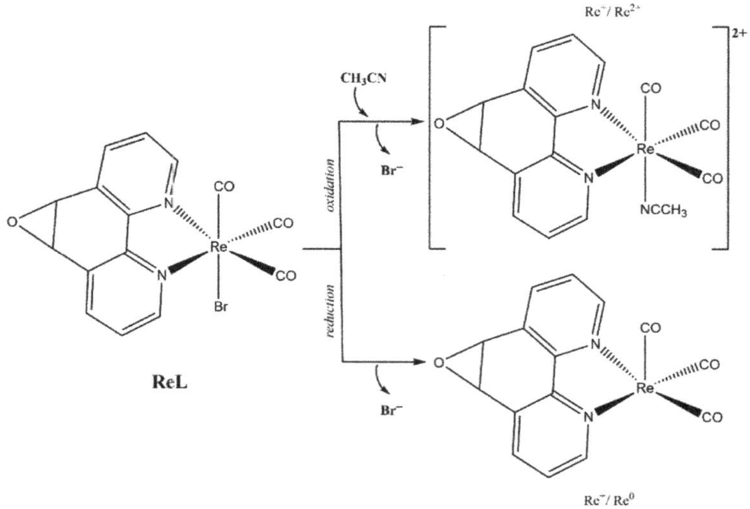

Figure 9. Electrochemical reaction scheme for ReL.

Table 1. Summary of the electrochemical processes.

Peak *	Potential, *E* (V vs. SCE)	Assignment	Diffusional Control?
L (5,6-epoxy-5,6-dihydro-1,10-phenanthroline)			
$Red^{I}_{(qrev)}$	−0.86	L reduction	No
$Ox^{I}_{(irr)}$	1.64	L oxidation	Yes
ReL (*fac*-[Re(CO)3(5,6-epoxy-5,6-dihydro-1,10-phenanthroline)Br])			
$Red^{I}_{(qrev)}$	−1.21	Ligand reduction	No
$Red^{II}_{(qrev)}$	−1.50	$Re^{I\rightarrow 0}$ reduction (Followed by bromide elimination)	Yes
$Ox^{I}_{(irr)}$	1.56	$Re^{I\rightarrow II}$oxidation (Followed by intramolecular $Re^{II\rightarrow I}$reduction)	Yes
$Ox^{II}_{(rev)}$	1.93	$Re^{I\rightarrow II}$ oxidation (After Br^- replacement with ACN)	No

* qrev = quasi-reversible; irr = irreversible.

Complementarily, a voltammetry experiment was conducted in a CO_2-saturated solution (Figure S20) to preliminarily check a possible electrocatalytical potential of the complex towards CO_2 reduction process. An increase in the current density was observed, which may be associated with the CO_2 reduction processes [62,63]. These results suggest a possible influence of the complex on the process, which must be rigorously evaluated through further studies. Future studies will involve modifying electrodes with these complexes to systematically investigate their catalytic behavior during the reduction $Re^{I\rightarrow 0}$ processes [4,52]. To better understand the metallic center reduction process $Re^{I\rightarrow 0}$, the half-wave potentials for a family of compounds measured in identical conditions are presented in Figure S21 [4,57]. As expected, some trends can be observed on the $Re^{I\rightarrow 0}$ potential depending on the size of the ligands and/or nature of the moieties present. On the one hand, the presence of the ancillary ligand corresponding to a Schiff base (red dashed enclosure) showed the most negative values. In previous studies [53,57], the ancillary ligand exhibited a withdrawing effect to the rhenium core, as evidenced by UV–Vis spectroscopy, theoretical

https://doi.org/10.3390/inorganics14010003

calculations, and other methods. These reports have pointed out that this kind of ancillary ligand leads to a less intense electron density on the rhenium core, which could explain why the reduction processes require higher potential values. In the middle, the complexes with a halide ion as a ligand (blue line) were observed. Finally, the most positive potential value for $Re^{I \to 0}$ belongs to the complex with a benzimidazole ligand. This ligand appears to influence the homogeneous distribution of electron density in the rhenium core. Thus, the reduction in the observed potential values in this Re complex family is strongly influenced by the aromaticity due to the nature of the substituents in the ligands' moiety. Reduction processes for this type of compound are important in the study of potential electrocatalytic capabilities for the carbon dioxide reduction process [64–66].

2.5. Computed Oxidation and Reduction Potentials of ReL

To better understand the redox processes described in Section 2.4, we employed a computational methodology using a thermodynamic cycle to calculate free energies in both the gas-phase and solvated states. The CPCM model with acetonitrile as the solvent was used to estimate the solvation energies. We found oxidation and reduction potentials at +2.04 V and −0.92 V, respectively, for L, which agree with the experimental trends (Table 2).

Table 2. Electrochemical values of L and ReL complex (experimental data and theoretical calculations).

Compound	Process	Potential (V)	
		Experimental	Theoretical Calculations
L	Ox^{I}	+1.64	+2.04
	Red^{I}	−0.86	−0.92
ReL	Ox^{I}	+1.56	+1.45
	Ox^{II}	+1.93	+2.11
	Red^{I}	−1.21	−1.31
	Red^{II}	−1.50	−1.95

As observed after ^{1}H-NMR analysis (Figure S2a,b), both isomers are present in the solution. For the computational calculations, only one of the isomeric structures (*trans*) was considered, as the other isomer is expected to exhibit similar properties based on previous computational studies. For ReL, we identified two oxidation processes at +1.45 V and +2.11 V, which agree with experimental data at +1.56 V and +1.93 V, respectively. These processes were attributed to rhenium oxidation following an electrochemical–chemical–electrochemical mechanism (see discussion above). Regarding the reduction, two processes are calculated at −1.31 and −1.95 V (see Table 2), consistent with the reduction of the L equatorial ligand at −1.21 V and the reduction of $Re^{I \to 0}$ at −1.50 V, respectively.

To complement these findings, electron density plots were calculated for the L and the ReL complex (see Table 2). The analysis focused on the Highest Occupied Molecular Orbital (HOMO) to identify the most probable oxidation site. For the L, the HOMO is primarily localized in the 1,10-phenanthroline ring, with a minor contribution from the oxygen atom of the epoxy moiety (see Table 3). The significant potential difference in oxidation peaks observed in the calculations can be attributed to the electron-withdrawing effect of the 1,10-phenanthroline moiety, which increases the energy required to oxidize the epoxy group in L. In contrast, the Lowest Unoccupied Molecular Orbital (LUMO) is predominantly distributed across the fused aromatic rings.

https://doi.org/10.3390/inorganics14010003

Table 3. Electron density of molecular orbitals HOMO and LUMO involved in electrochemical steps of redox mechanisms for L, ReL and ReL-ACN complexes.

Compound	Electron Density		ΔE (eV)
	HOMO	**LUMO**	
L	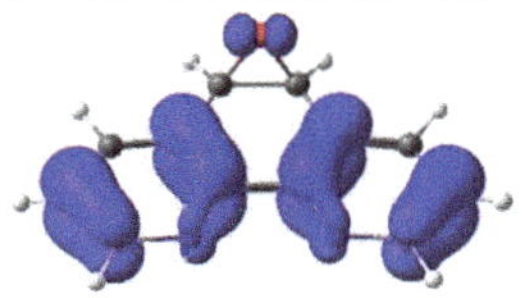 E = −6.77 eV	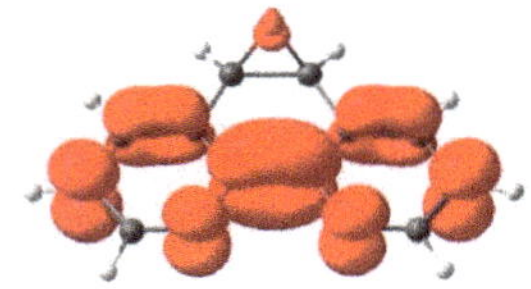 E = −1.90 eV	4.87
ReL	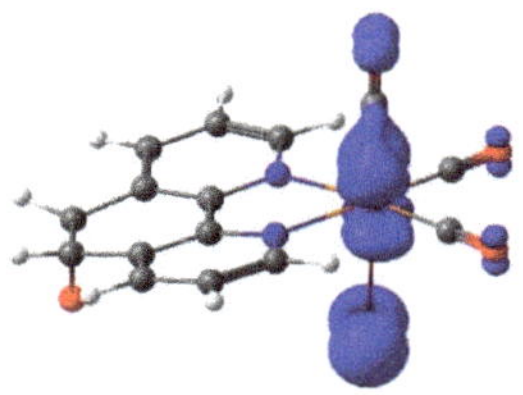 E = −5.59 eV	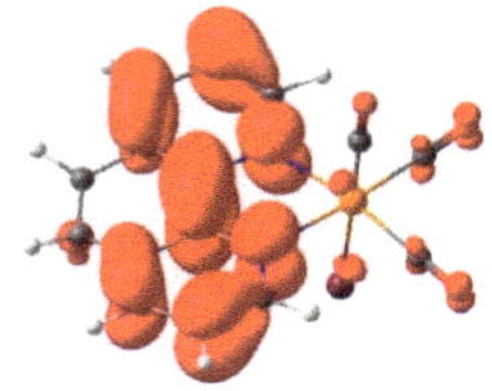 E = −3.51 eV	2.08
ReL-ACN *	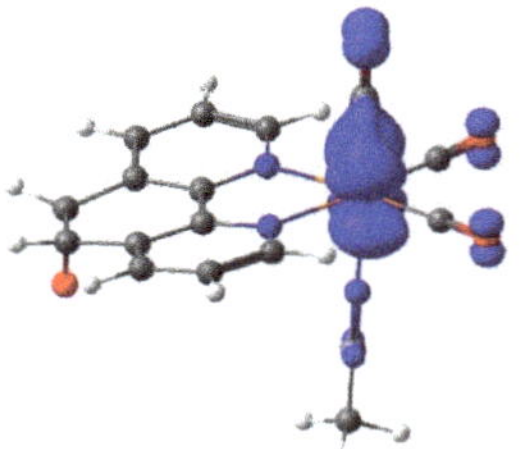 E = −6.18 eV	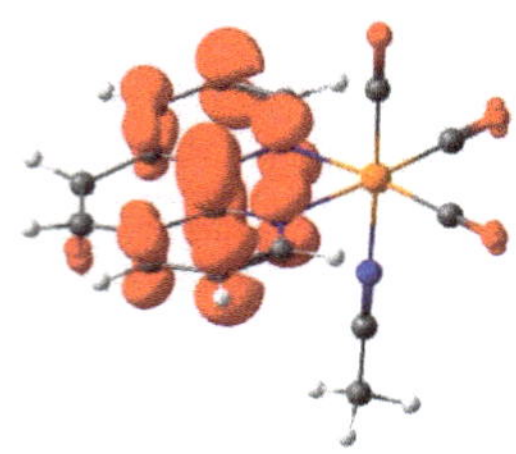 E = −3.68 eV	2.50

* (ReL-ACN = *fac*-$[Re(CO)_3$(5,6-epoxy-5,6-dihydro-1,10-phenanthroline)(ACN)$]^+$.

For the ReL complex, redox calculations indicated that the spatial modification of the epoxy group relative to the bromine ligand in the octahedral arrangement did not significantly alter the electronic properties. The HOMO and LUMO plots revealed that the electron density was primarily localized either on the metal core or within the aromatic ring system (Table 3). Specifically, the HOMO was concentrated in the rhenium tricarbonyl core rather than in the 1,10-phenanthroline ring, suggesting that oxidation is most likely to occur at the rhenium center ($Re^{I \to II}$). The experimental results revealed a second reversible oxidation at +1.93 V, suggesting the formation of a dicationic species. The ReL-ACN complex exhibited a HOMO distribution centered on the metal core, with a stabilization energy of ΔE = 2.50 eV. This value is lower than that obtained for the L (ΔE = 4.87 eV), confirming that oxidation is more likely to occur at the metal center. Regarding the LUMO, the electron density is primarily distributed in the 1,10-phenanthroline ring, indicating that reduction is more likely to occur in this region. Based on these analyses, ReL exhibited a characteristic cyclic voltammetry profile, beginning with an initial irreversible oxidation, followed by a second reversible oxidation that followed an electrochemical–chemical–electrochemical (ECE) mechanism. Additionally, the complex undergoes two reversible reductions processes consistent with the electrochemical–electrochemical–chemical (EEC) mechanism.

https://doi.org/10.3390/inorganics14010003

2.6. Cell Viability Assays

Rhenium(I) tricarbonyl complexes featuring dinitrogenated N,N′ ligands, such as 1,10-phenanthroline derivatives, combined with chloride or bromide as ancillary ligands have gained attention as promising candidates in anticancer research [67–69]. The incorporation of halides confers neutrality to these complexes, improving lipophilicity, cellular uptake, and physiological stability, critical attributes for effective drug delivery. These structural benefits correlate with significant cytotoxic activity against human cancer cell lines, notably HeLa (cervical cancer) [67]. Representative examples, including $Re(CO)_3$(1,10-phenanthroline)Cl and its functionalized analogues, exhibit strong potential for development as chemotherapeutic agents [29,70]. Based on this, the proposed complex *fac*-[$Re(CO)_3$(5,6-epoxy-5,6-dihydro-1,10-phenanthroline)Br] (ReL) was evaluated for its cytotoxic potential in HeLa cervical cancer cells. The cell viability was assessed using the MTT assay, which revealed a measurable reduction in metabolic activity, suggesting that ReL may exert cytotoxic effects under the tested conditions [71,72], Figure 10.

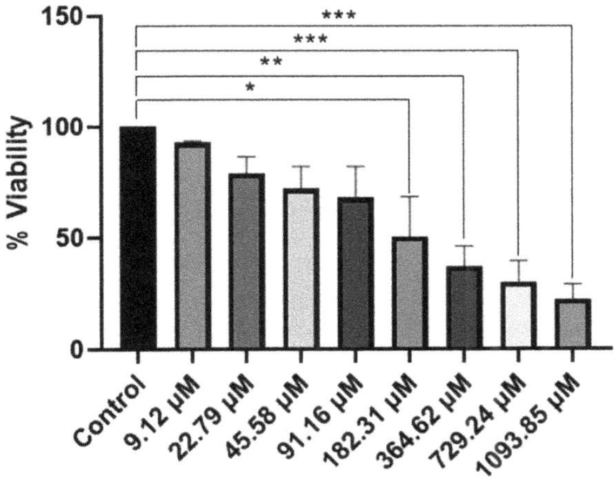

Figure 10. Evaluation of cell viability of ReL complex by a MTT assay in HeLa cells. A An MTT assay was performed to evaluate cell viability in the presence of the complex, dissolved in DMSO to obtain final concentrations of 9.12 µM, 22.79 µM, 45.58 µM, 91.16 µM, 182.31 µM, 364.62 µM, 729.24 µM and 1093.85 µM. Cell viability was expressed as a percentage relative to that of the control group. Analyses were performed with n = 3 and evaluated using one-way ANOVA, followed by Dunnett's post hoc test. Bars represent the mean ± standard error of the mean (SEM). Symbol (*) indicates a statistically significant difference with $p < 0.05$. Group differences versus control were evaluated by one-way ANOVA with Dunnett's post-hoc test. Significance codes: * $p < 0.05$; ** $p < 0.01$; *** $p < 0.001$ (two-sided; posthoc– adjusted where applicable).

The MTT assay provided comprehensive information on the cytotoxic effects of the ReL complex. A reduction in cell viability was observed, which depended on the concentration, indicating that ReL demonstrates cytotoxic activity under the conditions tested. The lowest concentration of the compound exhibited behavior like that of the control. Subsequently, at concentrations of 22.79 µM, 45.58 µM, and 91.16 µM, a decrease in cell viability was observed, suggesting an early onset of biological activity; however, these differences were not statistically significant. This trend was accentuated at concentrations of 182.31 µM and 364.62 µM, where significant cytotoxic effects were detected ($p < 0.01$ and $p < 0.001$, respectively). At higher concentrations (729.24 µM and 1093.85 µM), cell viability was reduced to below 40% compared to control, with highly significant differences ($p < 0.0001$).

https://doi.org/10.3390/inorganics14010003

Therefore, ReL complex demonstrated a clear concentration-dependent cytotoxic effect, characterized by a progressive reduction in cell viability as the dose increased.

Cell viability of *fac*-[Re(CO)$_3$(5,6-epoxy-5,6-dihydro-1,10-phenanthroline)Br] decreased in a dose-dependent manner, with significant reductions observed at 182.31 μM and above, yielding an IC$_{50}$ of 182.31 μM. In contrast, neutral rhenium(I) tricarbonyl complexes bearing unmodified phenanthroline ligands typically display much higher potency, with reported IC$_{50}$ values of 1.2–1.8 μM, surpassing cisplatin (6.6 μM) [25,29,67,73]. Although the bromide ligand may enhance lipophilicity, this effect does not offset the structural constraints imposed by the epoxide modification, highlighting the critical role of rational ligand design in achieving optimal biological performance [29,67]. Importantly, replacing the bromide ligand with a neutral donor that enables formation of cationic species could improve aqueous solubility and facilitate endocytotic uptake, potentially lowering IC$_{50}$ values to levels comparable with other highly active rhenium tricarbonyl analogues reported in the literature [74]. Complexes bearing benzimidazole or bipyridine derivatives typically display IC$_{50}$ values in the 50–100 μM range, depending on ligand lipophilicity and substitution patterns [67]. Neutral complexes of the type fac-[Re(CO)$_3$(Phenathroline)Cl or Br] are attractive as activatable precursors [29,75]. These features suggest their potential as prodrug candidates that could be activated in biological environments, although further optimization is required to improve stability and lipophilicity. These considerations highlight the need for rational ligand modification to optimize both physicochemical properties and anticancer activity [76].

In brief, the ReL complex exhibited a clear dose-dependent cytotoxic effect on HeLa cells, with an IC$_{50}$ significantly higher than cisplatin. This relatively low toxicity indicates that ReL could potentially meet key requirements for an effective fluorophore, including minimal cytotoxicity, high photostability, and efficient cellular uptake without permeabilizing agents, ensuring reliable bioimaging in epithelial (non-walled) cell models while preserving cell viability [26]. Interestingly, previous studies reported that ReL interacts effectively with walled cells such as Gram-negative bacteria and yeasts, showing minimal or no cytotoxicity, while significant effects were observed against Gram-positive bacteria, suggesting a selective biological response. Moreover, its intrinsic luminescent properties enable direct application as a fluorescent probe for microbial imaging, particularly in *Salmonella enterica* and *Candida albicans* [33]. These findings highlight the multifunctional nature of ReL, combining imaging capability with selective bioactivity and underscoring its potential as a versatile tool for microbiological applications.

3. Experimental and Theoretical Details

3.1. Materials and Instruments

The initial reagents, sourced from Merck and Aldrich (St. Louis, MO, USA), were used as received, without additional purification. Acetonitrile (CH$_3$CN) was dried using molecular sieves and purged with argon gas in preparation for the electrochemical experiments. The synthesis and comprehensive characterization of the rhenium(I) tricarbonyl complex are described in detail. Additional methodological, crystallographic, and computational details are provided in Appendix A

3.2. Synthesis of fac-[Re(CO)$_3$(5,6-epoxy-5,6-dihydro-1,10-phenanthroline)Br] (ReL)

The complex was prepared following a previously documented method, which involved the reaction of L with the [Re(CO)$_3$Br(THF)]$_2$ dimer in a 2:1 ratio in toluene. The mixture was stirred for 15 min at room temperature[2]. Reaction scheme is presented on Figure 11.

https://doi.org/10.3390/inorganics14010003

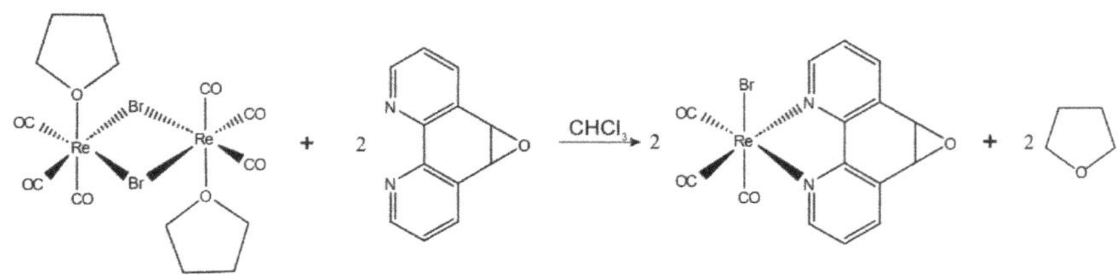

Figure 11. Reaction scheme for the synthesis of ReL.

FTIR (KBr, cm^{-1}): 3433 (νNH); 2021 (symmetric νCO); 1909 (antisymmetric νCO); and 1435 (νCC). ^{1}HNMR (400 MHz, DMSO-$_{d6}$, ppm): δ = 9.13–9.07 [m, 2H, C1,C10], 8.76–8.65 [m, 2H, C3,C8], 7.94–7.82 [m, 2H, C2,C9], 5.10 [d, J = 1.7 Hz, 1H,C6] and 5.08 [d, J = 1.6 Hz, 1H, C5]. ^{13}C-NMR (100 MHz, DMSO-$_{d6}$, ppm): δ = 197.40; 153.47; 150.93; 141.86; 133.89; 128.31; and 54.91.

For the characterization of ReL, we conducted FTIR spectra (ATR) using a Bruker Vector-22 FT-IR spectrophotometer (Billerica, MA, USA). Additionally, ^{1}H-NMR, ^{13}CNMR, and CHCOSY spectra were acquired on a Bruker AVANCE 400 spectrometer (400 MHz) and maintained at 25 °C. The compound was dissolved in deuterated DMSO.

3.3. Structure Determination

Crystals suitable for X-ray analysis of compound ReL were obtained as previously described and mounted using MiTeGen Micro Mounts (New York, NY, USA). Table S1 presents the experimental and crystallographic data for ReL, while selected bond distances and angles for all the studied compounds are shown in Table S2. The intramolecular interactions are outlined in Table S3. Detailed information on the crystal, X-ray data collection, and structure solution is provided in the Supporting Information. Intensity data were collected at room temperature using a Bruker Smart Apex diffractometer (Billerica, MA, USA) equipped with a bi-dimensional CMOS Photon100 detector and a graphite monochromator employing Mo-Kα radiation. A frame separation of 0.31 and a collection time of 10 s per frame were used at 296 K. APEX3 (Bruker AXS), SHELXL (Sheldrick), and the TONTO package integrated within CrystalExplorer 17.5 were used in their standard distributions. Data integration was performed using APEX3 software, with absorption corrections applied via SADABS. The solution and refinement for compound ReL were carried out with Olex2. The structure of ReL was solved using the Patterson method and refined with the SHELXL package through least squares minimization [77]. The refinement utilized full matrix least squares on reflection intensities (F^2), with all non-hydrogen atoms refined anisotropically and hydrogen atoms positioned in idealized locations.

The intermolecular interactions in the crystalline molecules were studied using Hirshfeld surface analysis. This surface is defined as the fraction of the electron density sphere of the atoms of the molecule of interest, "promolecule" that contributes to the total electron density of the crystalline molecule "procrystal" system, considering a normalized contact distance (d_{norm}), where r^{VdW} is the van der Waals (vdW) radius of the atom corresponding to the interior or exterior of the Surface (Equation (1)). For this reason, Hirshfeld surface analysis has gained prominence as a powerful tool to explore and describe various intermolecular interactions present in crystal packing. Furthermore, these surfaces are related to their corresponding 2D fingerprint plots [78–81]. This plot allows us to explore, quantify, and compare the contributions of the different interactions present in the crystal structure and helps to elucidate important information about close contacts and more distant interactions and areas where contacts are weak. Crystal Explorer 17.5 software [82] was used to calculate the Hirshfeld surface [78,79] and associated 2D-fingerprint plots [83,84] of the L1–4 compounds using the crystallographic information file (CIF) as input for the analysis.

https://doi.org/10.3390/inorganics14010003

The electrostatic potentials were mapped on the Hirshfeld surfaces using the 3–21G basis set at the level of Hartree-Fock theory over a range of ±0.002 au using the TONTO computational package integrated into the program Crystal Explorer [82]. For the generation of fingerprint plots, the bond lengths of hydrogen atoms involved in interactions were normalized to standard neutron values (C-H = 1.083 Å, N-H = 1.009 Å, O-H = 0.983 Å) [85].

$$d_{norm} = \frac{d_i - r_i^{VdW}}{r_i^{VdW}} + \frac{d_e - r_e^{VdW}}{r_e^{VdW}} \quad (1)$$

Equation (1): Normalized contact distance, defined in terms of de, d_i, and the VdW radii of the atoms. Red (distances shorter than the sum of VdW radii) through white to blue (distances longer than the sum of VdWradii).

3.4. Electrochemical Characterization

For all electrochemical studies, the working solution consisted of 0.01 mol/L of the respective compounds (L and ReL complex). This solution was supplemented with 0.1 mol/L of tetra-n-butylammonium hexafluorophosphate ($TBAPF_6$) as a supporting electrolyte. In this study, all CV experiments were conducted under standardized conditions using a platinum working electrode in anhydrous acetonitrile containing 10^{-3} mol·L^{-1} of the respective compound (L or ReL) and 0.1 mol·L^{-1} of tetrabutylammonium hexafluorophosphate ($TBAPF_6$) as the supporting electrolyte. Measurements were carried out under an argon atmosphere at a scan rate of 200 mV·s^{-1}. The compounds and electrolyte were prepared in high p.a. anhydrous acetonitrile (CH_3CN). Prior to each experiment, the working solution underwent purging with high-purity argon gas, with an argon atmosphere maintained throughout the experiment [4]. For carbon dioxide reduction experiments, the solution was purged with CO_2, and a constant pressure of this gas was maintained during measurements. A polycrystalline, non-annealed platinum disc with a 2 mm diameter was used as the working electrode. The working electrode was polished with alumina slurry (particle size 0.3 μm) on soft leather, then rinsed with bi-distilled water, acetone, and finally dried before and between measurements. A platinum gauze of substantial geometrical dimensions was employed for the counter electrode, segregated from the primary cell compartment by a fine sintered glass. All recorded potentials in this document are referenced to Ag/AgCl electrodes in tetramethylammonium chloride to align with the potential of a saturated calomel electrode (SCE) at room temperature [86]. Electrochemical cells, syringes, volumetric flasks, and the rest of the employed glassware were always washed with bi-distilled water, acetone, and oven-dried at 60 °C for at least 12 h prior to all experiments. All electrochemical experiments were executed at room temperature, using a CHI900B bipotentiostat CH Instruments (Austin, TX, USA) interfaced with a computer running CHI 9.12 software. This software facilitated experimental control and data acquisition.

3.5. Theoretical Details

All calculations in this study were performed within the Density Functional Theory (DFT) framework using ORCA 5.0.4 [87]. Molecular geometries were fully optimized employing the B3LYP hybrid functional with the def2-TZVPP basis set. The Tight SCF convergence criteria and the RIJCOSX approximation were applied for Coulomb integral calculations to enhance computational efficiency. The solvation effects were estimated using the Conductor-like Polarizable Continuum Model (CPCM) [88,89], with acetonitrile as the solvent. Frequency analyses were conducted to obtain the thermodynamic parameters necessary for determining the reduction potentials.

For the Fukui function analysis, an optimization calculation was performed using ORCA 5.0.4 to obtain the electronic density of the neutral species. Density Functional

https://doi.org/10.3390/inorganics14010003

Theory (DFT) was employed with the BP86 functional and the def2-TZVPP basis set [90–92]. For the rhenium atom, the SARC-ZORA-TZVPP basis set [92] was used to account for the relativistic effects associated with heavy metals. Adiabatic calculations were carried out to obtain the electronic density of the positively and negatively charged species. This approach assumes that the molecular geometry remains unchanged upon charge variation, which means that the geometric structures of the charged species are not re-optimized. Such an approximation is valid for systems in which the charge influence on geometry is minimal, as seen in radicals and other charged species where structural changes upon electron addition or removal are negligible. This adiabatic treatment simplifies the analysis and enables the study of the reactivity and electronic density of charged species without requiring geometric optimization for each charge state. To generate figures highlighting nucleophilic and electrophilic reactivity sites, Chemcraft 1.8 software was used to perform density subtraction operations.

The NCI approach relies on electron density and its derivatives, allowing the identification of noncovalent interactions based on the reduced density gradient (S) in low-density regions. Within this framework, the reduced density gradient is defined by the equation described for Trujillo et al. and Villegas-Escobar et al. [93,94]. The optimized structure from the Fukui function calculation of the neutral species was used for the NCI analysis. Multiwfn 3.8 software [95] was employed to generate .cube files to obtain density gradient data. The .cube files were then visualized using the VMD 1.9 software [96] to identify and display the different types of interactions within the structure. Additionally, Python programming language 3.12.12 script developed by our research group was utilized to generate RDG analysis figures.

The magnetically induced current density (MICD) was analyzed using the GIMIC program [97,98]. For a more detailed analysis of cyclic current Lows and to evaluate current profiles across selected integration planes, the planes were positioned perpendicular to the molecular plane. The ring currents and their intensities were then determined by integrating the current density across these planes. For this analysis, the Gauge Included Atomic Orbitals (GIAO) method [99] was used to study diatropic (aromatic) and paratropic (antiaromatic) ring currents circulating clockwise and counterclockwise, respectively. The two-dimensional Gauss-Lobatto algorithm was used to integrate the currents passing through an integration plane. The current density paths were done using Paraview5.10.0 [100]. Computations were performed by Gaussian 16.B015 with the BP86 (functional, in which a small-core fully relativistic effective core potential (ECP60MDF) [101] and associated segmented valence basis sets were adopted for Re and the def2-TZVP [102,103] basis set was used for the remaining atoms [101,104]. Also, aromaticity was evaluated by performing electron density of delocalized bonds (EDDB) calculations using EDDBRun scripts [105].

The methodology for computing the reduction and oxidation potentials was based on an established thermodynamic cycle approach [106–109]. This method calculates the free energy of the species in both gas-phase and solvated state. Separate thermodynamic cycles were used for each oxidation and reduction potential, as illustrated in Figures S22 and S23. To establish the computational protocol for estimating redox potentials, standard Gibbs free energies in acetonitrile were obtained from gas- and solution-phase calculations for the reduced and oxidized species. The gas-phase free energy was derived from single-point energy calculations. The standard absolute redox potential (E^0) was determined using the following equation:

$$E^0 = -\Delta G_s^0(\text{redox}, CH_3CN)/nF$$

In the given equation, n represents the number of electrons consumed or generated in the specific half-reaction, and F denotes the Faraday constant. This thermodynamic model establishes a direct connection between the changes in the free energy in the condensed

https://doi.org/10.3390/inorganics14010003

phase related to reduction or oxidation and the corresponding processes in the gas phase. Furthermore, to determine the free energy associated with ionization or electron attachment at 298 K, it is essential to calculate the thermal contributions to the free energy at 298 K. Considering this, the remaining link between the gas and the condensed phase is defined by the difference in solvation free energies between the oxidized and reduced species involved in the respective processes. The calculated absolute free energy of reduction or oxidation represents the generation or consumption of a free electron.

Additionally, the theoretical calculation of the free energy for the oxidation and reduction processes considers the complexity of the electrochemical mechanisms.

3.6. MTT Assay Protocol

HeLa cells were obtained from ATCC (Manassas, VA, USA), a recognized supplier of authenticated cell lines and microbial strains for biomedical research and diagnostics. The MTT assay is an established colorimetric method for assessing cell viability, based on the ability of living cells to reduce the tetrazolium (MTT) to a purple formazan [71,72]. This color change is indicative of cellular metabolic activity, as viable cells convert MTT into an insoluble product whose color intensity can be quantified to determine cell viability [72]. This assay was performed using a 96-well plate, in which 35,000 cells were seeded per well. Cells were cultured in Dulbecco's Modified Eagle Medium (DMEM) supplemented with 10% fetal bovine serum (FBS) and penicillin/streptomycin. . Cells were maintained in a humidified atmosphere with 5% CO_2 at 37 °C. 200 μL of each prepared solution (9.12, 22.79, 45.58, 91.16, 182.31, 364.62, 729.24, and 1093.85 μM) was added to the respective cell well. Stock solutions were obtained by dissolving precise amounts of the compound in DMSO. Appropriate controls were included to ensure the reliability of the assay: the positive control consisted of culture medium alone, the negative control was DMSO, and the vehicle control was a mixture of culture medium and DMSO. These controls provided a baseline for evaluating the specific effects of this compound on cell viability.

The plates were incubated at 37 °C in a 5% CO_2 atmosphere for 24 h. Subsequently, a 5 mg/mL MTT solution, diluted 1:10 in culture medium, was added to each well. The MTT incubation was continued for two hours to allow the conversion of MTT to formazan. After MTT incubation, 100 μL of DMSO was added to each well to dissolve the formazan complexes. The solution was carefully resuspended in each well to ensure complete dissolution of the formazan. Finally, absorbances at 570 nm and 630 nm were measured using a BioTek microplate reader. Cell viability was expressed as a percentage with respect to the control group. Analyses were performed with n = 3 and evaluated using one-way ANOVA followed by Dunnett's post hoc test [110,111].

4. Conclusions

In this study, we present a comprehensive structural, electronic, and biological characterization of a neutral rhenium(I) complex *fac*-[Re(CO)$_3$(5,6-epoxy-5,6-dihydro-1,10-phenanthroline)Br] (ReL). Single-crystal X-ray diffraction confirmed the *cis* configuration, and Hirshfeld surface analysis revealed dominant O···H interactions in the crystal packing. Variable-temperature NMR spectroscopy demonstrated the coexistence of *cis* and *trans* isomers in solution, with the *trans* form being slightly more stable than the *cis* form. DFT calculations, including Fukui functions, NCI indices, and aromaticity descriptors, showed that both isomers share nearly identical electronic and structural frameworks, with enhanced local aromaticity in the *trans* isomer, which explains its slight thermodynamic preference. This supports the treatment of the isomeric mixture as a unified species in electrochemical and biological studies. Electrochemical studies revealed a typical *fac*-[Re(CO)$_3$(N,N)X] profile, including a $Re^I \rightarrow Re^0$ transition, with theoretical redox potentials aligning well with the experimental data. Biologically, ReL demonstrated dose-dependent cytotoxicity in HeLa cells, in contrast to its previously reported low toxicity in microbial systems. Combined with its luminescent properties and antimicrobial activity, these findings underscore the multifunctional nature of ReL and its potential as a bioactive and imaging agent for therapeutic and microbiological applications.

Supplementary Materials: The following supporting information can be downloaded at: https://www.mdpi.com/article/10.3390/inorganics14010003/s1, Figure S1: FTIR (KBr) spectrum of ReL; Figure S2: a. ^{1}HNMR spectrum of ReL recorded at 400 MHz in DMSO-$_{d6}$ at 25 °C; Figure S2: b. Expanded aromatic zone of the ^{1}H-NMR spectrum of ReL at 400 MHz and 25 °C, with samples dissolved in deuterated DMSO; Figure S3: ^{1}HNMR spectrum of ReL recorded at 400 MHz in DMSO-$_{d6}$ at 40 °C; Figure S4: ^{1}HNMR spectrum of ReL recorded at 400 MHz in DMSO-$_{d6}$ at 50 °C; Figure S5: ^{1}HNMR spectrum of ReL recorded at 400 MHz in DMSO-$_{d6}$ at 60 °C; Figure S6: a. ^{13}C NMR spectrum of ReL recorded at 100 MHz in deuterated DMSO (DMSO-$_{d6}$) at 25 °C; Figure S6: b. Expanded aromatic zone of ^{13}C NMR spectrum of ReL recorded at 100 MHz in deuterated DMSO (DMSO-$_{d6}$) at 25 °C; Figure S7: ^{13}C NMR spectrum of ReL recorded at 100 MHz in deuterated DMSO (DMSO-$_{d6}$) at 40 °C; Figure S8: ^{13}C NMR spectrum of ReL recorded at 100 MHz in deuterated DMSO (DMSO-$_{d6}$) at 50 °C; Figure S9: ^{13}C NMR spectrum of ReL recorded at 100 MHz in deuterated DMSO (DMSO-$_{d6}$) at 60 °C; Figure S10: CH COSY spectrum of ReL recorded in deuterated DMSO (DMSO-$_{d6}$) at 25 °C; Figure S11: Comparison of the crystal packing patterns of ReL and *fac*-[Re(CO)$_3$(5,6-epoxy-5,6-dihydro-1,10-phenanthroline)Cl] (CCDC-239411)* complexes; Figure S12: Hirshfeld surfaces mapped over d_{norm} (top) and 2D fingerprint plots of ReL (bottom) show the principal percentage contribution of contacts between complexes; Figure S13: Computed FTIR of *cis* isomeric form of ReL, with respect to epoxy moiety 1,10-phenenthroline and bromide ancillary ligand; Figure S14: Computed FTIR of *the trans* isomeric form of ReL, with respect to epoxy moiety 1,10-phenenthroline and bromide ancillary ligand; Figure S15: Plot of the Fukui function of the ReL cis isomeric form: the left plot indicates a nucleophilic site (f^+), and the right plot indicates an electrophilic site (f^-); Figure S16: Plot of the Fukui function of the ReL trans isomeric form: the left plot indicates a nucleophilic site (f^+) and the right plot indicates an electrophilic site (f^-); Figure S17: Noncovalent interaction (NCI) analysis of the cis-isomeric form of ReL. Bottom: The NCI color scale is $-0.02 < \lambda H > 0.02$ a.u. These calculations were performed at the BP86 and def2-TZVPP level; Figure S18: Noncovalent interaction (NCI) analysis of the trans-isomeric form of ReL. Bottom: The NCI color scale is $-0.02 < \lambda H > 0.02$ a.u. These calculations were performed at the BP86 and def2-TZVPP level; Figure S19: Cyclic voltammetry scan-rate study of L and ReL. Interface: Pt | 10^{-3} mol/L of respective compound + 10^{-1} mol/L of $TBAPF_6$ in anhydrous CH_3CN under an argon atmosphere. Scan rate: 25–200 mV/s; Figure S20: Cyclic voltammetry for CO_2-electrocatalysis preliminary study. Interface: Pt | 10^{-3} mol/L of fac-Re(CO)$_3$(5,6-epoxy-5,6-dihydro-1,10-phenanthroline)Br (ReL) + 10^{-1} mol/L of $TBAPF_6$ in anhydrous CH_3CN under saturated Ar(—) or CO_2 (---). Scan rate: 100 mV/s; Figure S21: Half-wave potential values for Rhenium (I) tricarbonyl center reductions were determined for a family of *fac*-[Re(CO)$_3$(N,N)X]^n compounds; Figure S22: Thermodynamic free energy cycles used in the computation of equilibrium redox potentials for L: (A) L/L^- pair and (B) L^+/L pair; Figure S23: Thermodynamic free energy cycles used in the computation of equilibrium redox potentials for ReL complex: (A) FO/FR pair, (B) Different species (FO and FR) involved in electrochemical steps of redox mechanisms (ReL-ACN = *fac*-[Re(CO)$_3$(5,6-epoxy-5,6-dihydro-1,10-phenanthroline)(ACN)]$^+$); Table S1: Crystal data collection and structure refinement parameters for the ReL compound; Table S2: Bond distances (Å) and angles (°) for the ReL compound; Table S3: Intermolecular hydrogen-bonding interaction parameters for ReL compound. References [112,113] is cited in the supplementary materials.

Author Contributions: Conceptualization, A.C. and M.G.; methodology, A.C., R.M.-G., R.A., L.L.-P., C.V., M.C.O. and M.G.; software, E.A.-G.; validation, A.C. and M.G.; formal analysis, A.C., V.A., E.A.-G. and M.G.; investigation, A.C., V.A., E.A.-G., R.M.-G., R.A., L.L.-P., C.V., M.C.O. and M.G.; resources, A.C.; data curation, A.C. and M.G.; writing—original draft, A.C., E.A.-G., R.M.-G., R.A.,

L.L.-P., A.A.M., C.V., M.C.O. and M.G.; writing—review & editing, A.C., V.A., E.A.-G., R.M.-G., A.A.M., M.C.O. and M.G.; visualization, A.A.M. and M.G.; supervision, A.A.M. and M.G.; funding acquisition, A.C. All authors have read and agreed to the published version of the manuscript.

Funding: Funds were provided by ANID through grants numbers (FONDECYT) 1230917 and (FONDECYT) 1220272.

Institutional Review Board Statement: Not applicable.

Informed Consent Statement: Not applicable.

Data Availability Statement: The raw data supporting the conclusions of this article will be made available by the authors on request.

Acknowledgments: A. Carreño thanks Fondecyt 1230917 (ANID). M. Gacitúa thanks FONDECYT 1220272 (ANID). R. Arce thanks the Millennium Institute on Green Ammonia as Energy Vector MIGA, ANID/Millennium Science Initiative Program/ICN2021_023. A Carreño acknowledges Dayán Páez-Hernández (Center of Applied Nanosciences CANS, Facultad de Ciencias Exactas, Universidad Andres Bello, Chile) for software facilities and Bs. Juan Manuel Ortega Martinez (Fondecyt 1230917) for editing and the English translation. We also thank Poldie Oyarzún from the Laboratorio de Análisis de Sólidos (L.A.S.), Facultad de Ciencias Exactas, Universidad Andrés Bello, for helping with the crystal measurements.

Conflicts of Interest: The authors declare no conflict of interest.

Appendix A

Supplementary data CCDC reference number 2400807 contains the supplementary crystallographic data for this paper. These data can be obtained free of charge at www.ccdc.cam.ac.uk/conts/retrieving.html (accessed on 4 December 2025) [or from the Cambridge Crystallographic Data Centre, 12, Union Road, Cambridge CB2 1EZ, UK; Fax: (internat.) +44 1223/336 033; E-mail: deposit@ccdc.cam.ac.uk.

References

1. Ko, C.-C.; Cheung, A.W.-Y.; Lo, L.T.-L.; Siu, J.W.-K.; Ng, C.-O.; Yiu, S.-M. Syntheses and photophysical studies of new classes of luminescent isocyano rhenium(I) diimine complexes. *Coord. Chem. Rev.* **2012**, *256*, 1546–1555. [CrossRef]
2. Maldonado, T.; Flores, E.; Gómez, A.; Godoy, F.; Mascayano, C.; Martí, A.A.; Ferraudi, G. Kinetic study of the azo—Hydrazone photoinduced mechanism in complexes with a -Re(CO)$_3$L$^{0/+}$ core by flash photolysis. *J. Photochem. Photobiol. A Chem.* **2023**, *442*, 114802. [CrossRef]
3. Manicum, A.-L.E.; Louis, H.; Mathias, G.E.; Agwamba, E.C.; Malan, F.P.; Unimuke, T.O.; Nzondomyo, W.J.; Sithole, S.A.; Biswas, S.; Prince, S. Single crystal investigation, spectroscopic, DFT studies, and in-silico molecular docking of the anticancer activities of acetylacetone coordinated Re(I) tricarbonyl complexes. *Inorganica Chim. Acta* **2023**, *546*, 121335. [CrossRef]
4. Carreño, A.; Solís-Céspedes, E.; Zúñiga, C.; Nevermann, J.; Rivera-Zaldívar, M.M.; Gacitúa, M.; Ramírez-Osorio, A.; Páez-Hernández, D.; Arratia-Pérez, R.; Fuentes, J.A. Cyclic voltammetry, relativistic DFT calculations and biological test of cytotoxicity in walled-cell models of two classical rhenium (I) tricarbonyl complexes with 5-amine-1,10-phenanthroline. *Chem. Phys. Lett.* **2019**, *715*, 231–238. [CrossRef]
5. Wrighton, M. Photochemistry of metal carbonyls. *Chem. Rev.* **1974**, *74*, 401–430. [CrossRef]
6. Wrighton, M.; Morse, D.L. Nature of the lowest excited state in tricarbonylchloro-1,10-phenanthrolinerhenium(I) and related complexes. *J. Am. Chem. Soc.* **1974**, *96*, 998–1003. [CrossRef]
7. Dilworth, J.R. Rhenium chemistry—Then and Now. *Coord. Chem. Rev.* **2021**, *436*, 213822. [CrossRef]
8. Merillas, B.; Cuéllar, E.; Diez-Varga, A.; Torroba, T.; García-Herbosa, G.; Fernández, S.; Lloret-Fillol, J.; Martín-Alvarez, J.M.; Miguel, D.; Villafañe, F. Luminescent Rhenium(I)tricarbonyl Complexes Containing Different Pyrazoles and Their Successive Deprotonation Products: CO_2 Reduction Electrocatalysts. *Inorg. Chem.* **2020**, *59*, 11152–11165. [CrossRef]
9. Cuéllar, E.; Pastor, L.; García-Herbosa, G.; Nganga, J.; Angeles-Boza, A.M.; Diez-Varga, A.; Torroba, T.; Martín-Alvarez, J.M.; Miguel, D.; Villafañe, F. (1,2-Azole)bis(bipyridyl)ruthenium(II) Complexes: Electrochemistry, Luminescent Properties, And Electro- And Photocatalysts for CO_2 Reduction. *Inorg. Chem.* **2021**, *60*, 692–704. [CrossRef]

https://doi.org/10.3390/inorganics14010003

10. Gómez–Arteaga, B.; Gómez, A.; Flores, E.; Vega, A.; Cruz–Piñones, B.; Godoy, F.; Maldonado, T. Revealing the Unusual behavior of rhenium (I) tricarbonyl complex functionalized with Aza-Macrocycles in response to metal ions. *Inorganica Chim. Acta* **2025**, *574*, 122389. [CrossRef]
11. Choroba, K.; Penkala, M.; Palion-Gazda, J.; Malicka, E.; Machura, B. Pyrenyl-Substituted Imidazo[4,5-*f*][1,10]phenanthroline Rhenium(I) Complexes with Record-High Triplet Excited-State Lifetimes at Room Temperature: Steric Control of Photoinduced Processes in Bichromophoric Systems. *Inorg. Chem.* **2023**, *62*, 19256–19269. [CrossRef]
12. Thanasekaran, P.; Huang, J.-H.; Jhou, C.-R.; Tsao, H.-C.; Mendiratta, S.; Su, C.-H.; Liu, C.-P.; Liu, Y.-H.; Huang, J.-H.; Lu, K.-L. A neutral mononuclear rhenium(*I*) complex with a rare *in situ* -generated triazolyl ligand for the luminescence "turn-on" detection of histidine. *Dalton Trans.* **2023**, *52*, 703–709. [CrossRef]
13. Diksha; Somasundaram, M.; Ganeshan, M.; Samal, S.K.; Dharumadurai, D.; Madrahimov, S.; Deshwal, A.; Kaur, H.; Sinopoli, A.; Yempally, V. Synthesis and structural elucidation of cytotoxic mono and dinuclear rhenium carbonyl complexes bearing bis-{1,3-(imino pyrrolyl)-m-chloro phenyl)} ligand. *J. Mol. Struct.* **2025**, *1319*, 139506. [CrossRef]
14. Bertrand, H.C.; Clède, S.; Guillot, R.; Lambert, F.; Policar, C. Luminescence Modulations of Rhenium Tricarbonyl Complexes Induced by Structural Variations. *Inorg. Chem.* **2014**, *53*, 6204–6223. [CrossRef]
15. Palion-Gazda, J.; Choroba, K.; Maroń, A.M.; Malicka, E.; Machura, B. Structural and Photophysical Trends in Rhenium(I) Carbonyl Complexes with 2,2′:6′,2″-Terpyridines. *Molecules* **2024**, *29*, 1631. [CrossRef]
16. Szlapa-Kula, A.; Małecka, M.; Maroń, A.M.; Janeczek, H.; Siwy, M.; Schab-Balcerzak, E.; Szalkowski, M.; Maćkowski, S.; Pedzinski, T.; Erfurt, K.; et al. In-Depth Studies of Ground- and Excited-State Properties of Re(I) Carbonyl Complexes Bearing 2,2′:6′,2″-Terpyridine and 2,6-Bis(pyrazin-2-yl)pyridine Coupled with π-Conjugated Aryl Chromophores. *Inorg. Chem.* **2021**, *60*, 18726–18738. [CrossRef] [PubMed]
17. Alka, A.; Shetti, V.S.; Ravikanth, M. Coordination chemistry of expanded porphyrins. *Coord. Chem. Rev.* **2019**, *401*, 213063. [CrossRef]
18. Palion-Gazda, J.; Szłapa-Kula, A.; Penkala, M.; Erfurt, K.; Machura, B. Photoinduced Processes in Rhenium(I) Terpyridine Complexes Bearing Remote Amine Groups: New Insights from Transient Absorption Spectroscopy. *Molecules* **2022**, *27*, 7147. [CrossRef]
19. Schindler, K.; Horner, J.; Demirci, G.; Cortat, Y.; Crochet, A.; Mamula Steiner, O.; Zobi, F. In Vitro Biological Activity of α-Diimine Rhenium Dicarbonyl Complexes and Their Reactivity with Different Functional Groups. *Inorganics* **2023**, *11*, 139. [CrossRef]
20. Thorp-Greenwood, F.L. An Introduction to Organometallic Complexes in Fluorescence Cell Imaging: Current Applications and Future Prospects. *Organometallics* **2012**, *31*, 5686–5692. [CrossRef]
21. Sorsche, D.; Lima, M.A.L.; Meitinger, N.; Prasad, K.; Mandal, S.; Glusac, K.D.; Rau, S.; Pannwitz, A. Shifting the MLCT of d6 metal complexes to the red and NIR. *Coord. Chem. Rev.* **2025**, *530*, 216454. [CrossRef]
22. Diksha; Kaur, M.; Megha; Reenu; Kaur, H.; Yempally, V. Rhenium (I) tricarbonyl complex with thiosemicarbazone ligand derived from Indole-2-carboxaldehyde: Synthesis, crystal structure, computational investigations, antimicrobial activity, and molecular docking studies. *J. Mol. Struct.* **2024**, *1301*, 137319. [CrossRef]
23. Bakir, M.; Green, O. Rhenium tricarbonyl chloro of di-2-pyridylketone 4-aminobenzoyl hydrazone (dpk4abh), *fac*-[Re(CO)$_3$(κ^2-N,N-dpk4abh)Cl]: Synthesis, spectroscopic and electrochemical properties. *J. Mol. Struct.* **2011**, *996*, 24–30. [CrossRef]
24. Wenger, O.S. Proton-coupled electron transfer with photoexcited ruthenium(II), rhenium(I), and iridium(III) complexes. *Coord. Chem. Rev.* **2015**, *282–283*, 150–158. [CrossRef]
25. Murphy, B.L.; Marker, S.C.; Lambert, V.J.; Woods, J.J.; MacMillan, S.N.; Wilson, J.J. Synthesis, characterization, and biological properties of rhenium(I) tricarbonyl complexes bearing nitrogen-donor ligands. *J. Organomet. Chem.* **2020**, *907*, 121064. [CrossRef]
26. Otero, C.; Carreño, A.; Polanco, R.; Llancalahuen, F.M.; Arratia-Pérez, R.; Gacitúa, M.; Fuentes, J.A. Rhenium (I) Complexes as Probes for Prokaryotic and Fungal Cells by Fluorescence Microscopy: Do Ligands Matter? *Front. Chem.* **2019**, *7*, 461605. [CrossRef]
27. Capulín Flores, L.; Paul, L.A.; Siewert, I.; Havenith, R.; Zúñiga-Villarreal, N.; Otten, E. Neutral Formazan Ligands Bound to the *fac*-(CO)$_3$ Re(I) Fragment: Structural, Spectroscopic, and Computational Studies. *Inorg. Chem.* **2022**, *61*, 13532–13542. [CrossRef]
28. Manav, N.; Janaagal, A.; Gupta, I. Unveiling new horizons: Exploring rhenium and iridium dipyrrinato complexes as luminescent theranostic agents for phototherapy. *Coord. Chem. Rev.* **2024**, *511*, 215798. [CrossRef]
29. Enslin, L.E.; Purkait, K.; Pozza, M.D.; Saubamea, B.; Mesdom, P.; Visser, H.G.; Gasser, G.; Schutte-Smith, M. Rhenium(I) Tricarbonyl Complexes of 1,10-Phenanthroline Derivatives with Unexpectedly High Cytotoxicity. *Inorg. Chem.* **2023**, *62*, 12237–12251. [CrossRef]
30. Liew, H.S.; Mai, C.-W.; Zulkefeli, M.; Madheswaran, T.; Kiew, L.V.; Delsuc, N.; Low, M.L. Recent Emergence of Rhenium(I) Tricarbonyl Complexes as Photosensitisers for Cancer Therapy. *Molecules* **2020**, *25*, 4176. [CrossRef]
31. Rotundo, L.; Grills, D.C.; Gobetto, R.; Priola, E.; Nervi, C.; Polyansky, D.E.; Fujita, E. Photochemical CO_2 Reduction Using Rhenium(I) Tricarbonyl Complexes with Bipyridyl-Type Ligands with and without Second Coordination Sphere Effects. *ChemPhotoChem* **2021**, *5*, 526–537. [CrossRef]

32. Feng, Y.; Cheng, S.-C.; Ko, C.-C. Luminescent rhenium(I) complex with cyanopentafluorophosphate ligand: Synthesis, characterization, and photophysics. *J. Organomet. Chem.* **2023**, *1001*, 122882. [CrossRef]
33. Carreño, A.; Páez-Hernández, D.; Zúñiga, C.; Ramírez-Osorio, A.; Pizarro, N.; Vega, A.; Solis-Céspedes, E.; Rivera-Zaldívar, M.M.; Silva, A.; Fuentes, J.A. Exploring rhenium (I) complexes as potential fluorophores for walled-cells (yeasts and bacteria): Photophysics, biocompatibility, and confocal microscopy. *Dye. Pigment.* **2021**, *184*, 108876. [CrossRef]
34. Zamora, P.P.; Bieger, K.; Cuchillo, A.; Tello, A.; Muena, J.P. Theoretical determination of a reaction intermediate: Fukui function analysis, dual reactivity descriptor and activation energy. *J. Mol. Struct.* **2021**, *1227*, 129369. [CrossRef]
35. Novoa, T.; Laplaza, R.; Peccati, F.; Fuster, F.; Contreras-García, J. The NCIWEB Server: A Novel Implementation of the Noncovalent Interactions Index for Biomolecular Systems. *J. Chem. Inf. Model.* **2023**, *63*, 4483–4489. [CrossRef]
36. Carreño, A.; Solis-Céspedes, E.; Páez-Hernández, D.; Arratia-Pérez, R. Exploring the geometrical and optical properties of neutral rhenium (I) tricarbonyl complex of 1,10-phenanthroline-5,6-diol using relativistic methods. *Chem. Phys. Lett.* **2017**, *685*, 354–362. [CrossRef]
37. El Nahhas, A.; Consani, C.; Blanco-Rodríguez, A.M.a.; Lancaster, K.M.; Braem, O.; Cannizzo, A.; Towrie, M.; Clark, I.P.; Zális, S.; Chergui, M.; et al. Ultrafast Excited-State Dynamics of Rhenium(I) Photosensitizers [Re(Cl)(CO)$_3$(N,N)] and [Re(imidazole)(CO)$_3$(N,N)]$^+$: Diimine Effects. *Inorg. Chem.* **2011**, *50*, 2932–2943. [CrossRef] [PubMed]
38. Martí, A.A.; Mezei, G.; Maldonado, L.; Paralitici, G.; Raptis, R.G.; Colón, J.L. Structural and Photophysical Characterisation of *fac*-Tricarbonyl(chloro)(5,6-epoxy-1,10-phenanthroline)rhenium(I)]. *Eur. J. Inorg. Chem.* **2005**, *2005*, 118–124. [CrossRef]
39. Kearney, L.; Brandon, M.P.; Coleman, A.; Chippindale, A.M.; Hartl, F.; Lalrempuia, R.; Pižl, M.; Pryce, M.T. Ligand-Structure Effects on N-Heterocyclic Carbene Rhenium Photo- and Electrocatalysts of CO$_2$ Reduction. *Molecules* **2023**, *28*, 4149. [CrossRef]
40. Akman, F.; Demirpolat, A.; Kazachenko, A.S.; Kazachenko, A.S.; Issaoui, N.; Al-Dossary, O. Molecular Structure, Electronic Properties, Reactivity (ELF, LOL, and Fukui), and NCI-RDG Studies of the Binary Mixture of Water and Essential Oil of Phlomis bruguieri. *Molecules* **2023**, *28*, 2684. [CrossRef] [PubMed]
41. Bultinck, P.; Clarisse, D.; Ayers, P.W.; Carbo-Dorca, R. The Fukui matrix: A simple approach to the analysis of the Fukui function and its positive character. *Phys. Chem. Chem. Phys.* **2011**, *13*, 6110. [CrossRef]
42. Bultinck, P.; Van Neck, D.; Acke, G.; Ayers, P.W. Influence of electron correlation and degeneracy on the Fukui matrix and extension of frontier molecular orbital theory to correlated quantum chemical methods. *Phys. Chem. Chem. Phys.* **2012**, *14*, 2408. [CrossRef]
43. Senthilkumar, K.; Ramaswamy, M.; Kolandaivel, P. Studies of chemical hardness and Fukui function using the exact solution of the density functional theory. *Int. J. Quantum Chem.* **2001**, *81*, 4–10. [CrossRef]
44. Senthilkumar, L.; Kolandaivel, P. Study of effective hardness and condensed Fukui functions using AIM, *ab initio*, and DFT methods. *Mol. Phys.* **2005**, *103*, 547–556. [CrossRef]
45. Reddy, T.S.; Raja, K.; Mandapati, K.R.; Goli, S.R.; Babu, M.S.S. Efficient Approach for the Synthesis of Aryl Vinyl Ketones and Its Synthetic Application to Mimosifoliol with DFT and Autodocking Studies. *Molecules* **2023**, *28*, 6214. [CrossRef]
46. Sahoo, G.P.; Das, D.; Sheet, P.S.; Beg, H.; Salgado-Morán, G.; Misra, A. Morphology directing synthesis of 1-pyrene carboxaldehyde microstructures and their photo physical properties. *RSC Adv.* **2014**, *4*, 10903. [CrossRef]
47. Cornaton, Y.; Djukic, J.-P. Noncovalent Interactions in Organometallic Chemistry: From Cohesion to Reactivity, a New Chapter. *Acc. Chem. Res.* **2021**, *54*, 3828–3840. [CrossRef]
48. Fukin, G.K.; Cherkasov, A.V.; Zarovkina, N.Y.; Artemov, A.N. Experimental and Theoretical AIM and NCI Index Study of Substituted Arene Tricarbonyl Complexes of Chromium(0). *ChemistrySelect* **2016**, *1*, 5014–5018. [CrossRef]
49. Solà, M.; Ottosson, H.; Baranac, M.; Stojanovi'c, S.c. Aromaticity and Antiaromaticity: How to Define Them. *Chemistry* **2025**, *7*, 127. [CrossRef]
50. Sacanamboy, D.S.; García-Argote, W.; Pumachagua-Huertas, R.; Cárdenas, C.; Leyva-Parra, L.; Ruiz, L.; Tiznado, W. Interpreting Ring Currents from Hückel-Guided σ- and π-Electron Delocalization in Small Boron Rings. *Molecules* **2025**, *30*, 3566. [CrossRef] [PubMed]
51. Lin, Q.; Dawson, G.; Diao, T. Experimental Electrochemical Potentials of Nickel Complexes. *Synlett* **2021**, *32*, 1606–1620. [CrossRef]
52. Meitinger, N.; Mandal, S.; Sorsche, D.; Pannwitz, A.; Rau, S. Red Light Absorption of [ReI(CO)3(α-diimine)Cl] Complexes through Extension of the 4,4′-Bipyrimidine Ligand's π-System. *Molecules* **2023**, *28*, 1905. [CrossRef] [PubMed]
53. Carreño, A.; Gacitua, M.; Schott, E.; Zarate, X.; Manriquez, J.M.; Preite, M.; Ladeira, S.; Castel, A.; Pizarro, N.; Vega, A.; et al. Experimental and theoretical studies of the ancillary ligand (E)-2-((3-amino-pyridin-4-ylimino)-methyl)-4,6-di-tert-butylphenol in the rhenium(I) core. *New J. Chem.* **2015**, *39*, 5725–5734. [CrossRef]
54. Zhou, Q.-H.; Pan, M.-Y.; He, Q.; Tang, Q.; Chow, C.-F.; Gong, C.-B. Electrochromic behavior of *fac*-tricarbonyl rhenium complexes. *New J. Chem.* **2022**, *46*, 1072–1079. [CrossRef]
55. Rotundo, L.; Polyansky, D.E.; Gobetto, R.; Grills, D.C.; Fujita, E.; Nervi, C.; Manbeck, G.F. Molecular Catalysts with Intramolecular Re–O Bond for Electrochemical Reduction of Carbon Dioxide. *Inorg. Chem.* **2020**, *59*, 12187–12199. [CrossRef]

https://doi.org/10.3390/inorganics14010003

56. Lam, S.-T.; Zhu, N.; Au, V.K.-M.; Yam, V.W.-W. Synthesis, characterization, electrochemistry and photophysical studies of rhenium(I) tricarbonyl diimine complexes with carboxaldehyde alkynyl ligands. *Polyhedron* **2015**, *86*, 10–16. [CrossRef]
57. Carreño, A.; Gacitúa, M.; Fuentes, J.A.; Páez-Hernández, D.; Peñaloza, J.P.; Otero, C.; Preite, M.; Molins, E.; Swords, W.B.; Meyer, G.J.; et al. Fluorescence probes for prokaryotic and eukaryotic cells using $Re(CO)^{3+}$ complexes with an electron withdrawing ancillary ligand. *New J. Chem.* **2016**, *40*, 7687–7700. [CrossRef]
58. Carreño, A.; Gacitúa, M.; Molins, E.; Arratia-Pérez, R. X-ray diffraction and relativistic DFT studies on the molecular biomarker fac-Re(CO)3(4,4′-dimethyl-2,2′-bpy)(E-2-((3-amino-pyridin-4-ylimino)-methyl)-4,6-di-tert-butylphenol)(PF6). *Chem. Pap.* **2017**, *71*, 2011–2022. [CrossRef]
59. Czerwieniec, R.; Kapturkiewicz, A.; Lipkowski, J.; Nowacki, J. Re(I)(tricarbonyl)+ complexes with the 2-(2-pyridyl)-N-methyl-benzimidazole, 2-(2-pyridyl)benzoxazole and 2-(2-pyridyl)benzothiazole ligands—Syntheses, structures, electrochemical and spectroscopic studies. *Inorganica Chim. Acta* **2005**, *358*, 2701–2710. [CrossRef]
60. Świtlicka, A.; Klemens, T.; Machura, B.; Schab-Balcerzak, E.; Laba, K.; Lapkowski, M.; Grucela, M.; Nycz, J.; Szala, M.; Kania, M. Rhenium(I) complexes with phenanthrolines bearing electron-withdrawing Cl and electron-donating CH_3 substituents—Synthesis, photophysical, thermal, and electrochemical properties with electroluminescence ability. *RSC Adv.* **2016**, *6*, 112908–112918. [CrossRef]
61. Lo, K.K.-W.; Lau, J.S.-Y.; Fong, V.W.-Y.; Zhu, N. Electrochemical, Photophysical, and Anion-Binding Properties of a Luminescent Rhenium(I) Polypyridine Anthraquinone Complex with a Thiourea Receptor. *Organometallics* **2004**, *23*, 1098–1106. [CrossRef]
62. Costentin, C.; Limoges, B.; Robert, M.; Tard, C. A Pioneering Career in Electrochemistry: Jean-Michel Savéant. *ACS Catal.* **2021**, *11*, 3224–3238. [CrossRef]
63. Sun, Z.; Ma, T.; Tao, H.; Fan, Q.; Han, B. Fundamentals and Challenges of Electrochemical CO_2 Reduction Using Two-Dimensional Materials. *Chem* **2017**, *3*, 560–587. [CrossRef]
64. Chen, J.; Du, X.; Yu, T.; Zeng, Y.; Zhang, X.; Li, Y. Ligand Substituent Effects on Rhenium Tricarbonyl Catalysts in CO_2 Reduction. *Acta Chim. Sin.* **2016**, *74*, 523. [CrossRef]
65. Liang, Y.; Nguyen, M.T.; Holliday, B.J.; Jones, R.A. Electrocatalytic reduction of CO_2 using rhenium complexes with dipyrido[3,2-a:2′,3′-c]phenazine ligands. *Inorg. Chem. Commun.* **2017**, *84*, 113–117. [CrossRef]
66. Talukdar, K.; Sinha Roy, S.; Amatya, E.; Sleeper, E.A.; Le Magueres, P.; Jurss, J.W. Enhanced Electrochemical CO_2 Reduction by a Series of Molecular Rhenium Catalysts Decorated with Second-Sphere Hydrogen-Bond Donors. *Inorg. Chem.* **2020**, *59*, 6087–6099. [CrossRef]
67. Towett, E.K.; Tembu, V.J.; Kemboi, D.; Langat, M.K.; Manicum, A.-L.E. Review of Recent Medicinal Applications of Rhenium(I) Tricarbonyl Complexes. *Int. J. Mol. Sci.* **2025**, *26*, 7005. [CrossRef]
68. Schindler, K.; Zobi, F. Anticancer and Antibiotic Rhenium Tri- and Dicarbonyl Complexes: Current Research and Future Perspectives. *Molecules* **2022**, *27*, 539. [CrossRef]
69. Ajay Sharma, S.; Vaibhavi, N.; Binoy, K.; Utpal, D.; Priyankar, P. Target-specific mononuclear and binuclear rhenium(i) tricarbonyl complexes as upcoming anticancer drugs. *RSC Adv.* **2022**, *12*, 20264–20295. [CrossRef]
70. Ramoba, L.V.; Macharia, L.W.; Chakraborty, S.; Prince, S.; Malan, F.P.; Alexander, O.T.; Manicum, A.-L.E. Target-specific Rhenium(I) Tricarbonyl Complexes as Prospective Pharmacological Agents: Synthesis, X-ray Crystallography, and In Vitro Anticancer Evaluation. *J. Mol. Struct.* **2025**, *1350*, 144045. [CrossRef]
71. Qurishi, Y.; Seshadri, V.D.; Poyil, M.M.; Franklin, J.B.; Apte, D.A.; Alsharif, M.H.K.; Rajesh, R.P.; Zaki, R.M.; Qurishi, Y.; Seshadri, V.D.; et al. Anticancer activity in HeLa and MCF-7 cells via apoptopic cell death by a sterol molecule Cholesta-4,6-dien-3-ol (EK-7), from the marine ascidian Eudistoma kaverium. *J. King Saud Univ.—Sci.* **2021**, *33*, 101418. [CrossRef]
72. Ghasemi, M.; Turnbull, T.; Sebastian, S.; Kempson, I. The mtt assay: Utility, limitations, pitfalls, and interpretation in bulk and single-cell analysis. *Int. J. Mol. Sci.* **2021**, *22*, 12827. [CrossRef]
73. Knopf, K.M.; Murphy, B.L.; MacMillan, S.N.; Baskin, J.M.; Barr, M.P.; Boros, E.; Wilson, J.J. In Vitro Anticancer Activity and in Vivo Biodistribution of Rhenium(I) Tricarbonyl Aqua Complexes. *J. Am. Chem. Soc.* **2017**, *139*, 14302–14314. [CrossRef]
74. Mansour, A.M.; Ibrahim, N.M.; Farag, A.M.; Abo-Elfadl, M.T. Evaluation of cytotoxic properties of two fluorescent *fac*-$Re(CO)_3$ complexes bearing an *N,N*-bidentate benzimidazole coligand. *RSC Adv.* **2022**, *12*, 30829–30837. [CrossRef] [PubMed]
75. Marco, A.; Ashoo, P.; Hernández-García, S.; Martínez-Rodríguez, P.; Cutillas, N.; Vollrath, A.; Jordan, D.; Janiak, C.; Gandía-Herrero, F.; Ruiz, J. Novel Re(I) Complexes as Potential Selective Theranostic Agents in Cancer Cells and <i>In Vivo</i> in *Caenorhabditis elegans* Tumoral Strains. *J. Med. Chem.* **2024**, *67*, 7891–7910. [CrossRef] [PubMed]
76. Pivarcsik, T.; Kljun, J.; Clemente Rodriguez, S.; Cortéz Alcaraz, D.; Rapuš, U.; Nové, M.; Várkonyi, E.F.; Nyári, J.; Bogdanov, A.; Spengler, G.; et al. Structural and Solution Speciation Studies on fac-Tricarbonylrhenium(I) Complexes of 2,2′-Bipyridine Analogues. *ACS Omega* **2024**, *9*, 44601–44615. [CrossRef]
77. Sheldrick, G.M. A short history of *SHELX*. *Acta Crystallogr. Sect. A Found. Crystallogr.* **2008**, *64*, 112–122. [CrossRef]
78. Wood, P.A.; McKinnon, J.J.; Parsons, S.; Pidcock, E.; Spackman, M.A. Analysis of the compression of molecular crystal structures using Hirshfeld surfaces. *CrystEngComm* **2008**, *10*, 368–376. [CrossRef]

https://doi.org/10.3390/inorganics14010003

79. Bakavoli, M.; Rahimizadeh, M.; Feizyzadeh, B.; Kaju, A.A.; Takjoo, R. 3,6-Di(p-chlorophenyl)-2,7-dihydro-1,4,5-thiadiazepine: Crystal Structure and Decoding Intermolecular Interactions with Hirshfeld Surface Analysis. *J. Chem. Crystallogr.* **2010**, *40*, 746–752. [CrossRef]
80. McKinnon, J.J.; Mitchell, A.S.; Spackman, M.A. Hirshfeld Surfaces: A New Tool for Visualising and Exploring Molecular Crystals. *Chem.—A Eur. J.* **1998**, *4*, 2136–2141. [CrossRef]
81. McKinnon, J.J.; Spackman, M.A.; Mitchell, A.S. Novel tools for visualizing and exploring intermolecular interactions in molecular crystals. *Acta Crystallogr. Sect. B Struct. Sci.* **2004**, *60*, 627–668. [CrossRef]
82. Spackman, P.R.; Turner, M.J.; McKinnon, J.J.; Wolff, S.K.; Grimwood, D.J.; Jayatilaka, D.; Spackman, M.A. *CrystalExplorer*: A program for Hirshfeld surface analysis, visualization and quantitative analysis of molecular crystals. *J. Appl. Crystallogr.* **2021**, *54*, 1006–1011. [CrossRef] [PubMed]
83. Parkin, A.; Barr, G.; Dong, W.; Gilmore, C.J.; Jayatilaka, D.; McKinnon, J.J.; Spackman, M.A.; Wilson, C.C. Comparing entire crystal structures: Structural genetic fingerprinting. *CrystEngComm* **2007**, *9*, 648. [CrossRef]
84. Spackman, M.A.; McKinnon, J.J. Fingerprinting intermolecular interactions in molecular crystals. *CrystEngComm* **2002**, *4*, 378–392. [CrossRef]
85. Allen, F.H.; Kennard, O.; Watson, D.G.; Brammer, L.; Orpen, A.G.; Taylor, R. Tables of bond lengths determined by X-ray and neutron diffraction. Part 1. Bond lengths in organic compounds. *J. Chem. Soc. Perkin Trans. II* **1987**, S1–S19. [CrossRef]
86. East, G.A.; del Valle, M.A. Easy-to-Make Ag/AgCl Reference Electrode. *J. Chem. Educ.* **2000**, *77*, 97. [CrossRef]
87. Neese, F. The ORCA program system. *WIREs Comput. Mol. Sci.* **2012**, *2*, 73–78. [CrossRef]
88. Klamt, A.; Jonas, V.; Bürger, T.; Lohrenz, J.C.W. Refinement and Parametrization of COSMO-RS. *J. Phys. Chem. A* **1998**, *102*, 5074–5085. [CrossRef]
89. Lange, A.W.; Herbert, J.M. A simple polarizable continuum solvation model for electrolyte solutions. *J. Chem. Phys.* **2011**, *134*, 204110. [CrossRef]
90. Orio, M.; Pantazis, D.A.; Neese, F. Density functional theory. *Photosynth. Res.* **2009**, *102*, 443–453. [CrossRef]
91. Hirao, H. Which DFT Functional Performs Well in the Calculation of Methylcobalamin? Comparison of the B3LYP and BP86 Functionals and Evaluation of the Impact of Empirical Dispersion Correction. *J. Phys. Chem. A* **2011**, *115*, 9308–9313. [CrossRef] [PubMed]
92. Rolfes, J.D.; Neese, F.; Pantazis, D.A. All-electron scalar relativistic basis sets for the elements Rb–Xe. *J. Comput. Chem.* **2020**, *41*, 1842–1849. [CrossRef]
93. Trujillo, C.; Flood, A.; Sánchez-Sanz, G.; Twamley, B.; Rozas, I. Planarity or Nonplanarity: Modulating Guanidine Derivatives as α_2-Adrenoceptors Ligands. *J. Chem. Inf. Model.* **2019**, *59*, 2479–2486. [CrossRef]
94. Villegas-Escobar, N.; Schaefer, H.F.; Toro-Labbé, A. Formation of Formic Acid Derivatives through Activation and Hydroboration of CO_2 by Low-Valent Group 14 (Si, Ge, Sn, Pb) Catalysts. *J. Phys. Chem. A* **2020**, *124*, 1121–1133. [CrossRef]
95. Lu, T.; Chen, F. Multiwfn: A multifunctional wavefunction analyzer. *J. Comput. Chem.* **2012**, *33*, 580–592. [CrossRef] [PubMed]
96. Humphrey, W.; Dalke, A.; Schulten, K. VMD: Visual molecular dynamics. *J. Mol. Graph.* **1996**, *14*, 33–38. [CrossRef]
97. Jusélius, J.; Sundholm, D.; Gauss, J. Calculation of current densities using gauge-including atomic orbitals. *J. Chem. Phys.* **2004**, *121*, 3952–3963. [CrossRef]
98. Fliegl, H.; Taubert, S.; Lehtonen, O.; Sundholm, D. The gauge including magnetically induced current method. *Phys. Chem. Chem. Phys.* **2011**, *13*, 20500–20518. [CrossRef]
99. Wolinski, K.; Hinton, J.F.; Pulay, P. Efficient implementation of the gauge-independent atomic orbital method for NMR chemical shift calculations. *J. Am. Chem. Soc.* **2002**, *112*, 8251–8260. [CrossRef]
100. Ahrens, J.; Geveci, B.; Law, C. ParaView: An End-User Tool for Large-Data Visualization. In *Visualization Handbook*; Elsevier: Amsterdam, The Netherlands, 2005; pp. 717–731. [CrossRef]
101. Becke, A.D. A new mixing of Hartree–Fock and local density-functional theories. *J. Chem. Phys.* **1993**, *98*, 1372–1377. [CrossRef]
102. Weigand, A.; Cao, X.; Hangele, T.; Dolg, M. Relativistic Small-Core Pseudopotentials for Actinium, Thorium, and Protactinium. *J. Phys. Chem. A* **2014**, *118*, 2519–2530. [CrossRef]
103. Weigend, F.; Ahlrichs, R. Balanced basis sets of split valence, triple zeta valence and quadruple zeta valence quality for H to Rn: Design and assessment of accuracy. *Phys. Chem. Chem. Phys.* **2005**, *7*, 3297–3305. [CrossRef]
104. Lehtola, S.; Dimitrova, M.; Fliegl, H.; Sundholm, D. Benchmarking Magnetizabilities with Recent Density Functionals. *J. Chem. Theory Comput.* **2021**, *17*, 1457–1468. [CrossRef]
105. Szczepanik, D.W. A new perspective on quantifying electron localization and delocalization in molecular systems. *Comput. Theor. Chem.* **2016**, *1080*, 33–37. [CrossRef]
106. Rojas-Poblete, M.; Carreño, A.; Gacitúa, M.; Páez-Hernández, D.; Rabanal-León, W.A.; Arratia-Pérez, R. Electrochemical behaviors and relativistic DFT calculations to understand the terminal ligand influence on the $[Re_6(\mu_3\text{-}Q)_8X_6]^{4-}$ clusters. *New J. Chem.* **2018**, *42*, 5471–5478. [CrossRef]

https://doi.org/10.3390/inorganics14010003

107. Sebens, C.T. Electron Charge Density: A Clue from Quantum Chemistry for Quantum Foundations. *Found. Phys.* **2021**, *51*, 75. [CrossRef]
108. Psciuk, B.T.; Schlegel, H.B. Computational Prediction of One-Electron Reduction Potentials and Acid Dissociation Constants for Guanine Oxidation Intermediates and Products. *J. Phys. Chem. B* **2013**, *117*, 9518–9531. [CrossRef] [PubMed]
109. Sundstrom, E.J.; Yang, X.; Thoi, V.S.; Karunadasa, H.I.; Chang, C.J.; Long, J.R.; Head-Gordon, M. Computational and Experimental Study of the Mechanism of Hydrogen Generation from Water by a Molecular Molybdenum-Oxo Electrocatalyst. *J. Am. Chem. Soc.* **2012**, *134*, 5233–5242. [CrossRef]
110. Agbangba, C.E.; Aide, E.S.; Honfo, H.; Kakai, R.G. On the use of post-hoc tests in environmental and biological sciences: A critical review. *Heliyon* **2024**, *10*, e25131. [CrossRef] [PubMed]
111. Juarros-Basterretxea, J.; Aonso-Diego, G.; Postigo, Á.; Montes-Álvarez, P.; Menéndez-Aller, Á.; García-Cueto, E. Post-Hoc Tests in One-Way ANOVA: The Case for Normal Distribution. *Methodology* **2024**, *20*, 84–99. [CrossRef]
112. Spackman, M.A.; Jayatilaka, D. Hirshfeld Surface Analysis. *CrystEngComm* **2009**, *11*, 19–32. [CrossRef]
113. Carreño, A.; Gacitúa, M.; Solis-Céspedes, E.; Páez-Hernández, D.; Swords, W.; Meyer, G.; Preite, M.; Chávez, I.; Vega, A.; Fuentes, J. New Cationic fac-$[Re(CO)_3(deeb)B2]^+$ Complex, Where B2 Is aBenzimidazole Derivative, as aPotential New Luminescent Dye forProteins Separated by SDS-PAGE. *Front. Chem.* **2021**, *9*, 1–13. [CrossRef]

https://doi.org/10.3390/inorganics14010003

inorganics

MDPI

Article

Degradation of the Vaccine Additive Thimerosal by L-Glutathione and L-Cysteine at Physiological pH

Manon Fanny Degorge, Silas Mertz and Jürgen Gailer *

Department of Chemistry, University of Calgary, 2500 University Drive NW, Calgary, AB T2N 1N4, Canada
* Correspondence: jgailer@ucalgary.ca

Abstract

Humans are being exposed to a variety of potentially toxic metal compounds through the diet and/or the intravenous administration of metal-containing medicinal drugs. The organomercurial thimerosal (THI) is a bactericidal that is present in vaccines, but its potential degradation by biomolecules in vivo is incompletely understood. To probe its interaction with low-molecular-weight thiols that are highly abundant within cells, we have employed an LC-based analytical approach in conjunction with a mercury-specific detector. The injection of THI into a C_{18}-HPLC column equilibrated with mobile phases that contained increasing concentrations of up to 15 mM of glutathione (GSH) and 30% acetonitrile revealed the elution of a GS-EtHg adduct in conjunction with THI, as evidenced by electrospray ionization mass spectrometry. These results were confirmed by ^{199}Hg-NMR spectroscopy. While these results imply a rapid degradation of THI by GSH at physiological pH, it is important to point out that our results were obtained in aqueous solutions containing 30% (v:v) acetonitrile. Further studies need to confirm if the GS-EtHg adduct is also formed in biological fluids. Our results nevertheless demonstrate that GSH and L-cysteine (Cys) are potential targets of THI at physiological pH, which is relevant to better understand its side effects, including previously reported effects on Ca^{2+} channels.

Keywords: bioinorganic chemistry; reversed-phase-HPLC; flame atomic absorption spectrometry; electrospray ionization mass spectrometry; ^{199}Hg NMR spectroscopy

1. Introduction

The exposure of mammals to metal (loid) species from the environment [1–3] as well as medicinal drugs that contain metals [4], such as the organomercury compound thimerosal (THI) (Figure 1A) is being increasingly recognized as a major public health concern due to the associated potential short and long-term toxic effects [4]. Since potentially toxic metal species will invade the bloodstream, bioinorganic reactions therein play a fundamental role in the context of better understanding their toxicological chemistry. The interaction of toxic metal species with biomolecules in the bloodstream and/or organs may involve their binding to proteins as well as their interaction with cytosolic biomolecules [4]. As these latter interactions can contribute to their long-term adverse health effects at the organ level, they need to be better understood in order to evaluate their involvement in the etiology of adverse human health effects. THI has been shown to remain in the bloodstream for days after its administration to mice [5]. Since the red blood cell (RBC) membrane can be easily penetrated by a variety of toxic metal species [6] and since RBC cytosol contains ~2.5 mM of glutathione (GSH, Figure 1C), the interaction of THI with GSH therein is among the first interactions that are of potential human health relevance [7].

https://doi.org/10.3390/inorganics13090280

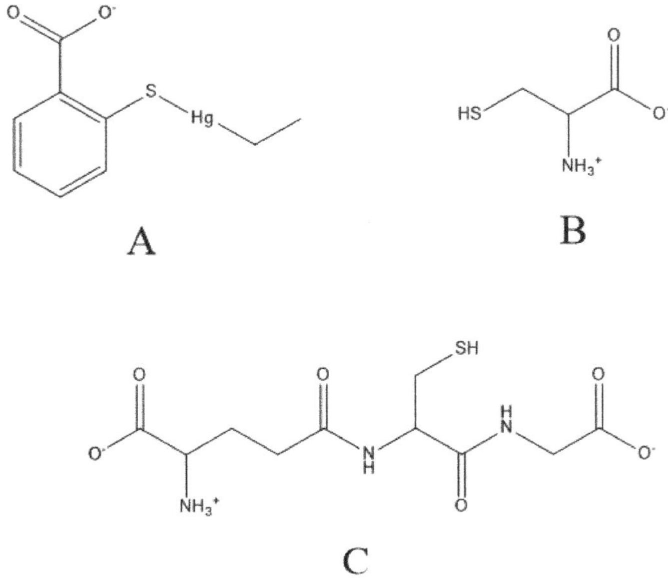

Figure 1. Chemical structure of thimerosal (**A**), L-cysteine (**B**), and L-glutathione (**C**) at physiological pH.

THI has been widely used as a preservative in many biological and pharmaceutical products, including vaccines, since the 1930s [8,9]. THI is a potent bactericidal, has antifungal activity [10], and effectively inhibits the SARS-CoV2 main protease [11]. While THI remains one of the most widely used preservatives in multidose vaccines, particularly in low-resource countries, this organomercury compound has been the subject of controversy because of concerns about a possible link between human exposure to thimerosal in vaccines and autism [12]. From a bioinorganic chemistry point of view, rather few studies have investigated the interaction of THI with thiols. While some studies have demonstrated the binging of a THI-degradation product to hemoglobin using mass spectrometry [13] and metallomics approaches [7], no studies have investigated its reaction with GSH at physiological pH. The related organomercury compound phenylmercuric acetate has been recently shown to be rapidly degraded by L-cysteine at physiological pH (Cys, Figure 1B) using a LC-based approach. We therefore decided to investigate the fate of THI at pH 7.4 in the presence of physiologically relevant GSH-concentrations as it is the most abundant small molecular weight thiol in mammalian cells at concentrations up to 19 mM [14]. The employed LC-based approach allows the specific detection of Hg-species and is therefore ideally suited to observe the degradation of THI provided that the degradation products can be chromatographically separated from the parent compound [15] and structurally characterized using analytical techniques, such as electrospray ionization mass spectrometry (ESI-MS) and ^{199}Hg NMR spectroscopy [16]. Our investigations are solely aimed at understanding the biochemical reaction between THI and GSH at physiological pH to gain insight into the corresponding bioinorganic chemistry. As such the anticipated results are insufficient to causally link human exposure to THI with the etiology of autism in children [17], but the results may nevertheless provide an important first step to uncover bioinorganic processes that may also unfold in mammalian cells and explain adverse effects therein. To the best of our knowledge our investigations appear to be the first of their kind to address the bioinorganic chemical reaction between THI and GSH at physiological pH in the presence of 30% of acetonitrile (ACN).

2. Results and Discussion

Since THI is widely used in a variety of pharmaceutical products numerous previous studies have investigated its stability in aqueous solutions, such as vaccines to determine its shelf life [18,19]. While the results from these studies provided insight into the integrity of THI under storage conditions, much less is known about the potential degradation of THI by biomolecules after its intramuscular injection into people. As soon as THI enters cells it will encounter small molecular weight thiols, such as Cys and GSH (Figure 1B,C) which are present at mM concentrations. Indeed, recent studies have allowed to observe the degradation of the organomercurial compound phenylmercuric acetate (PMA) by Cys at physiological pH [15]. To understand the degradation of another organomercury compound that is of pharmaceutical interest, we have employed a LC-approach which encompassed a Hg-specific detector, namely a flame atomic absorption spectrometer (FAAS) to study the degradation of THI. THI was injected into a C_{18}-HPLC column and eluted using mobile phases that contained different GSH concentrations at physiological pH.

The results obtained for the injection of THI using a GSH-free mobile phase revealed the elution of a single Hg peak (t_r = 137 s; Figure 2, red line), which exhibited a Hg recovery of 82.2%.

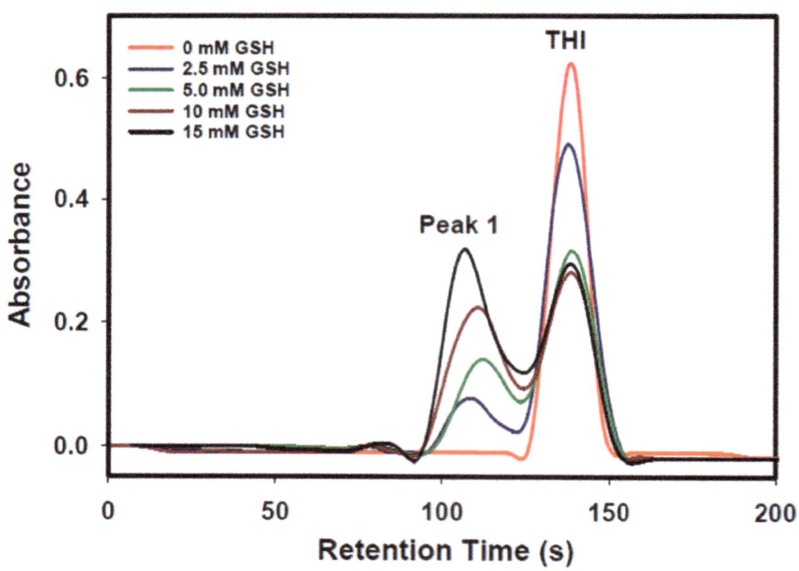

Figure 2. Representative Hg-specific chromatograms obtained for THI as a function of the GSH concentration in the 100 mM phosphate-buffer (pH 7.4):ACN 70:30 (v:v) mobile phase. Stationary phase: Gemini C_{18}-HPLC column (150 × 4.6 mm I.D., 5 μm particles); Flow rate: 1.0 mL/min; Detector: FAAS at 253.7 nm; Injection volume: 50 μL (1000 μg Hg).

The addition of up to 15 mM GSH concentrations to the mobile phase resulted in a concentration dependent decrease in the intensity of the Hg-peak that eluted at 137 s and the elution of a new Hg-peak that was progressively more intense with a 35 s reduced retention time compared to THI (Figure 2, blue, green, brown and black lines). With regard to the Hg peak areas of the Hg-specific chromatograms, Hg recoveries between 87.5 and 114.6% were obtained for the GSH-containing mobile phases (Table 1). Using the 15 mM GSH mobile phase the two Hg-containing fractions were collected and ESI-MS analysis revealed that the first Hg-peak (t_r = 105 s) corresponds to a GS-ethylmercury (EtHg) adduct with a m/z ratio of 536.0787 (Figure 3), while the second Hg-peak (t_r = 137 s) produced a peak in the ESI-MS mass spectrum of 383.0041 m/z ratio which is congruent with the elution of 'intact' THI (Figure 4). These results imply that increasing physiologically relevant GSH concentrations in the mobile phase progressively degraded THI on the column. With the 15 mM GSH mobile phase, 54% of THI was eluted intact (Table 1), while the remaining Hg corresponded to the GS-EtHg adduct.

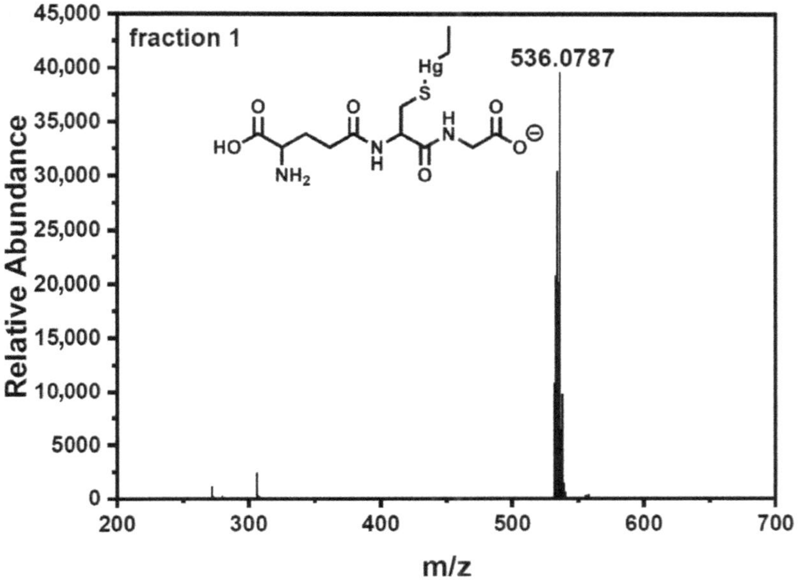

Figure 3. ESI-MS results (negative mode) obtained for Hg-containing fraction 1; obtained using the 15.0 mM GSH mobile phase.

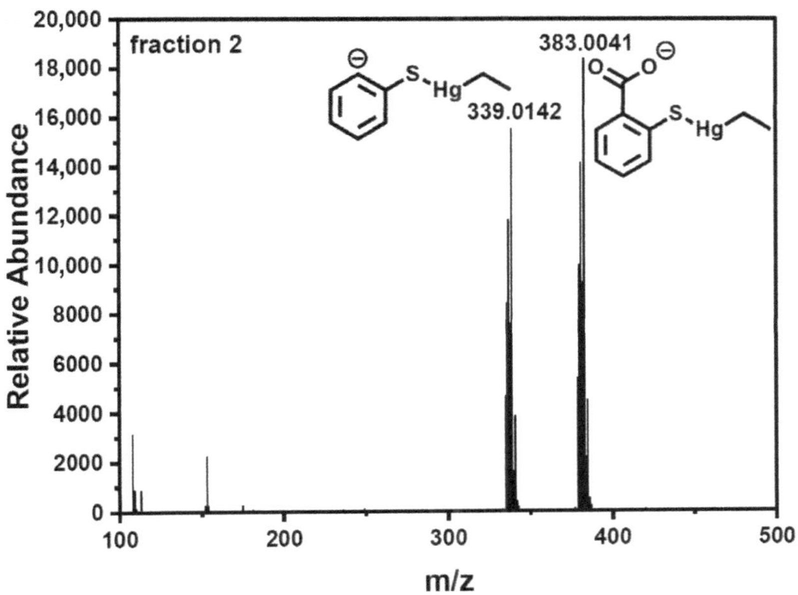

Figure 4. ESI-MS results (negative mode) obtained for Hg-containing fraction 2 obtained using the 15.0 mM GSH mobile phase.

Table 1. Peak areas of the Hg peaks obtained after the analysis of THI with different GSH concentrations in the mobile phase.

Concentration of GSH (mM) in Mobile Phase	Retention Time (s)		Peak Area (Area Units) (Area Percentage [%])		Sum Area (Area Units) (Sum Area Percentage [%])
	Peak 1	THI	Peak 1	THI	
0	-	137 ± 1 *	-	8.43 ± 1.07 *(82.2 ± 10.7%)*	8.43 ± 1.07 *(82.2 ± 10.7%)*
2.5	107 ± 2	137 ± 1	1.71 ± 0.07 *(16.7 ± 0.8%)*	8.11 ± 0.37 *(79.0 ± 4.3%)*	9.82 ± 0.38 *(95.7 ± 4.4%)*
5	111 ± 1	138 ± 1	2.79 ± 0.06 *(27.2 ± 1.0%)*	6.19 ± 0.20 *(60.3 ± 2.6%)*	8.98 ± 0.21 *(87.5 ± 2.8%)*
10	103 ± 1	135 ± 1	4.75 ± 0.06 *(46.3 ± 1.5%)*	5.73 ± 0.37 *(55.8 ± 4.0%)*	10.48 ± 0.37 *(102.1 ± 4.3%)*
15	105 ± 1	136 ± 1	6.23 ± 0.04 *(60.7 ± 1.8%)*	5.53 ± 0.02 *(53.9 ± 1.6%)*	11.98 ± 0.29 *(114.6 ± 2.4%)*

* n = 3.

Since Cys is also present in various cell cytosols at comparatively lower concentrations than GSH (e.g., hepatocytes contain ~0.5 mM Cys [15]), we replicated the LC-experiments with Cys-containing mobile phases. While the employed Cys concentrations are supra physiological, the obtained results nevertheless resembled those obtained with equimolar concentrations of GSH in the mobile phase (Figure 5). With regard to the obtained Hg peak areas of the obtained Hg-specific chromatograms, Hg recoveries between 78.9 and 95.7% were obtained for the Cys-containing mobile phases and 85.1% for the Cys-free mobile phase (Table 2). Similarly to the GSH LC-results we observed the parent THI Hg-peak at the same retention time (t_r = 137 s) as well as the elution of a progressively more intense Hg-peak (t_r = 115 s), which consistently eluted at a 10 s increased retention time compared to the corresponding GS-EtHg adduct (Figure 2). While a fraction containing the putative Cys-EtHg adduct was submitted to ESI-MS analysis, the results were not as clear as those obtained for the GS-EtHg adduct. Nevertheless, these Cys HPLC results suggest that the putative Cys-EtHg adduct that was formed on the column is slightly more hydrophobic than the GS-EtHg adduct. Compared to the 15 mM GSH mobile phase, 42.5% of Hg eluted in the form of intact THI, while 53.2% eluted as the putative Cys-EtHg adduct. These results imply a fast degradation of THI by Cys as compared to GSH at comparable mobile phase concentrations.

Table 2. Peak areas of the Hg peaks obtained after the analysis of THI with different Cys concentrations in the mobile phase.

Concentration of Cys (mM) in Mobile Phase	Retention Time (s)		Peak Area (Area Units) (Area Percentage [%])		Sum Area (Area Units) (Sum Area Percentage [%])
	Peak 1	Peak 2	Peak 1	Peak 1	
0	-	140 ± 1 *	-	8.73 ± 0.30 *(85.1 ± 3.8%)*	8.73 ± 0.30 *(85.1 ± 3.8%)*
2.5	115 ± 2	140 ± 1	0.51 ± 0.01 *(4.9 ± 0.2%)*	8.55 ± 0.36 *(83.3 ± 4.3%)*	9.06 ± 0.37 *(88.3 ± 4.4%)*
5	111 ± 1	139 ± 1	0.58 ± 0.06 *(5.7 ± 0.6%)*	8.44 ± 0.14 *(82.2 ± 2.8%)*	9.02 ± 0.20 *(87.9 ± 3.2%)*
10	121 ± 1	132 ± 1	3.84 ± 0.31 *(37.4 ± 3.2%)*	4.83 ± 0.49 *(47.1 ± 4.9%)*	8.67 ± 0.80 *(84.5 ± 8.2%)*
15	116 ± 1	133 ± 0	5.46 ± 0.39 *(53.2 ± 4.1%)*	4.36 ± 0.25 *(42.5 ± 2.7%)*	9.82 ± 0.64 *(95.7 ± 6.8%)*
20	117 ± 1	134 ± 1	5.95 ± 0.17 *(57.9 ± 2.4%)*	2.15 ± 0.19 *(20.9 ± 1.9%)*	8.10 ± 0.36 *(78.9 ± 4.2%)*

* n = 4.

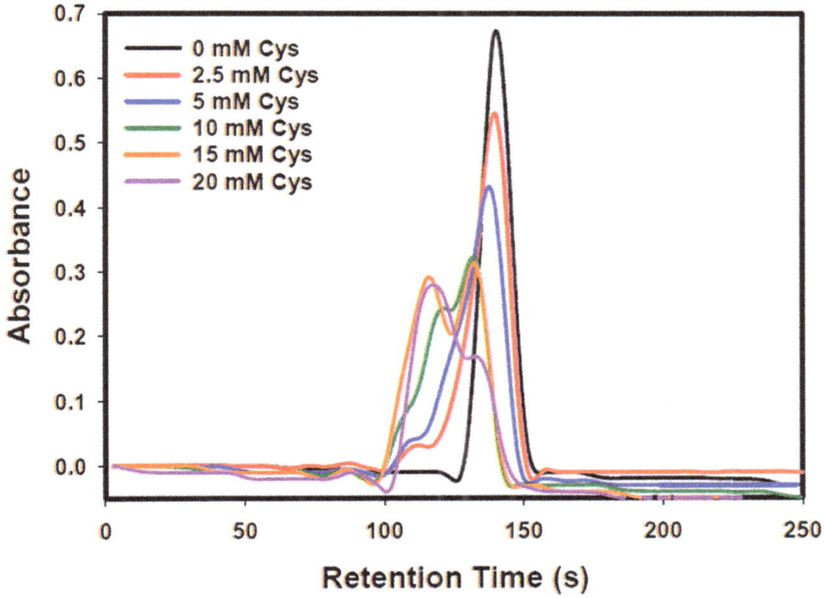

Figure 5. Representative Hg-specific chromatograms obtained for THI as a function of the Cys concentration in the 100 mM phosphate-buffer (pH 7.4):ACN 70:30 (v:v) mobile phase. Stationary phase: Gemini C_{18}-HPLC column (150 × 4.6 mm I.D., 5 μm particles); Flow rate: 1.0 mL/min; Detector: FAAS at 253.7 nm; Injection volume: 50 μL (1000 μg Hg).

To substantiate these HPLC/ESI-MS results for THI we conducted ^{199}Hg-NMR experiments, the results of which are depicted in Figure 6. The ^{199}Hg-NMR spectrum of a THI solution in a phosphate buffer that contained 30% of ACN revealed a well-defined complex multiplet at −743 ppm (Figure 6A). When GSH was added to a THI solution to achieve the same mM concentration ratio as in the LC-experiments, the observed multiplet (Figure 6A) shifted slightly to −740 ppm and it collapsed to a broad singlet (Figure 6B). This observation is indicative of a more complex coupling environment and likely conformational isomerism while conclusively confirming a chemical reaction between the Hg-center of THI with GSH by ^{199}Hg-NMR spectroscopy.

The results that were obtained for the interaction of THI with GSH at physiological pH in the presence of 30% ACN are all congruent with the formation of a GS-EtHg adduct at these conditions (Figure 7). Although our results were not obtained in 100% aqueous solution, previous studies have conclusively demonstrated that THI—after its incubation with RBC cytosol at 37° Celsius—degraded over a 6 h period and progressively co-eluted with hemoglobin (Hb) [7]. The binding of the EtHg-breakdown product of THI to Hb was confirmed by analyzing a fraction containing the putative EtHg-Hb adduct by X-ray absorption spectroscopy, which revealed a Hg-S bond length that was congruent with literature results [7]. Based on these previous studies obtained with RBC cytosol it appears that THI degrades within cells and the formed degradation product EtHg will then competitively bind to biological thiols, such as GSH and/or Hb based on steric factors. Future studies should investigate whether the GS-EtHg adduct can bind to cytosolic proteins, such as Hb using established metallomics techniques [7]. Based on previous results by others another promising future investigation is to evaluate if the GS-EtHg adduct will bind to calcium channels as there appears to be a link between THI and the perturbation of the Ca-metabolism in certain cell types [20,21].

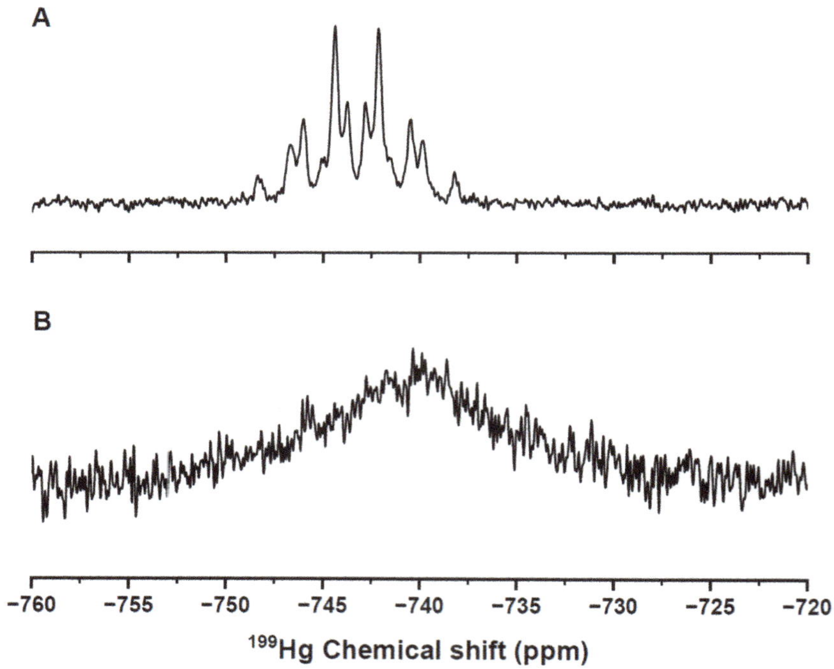

Figure 6. Hg-NMR spectra obtained for 0.2 M THI in 100 mM phosphate buffer (in D_2O):ACN 70:30 (v:v) (**A**) and 0.2 M THI and 30 mM GSH in 100 mM phosphate buffer (in D_2O):ACN 70:30 (v:v) (**B**).

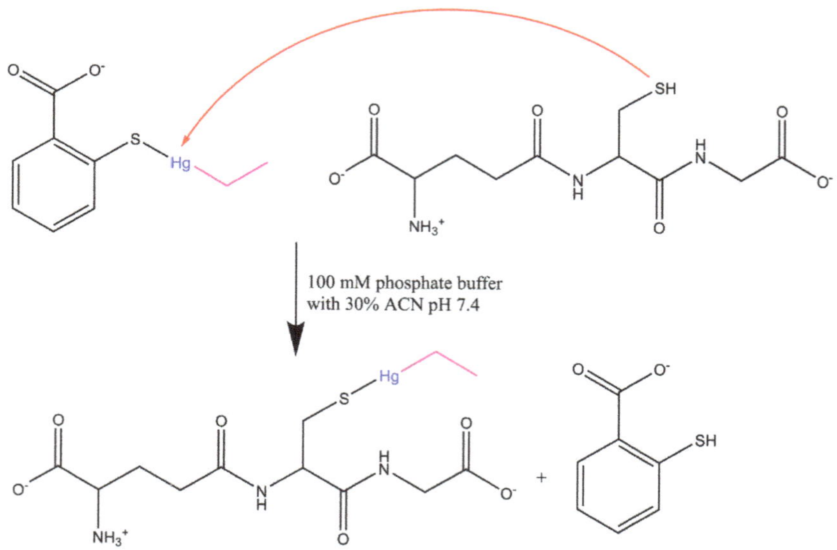

Figure 7. Proposed biomolecular mechanism by which THI reacts with GSH to form a GS-EtHg adduct at pH 7.4.

3. Experimental

3.1. Chemicals and Solutions

Thimerosal (97–101%), L-glutathione reduced (>98%), L-cysteine (>98%), sodium phosphate dibasic (Na_2HPO_4, >98.5%), sodium phosphate monobasic (NaH_2PO_4, >98%), sodium chloride (NaCl, >99.5%), acetonitrile (ACN, Chromasolv, HPLC grade, >99.9%) and deuterium oxide (D_2O, 99.9 atom % D) were purchased from Sigma Aldrich (St. Louis, MO, USA).

A 0.1 M solution of THI was made every day by dissolving 10.1 mg in 250 µL 0.8% NaCl, prepared by dissolving 0.8 g of NaCl in 100 mL deionized (dI) water from a Simplicity UV water purification system (Millipore, Billerica, MA, USA). Mobile phases were prepared by mixing 100 mM of Na_2HPO_4 and 100 mM of NaH_2PO_4 solutions to obtain pH 7.4 using a VWR Symphony SB20 pH meter (Thermo Electron Corporation, Beverly, MA, USA). Then ACN was added to achieve a 30% final concentration (v:v). Mobile phases that contained GSH (0–20 mM) were prepared by adding the corresponding amount of GSH to the ACN containing phosphate buffer to give 150 mL of GSH-containing buffer (0.6915 g of GSH to obtain the 15 mM GSH buffer). Mobile phases containing up to 20 mM Cys were prepared in an analogous manner. Then, the pH of the final solution was adjusted to 7.4 with 0.4 M HCl and all mobile phases were filtered through 0.45 µm MCE nitrocellulose membrane filters (Millipore, Mandel Scientific, Guelph, ON, Canada).

3.2. Instrumentation

The HPLC system comprised an Azura P2.1S pump equipped with a ceramic pump head, a Rheodyne 9725 injector (Rheodyne, LLC, 600 Park Court. Rohnert Park, CA, USA), a 50 µL sample loop and a Gemini 5 µm C_{18} HPLC column (Phenomenex, 411 Madrid Avenue Torrance, CA, USA) 150 × 4.6 mm, 5 µm particle size). The flow rate was 1.0 mL min^{-1} and 0.1 M solutions of THI were injected. All separations were conducted at room temperature (22 °C) and Hg was detected using a Buck Model 200A flame atomic absorption spectrometer (FAAS, Buck Scientific, East Norwalk, CT, USA) at 253.7 nm. The HPLC column exit was connected to the FAAS inlet with polyethylene tubing (I.D. 0.13 mm, length 8 cm) and its void volume (t_0 = 92 s) was determined by injecting a NaCl solution and observing the yellow color of the flame using dI water as the mobile phase. To determine the recovery of Hg in the HPLC experiments, we injected 1000 µg of Hg into a hollow plastic tube (0.15 mm × 21 cm) and defined the area that was obtained from the FAAS (10.26 ± 0.3 area units) as 100%. The total Hg peak areas determined by integrating all Hg peaks were then expressed as a percentage of this theoretical peak area of 100%.

The ESI-MS experiments to identify the Hg-species in the column effluent were conducted using previously established parameters [15]. To identify the Hg-degradation product that was formed on the column when THI was injected with the 15 mM GSH containing mobile phase, the corresponding Hg-containing fractions were collected and analyzed by LC-ESI MS with an Agilent 1200 series HPLC coupled to an Agilent 6520 Q-TOF mass spectrometer (Agilent, Santa Clara, CA, USA). Separation of the Hg complexes from the buffer matrix was achieved with an Agilent Eclipse Plus C_{18} column (2.1 mm × 100 mm, 3.5 µm particle size) and a 20 mM NH_4HCO_3/methanol gradient (0–1 min 20% MeOH, 1–9 min gradient to 90% MeOH, 9–10 min 90% MeOH). Flow rate, injection volume, and column temperature were 0.4 mL/min, 1 µL, and 45 C, respectively. Fraction 1 (~105 s) was analyzed in positive and negative ESI modes, and fraction 2 (~139 s) was analyzed in negative ESI mode only. Extracted ion chromatograms (EIC) of the Hg complexes were obtained from the TIC using the calculated theoretical mass ± 10 ppm.

The ^{199}Hg NMR experiments were recorded on a Avance III 600 spectrometer (Bruker, Karlsruhe, Germany) at a resonance frequency of 107.36 MHz using a broadband probe. The chemical shift was externally referenced to a saturated solution of $HgCl_2$ in D_2O which resulted in a signal at −1497 ppm. Spectra were acquired using a 90° pulse, a sweep width of 217.4 kHz, and 32,000 data points. A 0.3 s delay was used between the scans, and a total of 8000 scans were accumulated at 298 K. To conduct the actual measurement, we needed to make a 0.2 M solution of THI in a solution prepared by mixing 100 mM of Na_2HPO_4 and 100 mM of NaH_2PO_4 solutions each prepared with D_2O to obtain pH 7.4 followed by the addition of ACN to obtain 30% (v:v). We dissolved 80.8 mg of THI in 1.0 mL of the ACN containing 100 mM phosphate buffer. This solution was then filtered using a Pasteur pipet (VWR International, LLC, Radnor, PA, USA), which had been equipped with cotton on the inside into the 5 mm NMR tube and the ^{199}Hg-NMR spectrum was measured. Thereafter, we made a 1.0 mL solution of 0.2 M THI in 100 mM of phosphate-buffered solution containing 30% of ACN (v:v) of pH 7.4 prepared with D_2O and then added the amount of GSH to achieve a 30 mM solution (9.2 mg).

Raw LC-FAAS data were imported into Sigma Plot 15 and smoothed using the bisquare algorithm. Retention times (t_r) and Hg peak areas were determined using OriginPro software (Version 2025). All experiments were performed in quadruplets. Raw ESI-MS and ^{199}Hg NMR data files were imported into OriginPro software for spectral visualization.

4. Conclusions

Previously reported results pertaining to the Cys-mediated degradation of the organomercurial phenylmercuric acetate (PMA) at physiological pH to form a phenylmercury-cysteineate species prompted us to investigate the interaction of THI with the most abundant cytosolic low-molecular-weight thiol (GSH) at pH 7.4. Employing an LC-based approach allowed us to observe a progressive GSH-mediated degradation of THI, which was associated with the elution of a Hg-containing degradation product, which was identified as a GS-EtHg adduct by ESI-MS. The confirmation of this chemical reaction by ^{199}Hg-NMR experiments implies that under our chosen experimental conditions, THI is effectively degraded by GSH. To corroborate the LC-results that were obtained with GSH-containing mobile phases, we replicated these experiments with Cys-containing mobile phases and the same concentration range, obtaining results that were comparable, suggesting the formation of a Cys-EtHg adduct. Taken together, our results serve as a starting point to investigate if THI will similarly interact with GSH and Cys in biochemically complex cell lysates to form GS-EtHg and/or Cys-EtHg adducts. These adducts may interfere with other biochemical processes within cells (e.g., calcium channels) to provide a mechanistic link to adverse biochemical toxicological effects. The results of our investigations are also relevant to the development of better antivirals, as PMA and THI have been demonstrated to be effective inhibitors of the main protease of SARS-CoV 2 [11], which may involve processes similar to the established mechanism depicted in Figure 7.

Author Contributions: Conceptualization, J.G.; Methodology, M.F.D. and S.M.; Software, M.F.D. and S.M.; Formal analysis, M.F.D. and S.M.; Data curation, M.F.D. and S.M.; Writing—original draft, J.G.; Writing—review & editing, M.F.D., S.M. and J.G.; Visualization, M.F.D.; Supervision, J.G.; Project administration, J.G.; Funding acquisition, J.G. All authors have read and agreed to the published version of the manuscript.

Funding: M.F.D. was funded by a scholarship from Sorbonne University (SU IDEX 2025). Work carried out in the lab of JG was funded by an NSERC-DG.

Institutional Review Board Statement: Not applicable.

Informed Consent Statement: Not applicable.

Data Availability Statement: The original contributions presented in this study are included in the article. Further inquiries can be directed to the corresponding author.

Acknowledgments: The ESI/LC-MS and the ^{199}Hg-NMR measurements were provided by Michelle Thibault from the Chemistry Instrumentation Facility.

Conflicts of Interest: The authors declare no conflict of interest.

References

1. Hou, D.; Jia, X.; Wang, L.; McGrath, S.P.; Zhu, Y.-G.; Hu, Q.; Zhao, F.J.; Bank, M.S.; O'Connor, D.; Nriagu, J. Global soil pollution by toxic metals threatens agriculture and human health. *Science* **2025**, *388*, 316–321.
2. Niyogi, S.; Shekh, K.; Amuno, S. Toxicology of trace metals in the environment: A current perspective. In *Environmental and Biochemical Toxicology, Concepts, Case Studies and Challenges*; Gailer, J., Turner, R.J., Eds.; De Gruyter: Boston, MA, USA, 2022; p. 335.
3. Charette, T.; Kamisnki, G.; Rosabal, M.; Amyot, M. Effects of speciation, cooking and changes in bioaccessibility on methylmercury exposure assessment for contrasting diets of fish and marine mammals. *Int. J. Environ. Res. Public Health* **2021**, *18*, 2565. [CrossRef] [PubMed]
4. Doroudian, M.; Gailer, J. Integrative metallomics studies of toxic metal(loid) substances at the blood plasma-red blood cell-organ/tumor nexus. *Inorganics* **2022**, *10*, 200. [CrossRef]
5. Carneiro, M.F.H.; Souza, J.M.O.; Grotto, D.; Batista, B.L.; de Oliveira Sousa, V.C.; Barnbosa, F.J. A systematic study of the disposition and metabolism of mercury species in mice after exposure to low levels of thimerosal (ethymercury). *Environ. Res.* **2014**, *134*, 218–227. [CrossRef] [PubMed]
6. Gailer, J. Probing the bioinorganic chemistry of toxic metals in the mammalian bloodstream to advance human health. *J. Inorg. Biochem.* **2012**, *108*, 128–132. [CrossRef] [PubMed]
7. Gibson, M.A.; Sarpong-Kumankomah, S.; Nehzati, S.; George, G.N.; Gailer, J. Remarkable differences in the biochemical fate of Cd^{2+}, Hg^{2+}, CH_3Hg^+ and thimerosal in red blood cell lysate. *Metallomics* **2017**, *9*, 1060–1072. [CrossRef] [PubMed]
8. de Magalhanes Silva, M.; de Araujo Dantas, M.D.; de Sila Filho, R.C.; dos Santos Sales, M.V.; de Almeida Xavier, J.; Leite, A.C.R.; Goulart, M.O.F.; Grillo, L.A.M.; de Barros, W.A.; de Fatima, A.; et al. Toxicity of thimerosal in biological systems: Conformational changes in human hemoglobin, decrease of oxygen binding, increase of protein glycation and amyloid formation. *Intern. J. Biol. Macromol.* **2020**, *154*, 661–671. [CrossRef] [PubMed]
9. Zareba, M.; Sanecki, P.T.; Rawski, R. Simultaneous determination of thimerosal and aluminium in vaccines and pharmaceuticals with the use of HPLC Method. *Acta Chromatogr.* **2016**, *28*, 299–311. [CrossRef]
10. Elferink, J.G.R. Thimerosal: A versatile sulfhydryl reagent, calcium mobilizer, and cell function-modulating agent. *Gen. Pharmacol.* **1999**, *33*, 1–6. [CrossRef] [PubMed]
11. Coelho, C.; Gallo, G.; Campos, C.B.; Hardy, L.; Wuertele, M. Biochemical screening for SARS-CoV-2 main protease inhibitors. *PLoS ONE* **2020**, *15*, e0240079. [CrossRef] [PubMed]
12. Kern, J.K.; Geier, D.A.; Sykes, L.K.; Haley, B.E.; Geier, M.R. The relationship between mercury and autism: A comprehensive review and discussion. *J. Trace Elem. Med. Biol.* **2016**, *37*, 8–24. [CrossRef] [PubMed]
13. Janzen, R.; Schwarzer, M.; Sperling, M.; Vogel, M.; Schwerdtle, T.; Karst, U. Adduct formation of thimerosal with human and rat hemoglobin: A study using liquid chromatography coupled to electrospray time-of-flight mass spectrometry (LC/ESI-TOF-MS). *Metallomics* **2011**, *3*, 847–852. [CrossRef] [PubMed]
14. Stahl, S.H.; Yates, J.W.; Nicholls, A.W.; Kenna, J.G.; Coen, M.; Ortega, F.; Nicholson, J.K.; Wilson, I.D. Systems toxicology: Modelling biomarkers of glutathione homeostasis and paracetamol metabolism. *Drug Discov. Today Technol.* **2015**, *15*, e9–e14. [CrossRef]
15. Attia, M.I.; Gailer, J. RP-HPLC reveales the L-cysteine induced degradation of phenylmercuric acetate. *J. Inorg. Biochem.* **2025**, *266*, 112851. [CrossRef] [PubMed]
16. Girault, L.; Boudou, A.; Dufourc, E.J. Methyl mercury interactions with phospholipid membranes as reported by fluorescence, ^{31}P and ^{199}Hg NMR. *Biochim. Biophys. Acta* **1997**, *1325*, 250–262. [CrossRef] [PubMed]
17. Leung, J.W.-H.; Loan, A.; Xu, Y.; Yang, G.; Wang, J.; Chan, H.M. Reduction of glyoxalate 1 expression links fetal methylmercury exposure to autism spectrum disorder pathogenesis. *Toxics* **2024**, *12*, 449.
18. Tan, M.; Parkin, J.E. Route of decomposition of thiomersal (thimerosal). *Intern. J. Pharm.* **2000**, *208*, 23–34. [CrossRef] [PubMed]
19. Reader, M.J.; Lines, C.B. Decomposition of thimerosal in aqueous solution and its determination by high-performance liquid chromatography. *J. Pharm. Sci.* **1983**, *72*, 1406–1409. [CrossRef] [PubMed]

20. Bootman, M.D.; Taylor, C.W.; Berridge, M.J. The thiol reagent, thimerosal, evokes Ca^{2+} spikes in Hela-cells by sensitizing the inositol 1,4,5-trisphosphate receptor. *J. Biol. Chem.* **1992**, *267*, 25113–25119. [CrossRef] [PubMed]
21. Abramson, J.J.; Zable, A.C.; Favero, T.G.; Salama, G. Thimerosal interacts with the Ca2+ release channel ryanodinereceptor from skeletal-muscle sarcoplasmatic-reticulum. *J. Biol. Chem.* **1995**, *270*, 29644–29647. [CrossRef] [PubMed]

inorganics

MDPI

Article

Synthesis of Some Novel Cr(III), Mn(II), and Pd(II) Complexes via the Sono-Chemical Route with a Chlorinated Quinolinyl-Imine Ligand: Structural Elucidation, Bioactivity Analysis, and Docking Simulations

Dalal Alhashmialameer

Chemistry Department, College of Science, Taif University, Taif 21944, Saudi Arabia; dsamer@tu.edu.sa

Abstract

The present study reports the sono-chemical synthesis of novel nanosized Cr(III), Mn(II), and Pd(II) complexes incorporating the chloro-2-(quinolin-8-yliminomethyl)-phenol imine ligand. The synthesized complexes were characterized using various spectroscopic and analytical techniques, including Fourier-transform infrared (FT-IR) spectroscopy, ultraviolet–visible (UV–Vis) spectroscopy, scanning electron microscopy (SEM), and thermal gravimetric analysis (TGA). The results confirmed the successful coordination of the ligand-to-metal centers, forming stable nanosized metal complexes with distinct physicochemical properties. Biological evaluations, including antimicrobial and antioxidant assays, revealed that the synthesized complexes exhibited enhanced biological activity compared to the free ligand, demonstrating potent antibacterial and antifungal properties against various pathogenic strains. The potential of the complexes to serve as efficient free-radical inhibitors was determined by employing DPPH radical scavenging assays, which underscored their significant antioxidant properties. Furthermore, molecular docking studies were conducted to elucidate the binding interactions of the metal complexes with biological targets, providing insights into their mechanism of action. The findings suggest that the synthesized nanosized Cr(III), Mn(II), and Pd(II) complexes possess promising biological properties, making them potential candidates for pharmaceutical and biomedical applications. The study also demonstrates the effectiveness of sono-chemical synthesis in producing nanosized metal complexes with enhanced physicochemical and biological characteristics.

Keywords: sono-chemical synthesis; nanosized metal complexes; quinolinyl-imine ligand; Cr(III); Mn(II); Pd(II); biological activity; molecular docking

1. Introduction

Pharmaceutical innovation frequently involves a chemical strategy that combines several molecules to produce new ones with unprecedented biological properties. The synthesis of compounds, particularly the versatile class of Schiff bases, is notable for their anticancer, antibacterial, and anti-inflammatory characteristics, as well as various catalytic, optical, and sensing properties [1–4]. These molecules readily interact with almost every metal ion, which is why they are pivotal components in the realm of coordination chemistry. The objective of creating metal complexes is to tailor their physicochemical properties due to the broad spectrum of coordination environments, oxidation states, redox potentials, and ligand arrangements. Consequently, the biological actions of diverse ligands are significantly influenced when they form complexes [5–7]. The domain of metal-complex-based anticancer therapy is in a constant state of advancement, with scientists investigating innovative metal complexes, combination treatments, targeted delivery mechanisms, and tactics to surmount resistance [8–10]. Personalized medical approaches, which consider the individual genetic profiles of patients and their tumor cells, show substantial promise for enhancing the effectiveness of these therapies and reducing side effects. This category of metal-based treatments represents a vast and promising avenue for cancer care, potentially leading to superior outcomes. Furthermore, antimicrobial metal complexes boast a multitude of potential uses across different sectors, including medical applications such as antimicrobial medications and wound coverings, agricultural purposes like pesticides and fungicides, and consumer goods like antimicrobial coatings for various surfaces. A significant hurdle in the development of these complexes lies in achieving selective toxicity towards pathogenic microbes while minimizing harm to human cells. The design focus is on complexes that can target microbial cells with precision, thereby reducing the impact on the host organism. This selectivity is essential for their successful application across various fields. The design and synthesis of transition metal complexes with Schiff base ligands have attracted significant interest due to their diverse structural features, chemical versatility, and broad spectrum of biological applications, including antimicrobial, anticancer, and antioxidant activities [11,12]. Schiff bases, derived from the condensation of primary amines with carbonyl compounds, possess azomethine (-C=N-) functional groups, which facilitate strong coordination with metal ions, thereby enhancing their stability and biological properties [13]. Among the various Schiff base ligands, those incorporating quinoline moieties have demonstrated remarkable pharmacological potential due to their π-conjugated system, which enhances electronic interactions and biological activity [14]. The incorporation of metal ions such as chromium Cr(III), manganese Mn(II), and palladium Pd (II) into these ligands is particularly interesting, as these metals exhibit diverse coordination geometries and oxidation states, influencing the electronic and biological behavior of the resulting complexes [15,16]. Ultrasound-assisted (sono-chemical) synthesis has emerged as a promising technique in the preparation of nanostructured metal complexes, offering advantages such as improved reaction rates, reduced particle size, and enhanced purity compared to conventional synthetic routes [17]. Sono-chemical methods induce cavitation effects, leading to localized high temperatures and pressures, which significantly impact nucleation and growth kinetics, favoring the formation of nanosized materials with improved physicochemical properties [18].

The investigated ligand, chloro-2-(quinolin-8-yliminomethyl)-phenol, was not chosen arbitrarily. It features a Schiff base framework renowned for its versatility in coordination chemistry, yet distinguished by a unique combination of functional groups that enhance its potential in biomedical and catalytic applications. The quinoline moiety serves as a well-established pharmacophore, widely recognized for its ability to intercalate with DNA and for exhibiting diverse biological activities, including antitumor, antibacterial, and antimalarial effects. The chloro-substituted phenol ring contributes electron-withdrawing effects that can significantly influence metal-binding affinity, lipophilicity, and cellular uptake—properties that are crucial to pharmacological performance. Meanwhile, the imine linkage facilitates strong chelation with metal centers, while also modulating redox behavior and biological reactivity, further enhancing the ligand's multifunctional character. The chosen metal ions—Cr(III), Mn(II), and Pd(II)—were selected for their distinct biological relevance and strong coordination compatibility with the tetra-dentate N,N,O,O-donor set of the investigated ligand. Their selection reflects both their diverse pharmacological potential and their ability to form stable, bioactive complexes. Cr(III) is an essential trace element involved in glucose and lipid metabolism, with reported insulin-enhancing and antioxidant properties. Mn(II) complexes are valued for their catalytic roles in redox processes, as well as their superoxide dismutase-mimetic activity, which is linked to antioxidant and neuroprotective effects. Pd(II) is a versatile transition metal with significant applications in medicinal chemistry, particularly due to its potent cytotoxic activity and structural analogy to platinum-based anticancer agents, making it a promising candidate for chemotherapeutic development.

From this point of view, I report the sono-chemical synthesis of novel nanosized Cr(III), Mn (II), and Pd (II) complexes incorporating a chloro-2-(quinolin-8-yliminomethyl)-phenol imine ligand. The synthesized complexes are characterized using various spectroscopic and analytical techniques, including Fourier-transform infrared (FT-IR) spectroscopy, UV–Vis spectroscopy, scanning electron microscopy (SEM), and thermal gravimetric analysis (TGA). Furthermore, their biological activities, such as antimicrobial, anticancer, and antioxidant properties, are investigated, supported by molecular docking studies to elucidate their interaction mechanisms at the molecular level. The combined physicochemical and computational approach aims to provide insights into the structure–activity relationships of these novel nanosized metal complexes, paving the way for potential biomedical applications.

2. Result and Discussion

2.1. Characterization of 4-Chloro-2-(Quinolin-8-Yliminomethyl)-Phenol Ligand and Its Complexes

The solubility of the synthesized complexes was evaluated in various solvents. All complexes exhibited good solubility in polar aprotic solvents such as dimethyl sulfoxide (DMSO) and dimethylformamide (DMF), which were used for spectral and biological studies. However, they showed limited or no solubility in non-polar solvents such as hexane and toluene and moderate solubility in ethanol and methanol.

2.2. FTIR Spectrum

Infrared spectroscopy in the form of FT-IR analysis was employed to study the characteristics of the 4-chloro-2-(quinolin-8-yliminomethyl)-phenol ligand and its corresponding metal complexes, as presented in Table 1, Figure S1. A notable feature of the **L** ligand is its distinct absorption peak at 1618 cm^{-1}, which is attributed to the νC-H=N) stretching vibration. Upon interaction with chromium, manganese, and palladium ions, this peak is observed to shift to lower frequencies of 1607 cm^{-1}, 1610 cm^{-1}, and 1602 cm^{-1} for the **L-Cr**, **L-Mn**, and **L-Pd** complexes, respectively. This spectral alteration is indicative of the metal ions' chelation with the azomethine group (-CH=N) [19,20]. The ligand's FT-IR spectrum also reveals characteristic absorption bands at 1552 cm^{-1}, corresponding to the ν(C=N) mode in the quinoline moiety. After complexation with the metal species, these bands undergo a shift to lower wavenumbers: 1526 cm^{-1} for the chromium complex, 1533 cm^{-1} for the manganese complex, and 1517 cm^{-1} for the palladium complex. This phenomenon suggests the active participation of the quinoline ring in the metal coordination process [21]. Moreover, the ν(OH) stretching vibration of the free ligand at 3428 cm^{-1} diminishes, and the ν(C-O) band initially at 1269 cm^{-1} moves to lower frequencies of 1249 cm^{-1}, 1258 cm^{-1}, and 1242 cm^{-1} for the **L-Cr**, **L-Mn**, and **L-Pd** complexes, respectively. This change implies that the phenolic oxygen atom engages in C-O-M bond formation following the loss of its hydrogen atom. The appearance of novel absorption bands in the Fourier transforms infrared (FT-IR) spectra of the complexes **L-Cr**, **L-Mn**, and **L-Pd** at wavenumbers of 457 cm^{-1}, 445 cm^{-1}, and 524 cm^{-1}, respectively, can be attributed to the presence of metal–oxygen (M-O) stretching vibrations. In a similar manner, the bands observed at 513 cm^{-1}, 517 cm^{-1}, and 478 cm^{-1} for **L-Cr**, **L-Mn**, and **L-Pd** correspond to metal–nitrogen (M-N) stretching vibrations [22,23]. The specific wavenumbers and intensities of these bands provide insights into the coordination geometry and the nature of interactions surrounding the metal ions, thereby substantiating the proposed coordination models for the ligands in the metal complexes. The distinct vibrational characteristics confirm the participation of both nitrogen and oxygen in the ligation process, which is essential for understanding the structural and bonding properties of these compounds

Table 1. The synthesized 4-chloro-2-(quinolin-8-yliminomethyl)-phenol ligand along with its metal complexes, chemical formulas, M.P, CHN, and IR frequencies.

Compounds	Empirical Formula (Formula Weight)	Color	(M. p.) and De-comp. Temp. (°C)	Λ_m (Ω^{-1} cm^2 mol^{-1})	μ_{eff} (B.M.)	Analysis (%) Found (Calc.)			IR, Cm^{-1}					
						C	H	N	(υOH)/H_2O	υ(CH=N)υ_{ph}	(C=N)py	υ(C-O)	υ(M-N)	υ(M-O)
L	$C_{16}H_{11}ClON_2$ (282.72)	Canary yellow	180	-	-	67.89 (67.96)	3.88 (3.92)	9.97 (9.91)	3428	1618	1552	1269	-	-
L-Cr	$C_{32}H_{22}Cl_2N_5O_6$ Cr (695.45)	Dark brown	>300	60.50	3.65	55.33 (55.27)	3.14 (3.19)	10.14 (10.07)	3359	1607	1526	1249	513	457
L-Mn	$C_{32}H_{24}Cl_2N_4O_4$ Mn (654.40)	brown	>300	8.50	5.32	58.79 (58.73)	3.76 (3.70)	8.51 (8.56)	3345	1610	1533	1258	517	445
L-Pd	$C_{18}H_{15}ClN_2O_4$ Pd (465.20)	Orange	>300	10.21	-	46.54 (46.47)	3.19 (3.25)	6.10 (6.02)	3365	1602	1517	1242	478	524

2.3. *^{1}H-NMR and ^{13}C-NMR Spectral Evaluations*

The ligand in question, denoted as 4-chloro-2-(quinolin-8-yliminomethyl)-phenol, underwent comprehensive analysis via ^{1}H-NMR spectroscopy, utilizing DMSO-d$_6$ as the solvent. This analysis is presented in Figure S2. The ^{1}H-NMR spectra revealed a singlet at 13.98 ppm, which can be ascribed to the OH group of the phenolic moiety. The presence of the CH=N group was discerned at a chemical shift of 9.26 ppm. Additionally, the spectra exhibited a doublet signal at 8.83–8.80 ppm, corresponding to the hydrogen atom adjacent to the nitrogen within the quinoline ring (d, 1H, CH$_{arm}$ adjacent nitrogen of quinoline ring). Another set of doublet signals at 8.61–8.07 ppm (d, J = 6.9 Hz, 5H, CH$_{arm}$) of pyridine ring in quinoline moiety and at 7.90–6.83 ppm (d, J = 7.1 Hz, 2H, CH$_{arm}$) and at 7.42–6.81 ppm (d, J = 7.1 Hz, 3H, CH$_{arm}$) was observed, attributable to the aromatic protons quinoline moiety. The ^{13}C-NMR data, obtained in DMSO-d6 medium (Figure S3), provided further structural insights with chemical shifts at 162.10 ppm for the CH=N carbons, 167.30 ppm for the CH=N carbons, 163.60 ppm for the (C-OH) carbon, 152.20 (C=N of pyridine ring in quinoline moiety), 140.50, 138.20, 134.50, 132.10, 128.70, 127.10, 125.50,124.70, 123.10, 121.10, 120.00, 115.20, and 109.50 ppm. This information collectively elucidates the molecular structure and confirms the synthesis of the ligand according to the expected chemical structure.

2.4. *Elemental Analysis and Molar Conductance*

The stoichiometry of the formed **L-Pd** complex with a 1:1 ratio and **L-Cr** and **L-Mn** complexes with a 2:1 ratio (L to Cr or Mn) was affirmed through elemental analysis, aligning with the anticipated proportions of carbon, hydrogen, and nitrogen, thereby confirming the adherence to their theoretical chemical formulas. Table 1 encapsulates the extensive outcomes of this research, substantiating the molecular structure of the synthesized complexes. For the L-Pd complex, this behavior can be attributed to the square-planar geometry preferred by Pd(II) ions, which coordinate with the inspected tri-dentate ligand to form a stable four-coordinate complex. The investigated Schiff base ligand acts as a tri-dentate donor, offering two nitrogen and one oxygen donor atoms, which can fully satisfy the coordination requirements of the Pd(II) center in a 1:1 molar ratio with acetate ions. Additionally, steric hindrance and electronic factors may limit the possibility of forming higher-order complexes.

The molar conductivities of the **L-Cr ([Cr(L)$_2$]NO$_3$.H$_2$O)**, **L-Mn ([Mn(L)$_2$].2H$_2$O)**, and **L-Pd** ([Pd(L)(CH$_3$COO)]·H$_2$O) complexes were determined in 10^{-3} M DMF solution. The recorded values for **L-Mn** and **L-Pd** were 8.50 Ω^{-1} cm^2 mol^{-1} and

10.21 8.50 Ω^{-1} cm^2 mol^{-1}, respectively, as indicated in Table 1. These results suggest that both **L-Mn** and **L-Pd** function as non-electrolytes [24], consistent with their low conductivity values. In contrast, the Cr(III) complex exhibits a higher conductivity of 60.50 8.50 Ω^{-1} cm^2 mol^{-1}, which is indicative of its nature as a mono-electrolyte [25].

2.5. SEM Analysis of the Prepared Nanosized Cr(III), Mn(II), and Pd(II) Metal Complexes

Scanning electron microscopy (SEM) was employed to examine the surface morphology, particle size, and structural features of the synthesized nanosized Cr(III), Mn(II), and Pd(II) metal complexes. The SEM images revealed that the complexes exhibit well-defined nanosized structures with varying morphologies influenced by the nature of the metal ion and synthesis conditions (Figure 1).

- The SEM images of the Cr(III) complex showed an aggregated but uniformly distributed morphology with spherical-like nanosized particles. The observed structures suggest a tendency for slight agglomeration due to intermolecular interactions.
- The Mn(II) complex displayed rod-like or irregular morphology, indicating variations in nucleation and growth mechanisms during sono-chemical synthesis.
- The Pd(II) complex exhibited a well-dispersed, granular, and slightly crystalline nanostructure. The smaller particle size observed for Pd(II) could be attributed to the strong coordination interactions between the Schiff base ligand and the palladium ion, stabilizing the nanostructures effectively.

The particle size of the metal complexes, estimated using ImageJ software 12 from the SEM micrographs, ranged between 25 and 45 nm with average of 36 nm for the L-Cr complex, 30–67 with average of 49 nm for the L-Mn complex and 50 and 110 nm with average 81 nm for the L-Pd complex confirming the nanoscale dimensions. The differences in morphology and size among the complexes highlight the impact of metal coordination chemistry on the self-assembly of nanosized structures. The results also indicate that the sono-chemical approach facilitated the formation of well-defined nanosized complexes with enhanced surface properties.

2.6. Electronic Absorption Spectra (EAS)

The detailed spectroscopic study of the **L** ligand, along with its complexes with **Cr(III)**, **Mn(II)**, and **Pd(II)**, was thoroughly recorded at a consistent concentration of 1×10^{-3} mol. dm^{-3} in DMF, as depicted in Figure 2 and outlined in Table S1. The ligand exhibited absorption maxima at 243 nm (ε_{max} = 1846 dm^3 mol^{-1} mm^{-1}), 256 nm (ε_{max} = 1771 dm^3 mol^{-1} mm^{-1}), and 346 nm (ε_{max} = 1836 dm^3 mol^{-1} mm^{-1}), which can be ascribed to the electronic transitions occurring between π–π* within the aromatic systems and n-π* within the heteroatom-containing aromatic structures. The absorption feature at 421 nm (ε_{max} = 1372 dm^3 mol^{-1} mm^{-1}) is indicative of an intra-ligand transition [26]. Upon complexation, new absorption bands appeared in the range of 331–325 nm, which can be attributed to π–π* transitions of the aromatic system and n–π* transitions associated with the C=N (azomethine), –CH=N–, and phenolic OH groups [27,28]. The spectral data suggests that the ligands are interacting with the respective metal ions, as evidenced by the observed shifts in band positions. The new absorption bands that appeared at 435 nm (ε_{max} = 1936 dm^3 mol^{-1} mm^{-1}) in the **L-Pd** and 383 nm (ε_{max} = 1182 dm^3 mol^{-1} mm^{-1}) in the **L-Mn** complexes are indicative of ligand-to-metal charge transfer (LMCT) transitions. These transitions likely originate from the p-orbitals of the ligands and are destined for the d-orbitals of the metal ions. Furthermore, the **L-Cr** and **L-Mn** complexes revealed an additional low-intensity broad peak at 475 nm (ε_{max} = 197 dm^3 mol^{-1} mm^{-1}) and 479 nm (ε_{max} = 1953 dm^3 mol^{-1} mm^{-1}), respectively, which was absent in the spectrum of the un-complexed ligand. This feature can be predominantly attributed to the d-d transitions within the metal chelates' structural frameworks, which are characteristic of the metal–ligand complexation process [29].

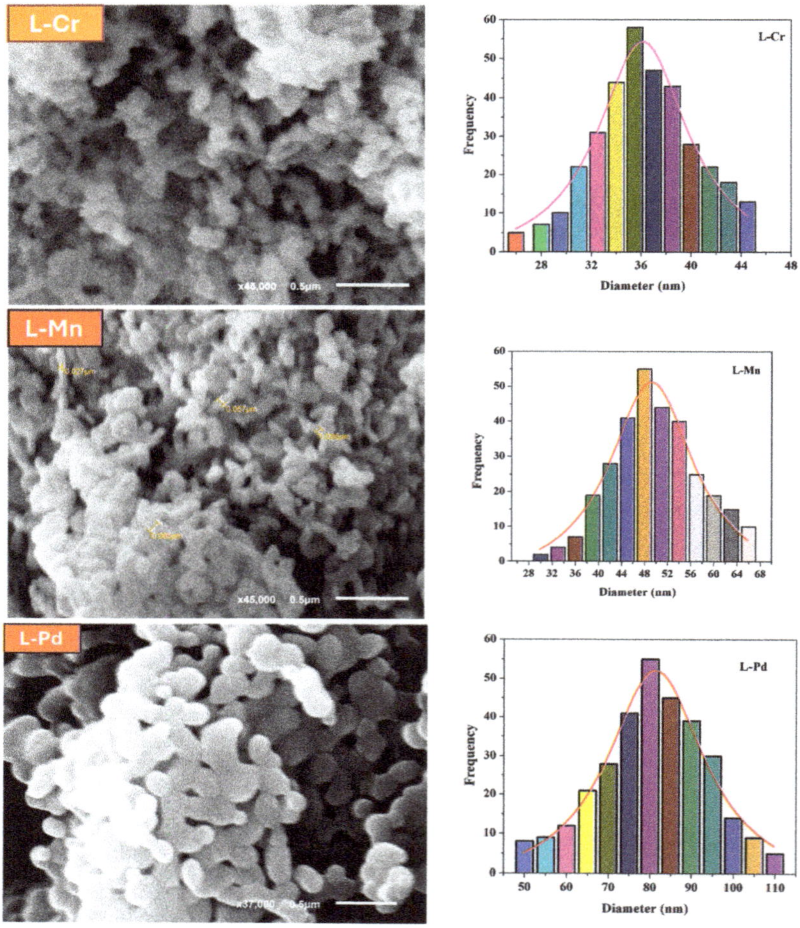

Figure 1. SEM images of the prepared nanosized metal chelates and particle size distribution for each complex.

2.7. Magnetic Moment

The magnetic moment of metal complexes provides valuable information about the electronic configuration and magnetic properties of the metal center and its surrounding ligands. The estimated magnetic susceptibility data obtained from this study closely align with the theoretical results derived from the equation $\mu_{eff} = [4S\ (S + 1)]^{0.5}$. The observed magnetic data for Cr (III), Mn (II), and Pd(II) chelates with the 4-chloro-2-(quinolin-8-yliminomethyl)-phenol ligand are presented in Table 1. The findings show that the magnetic moments for the L-Cr and L-Mn complexes are 3.65 and 5.32 B.M., respectively. These values are consistent with those expected for high-spin octahedral complexes of Cr (III) and Mn (II) [30]. While the synthesized **L-Pd** complex has tetra-coordinates, resulting in a square-planar structure with diamagnetic character [31].

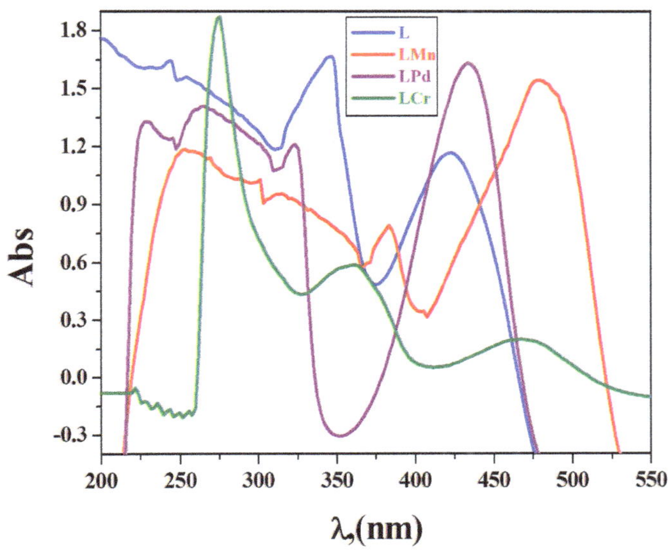

Figure 2. Electronic absorption spectra of all the prepared compounds in DMF at 298 K.

2.8. Thermal Analysis

Thermogravimetric analysis (TGA) was conducted to examine the thermodynamic stability and determine the presence of water within the crystal lattice of the synthesized compounds [32]. TGA/DTGA results for the 4-chloro-2-(quinolin-8-yliminomethyl)-phenol derived metal complexes are presented in Table 2 and Figure S4, with measurements taken between 30 °C and 800 °C. The [$C_{32}H_{22}Cl_2N_5O_6$-Cr] complex exhibits five distinct thermal degradation events. Initially, between 38 °C and 121 °C, a 2.62% mass loss occurs, which is indicative of the removal of a lattice water molecule. This is followed by an 8.96% weight reduction between 122 °C and 225 °C, corresponding to the loss of the nitrate (calc. 8.92%). The third event, at 230 °C to 305 °C, involves the elimination of a $C_9H_5NCl_2$ group with a mass loss of 28.40% (calc. 28.48%). A subsequent mass loss of 25.66% (calc. 25.59%) between 310 °C and 415 °C is attributed to the removal of a $C_{13}H_8N$ fragment. Lastly, from 420 °C to 595 °C, a $C_{10}H_7N_2O$ ligand is lost, resulting in an estimated weight decrease of 24.50% (calc. 24.58%). The ultimate residue following these degradation steps is chromium oxide. For the [$C_{32}H_{24}Cl_2N_4O_4$-Mn] complex, the TGA curve demonstrates five deterioration phases between 34 °C and 550 °C. The initial stage at 34 °C to 118 °C reveals the expulsion of two water molecules. The second phase, from 120 °C to 210 °C, corresponds to the loss of $C_7H_5Cl_2$, amounting to approximately 18.52% of the total mass (calc. 18.64%). Between 215 °C and 365 °C, a C_7H_4NO ligand is removed, contributing to a mass decrease of 19.13% (calc. 19.21%). At 370 °C to 460 °C, the TGA indicates the loss of a $C_{10}H_6N_2$ molecule, accounting for 23.60% of the mass (calc. 23.55%). Finally, the fifth stage at 465 °C to 545 °C corresponds to the degradation of a C_8H_5N unit with a mass loss of 20.70% (calc. 20.75%), leaving manganese oxide as the residue. The [$C_{18}H_{15}ClN_2O_4$-Pd] complex undergoes four weight loss stages. At the onset, from 38 °C to 125 °C, one water molecule is lost, equivalent to 3.80% of the mass (calc. 3.86%). The second stage, occurring between 125 °C and 232 °C, is associated with the removal of $C_2H_3O_2$, resulting in a 12.75% mass reduction (calc. 12.68%). The third stage, at 235 °C to 410 °C, involves the degradation of the C_7H_4NCl moiety, contributing to a 29.51% weight loss (calc. 29.55%). The last event, from 415 °C to 685 °C, corresponds to the elimination of a C_9H_6N group, with a calculated mass loss of 27.51%, ultimately resulting in palladium oxide as the remaining product after thermal decomposition.

Table 2. Thermal decomposition steps, mass loss (%), final residue, and thermo-kinetic activation parameters.

Complexes	Temp (°C)	Fragment Loss %		Weight Loss %		E* (KJmol^{-1})	A (S^{-1})	ΔH* (KJmol^{-1})	ΔG* (KJmol^{-1})	ΔS* (Jmol^{-1}K^{-1})
		M. Formula	M. Wt.	Found	Calc					
L-Cr 695.45 Residue	38–121	H_2O	18	2.62	2.58			30.06	51.73	−270.79
	122–225	NO_3	62	8.96	8.92			29.08	79.35	−275.25
	230–305	$C_9H_5NCl_2$	198.05	28.40	28.48	30.25	0.011	28.20	104.85	−277.84
	310–415	$C_{13}H_8N$	178	25.66	25.59			27.71	130.57	−283.36
	420–595	$C_{10}H_7N_2O$	171	24.50	24.58			26.51	171.88	−286.16
	>600	CrO	68	9.85	9.77			-	-	-
L-Mn 654.40 Residue	34–118	$2H_2O$	36	5.42	5.50			48.65	69.46	−273.73
	120–210	$C_7H_5Cl_2$	160	24.55	24.46			47.91	94.14	−280.18
	215–365	C_7H_4NO	118	17.95	18.04	49.28	0.008	46.87	129.48	−284.87
	370–460	$C_{10}H_6N_2$	154	23.60	23.54			45.83	165.29	−287.85
	465–545	C_8H_5N	115	17.55	17.58			45.08	191.27	−289.48
	>550	MnO	71	10.82	10.85			-	-	-
L-Pd 465.20 Residue	38–125	H_2O	18	3.80	3.86			25.66	48.015	−272.51
	125–232	$C_2H_3O_2$	59	12.75	12.68			24.86	74.80	−279.01
	235–410	C_7H_4NCl	137.5	29.51	29.55	26.35	0.01	23.66	115.36	−283.91
	415–685	C_9H_6N	128	27.55	27.51			21.77	180.36	−288.33
	>690	PdO	122.5	26.29	26.33			-	-	-

Kinetic Parameter

The thermodynamic parameters presented in Table 2 exhibit variations with respect to temperature. Specifically, the observed increase in G* values is indicative of a positive temperature coefficient, suggesting a rise in the rate of the degradation process as temperature escalates. This is consistent with the interpretation that higher temperatures are conducive to the activation of the system, thereby promoting the degradation reaction. The magnitude of the H* values, being positive, underscores the endothermic nature of the degradation. This implies that the reaction absorbs heat from its surroundings as it proceeds, which is characteristic of endothermic reactions that require additional energy to overcome the activation energy barrier. Concerning the entropy of activation (S*), most steps in the thermal degradation process exhibit negative values. This negative S* implies that the activated complex is more ordered than the reactants, which is indicative of a decrease in entropy during the reaction. This observation can be rationalized by considering that the degradation may occur through an abnormal mechanism, where the breakdown of molecular structures leads to a more ordered state at the intermediate stages. The negative values of S* can also be linked to the chemisorption of oxygen and the presence of other decomposition byproducts. The interaction of these species with the substrate may induce a more organized arrangement of molecules within the activated state, contributing to the reduced entropy. This ordered state is believed to facilitate the bond-breaking and bond-making processes necessary for degradation, thereby lowering the overall activation entropy. The nature of the activated state's increased order can be understood in terms of bond polarization. As the system transitions from reactants to the activated complex, there may be significant electronic rearrangements that lead to a more polarized state. These electronic transitions are essential for the cleavage of bonds and the formation of new ones, which are fundamental steps in the degradation process [33].

2.9. Stoichiometry of Complexes in Solution

Employing the techniques of continuous variation and molar ratio analysis, the stoichiometry of the synthesized complexes was ascertained [34–37]. The continuous variation graph revealed an absorbance peak at a ligand mole fraction range of 0.51–0.62, indicating the formation of complexes where Cr(III) and Mn(II) ions bind to the ligand in a 1:2 molar ratio and Pd(II) ion in a 1:1 molar ratio (Figure 3). This observation was corroborated by the molar ratio plot, which also affirmed these specific stoichiometric relationships between the metal ions and the ligand (Figure S5).

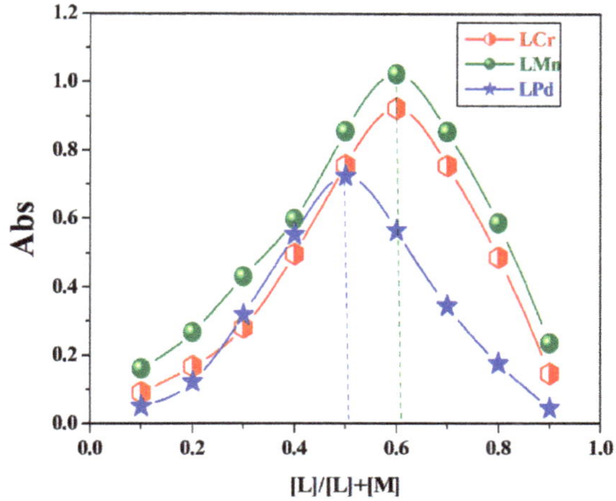

Figure 3. Continuous variation plots for the prepared complexes in DMF at [complex] = 1×10^{-3} M and 298 K.

2.10. The Apparent Formation Constants of the Synthesized Complexes

Evaluation of the Synthesized Complex Formation Constants: The synthesized complexes' formation constants (Kf) were derived from spectrophotometric data using the continuous variation technique, as presented in Table 3. The results indicate that the complexes exhibit substantial stability, with the order of stability being **L-Mn** > **L-Cr** > **L-Pd** complexes. Additionally, the stability constants (pK) and Gibbs free energy (ΔG*) values were determined for these complexes. The calculated ΔG* values are negative, suggesting that the complexation reactions occur spontaneously and are energetically favorable [35].

Table 3. The formation constant (K_f), stability constant (-log K_f), and Gibbs free energy (ΔG*) values of the prepared complexes in aqueous ethanol at 298 K.

Complexes	K_f	pK	$\Delta G^{\neq}$ kJ mol^{-1}
L-Cr	4.75×10^7	7.67	−43.99
L-Mn	6.18×10^7	7.79	−44.45
L-Pd	4.87×10^4	4.69	−26.67

2.11. pH Profile of the Investigated Complexes

The analysis of complex stability in relation to pH was conducted through spectrophotometry, which entailed measuring the absorbance of the complexes across an extensive pH spectrum from 2 to 12. The characteristic absorption bands corresponding to the metal–ligand charge transfer (LMCT) in the L-Pd complex and d-d transition in the L-Mn and L-Cr complexes were used to assess the pH dependence. This research aimed to establish the exact pH intervals wherein the complexes maintain their structural integrity, thereby ascertaining their operational effectiveness across a variety of environmental pH conditions, which is essential for their intended functionality. The stability curve presented in Figure 4 reveals that these complexes demonstrate persistent steadiness within a pH window of 4 to

11. Given this stability profile, it is anticipated that the complexes will function effectively in surroundings where the pH is contained within these bounds [38].

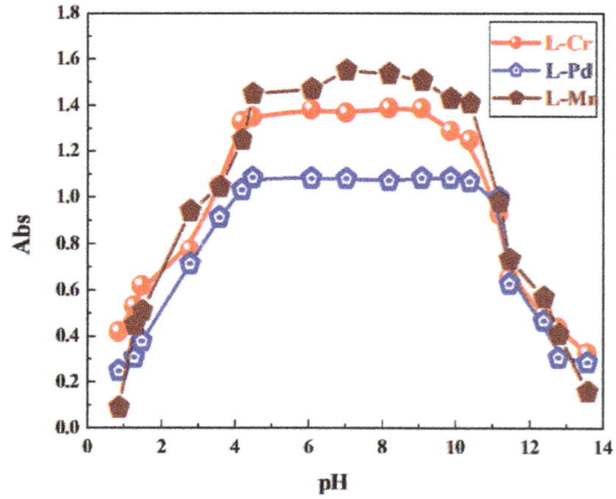

Figure 4. The effects of pH on the complexing elements in aqueous ethanolic media at 25 °C.

2.12. DFT Details

2.12.1. Geometry Optimization and Mulliken Charge

The geometry optimization of the L, L-Pd, L-Mn, and L-Cr compounds was performed using the DFT/B3LYP method in the gas phase. For this analysis, the 6-311g (d,p) basis set was applied to lighter atoms (C, H, N, Cl, and O), while the LANL2DZ basis set was used for the heavier atoms (Pd, Mn, and Cr). The optimized configurations for L, L-Pd, L-Mn, and L-Cr yielded minimum energies of −1261.48, −1616.22, −2625.44, and −2607.80 Hartree, respectively. These findings suggest that the metal complexes exhibit better stability energy than the free L. The calculated dipole moment values for L-Pd, L-Mn, and L-Cr are 9.45, 5.72, and 5.70 Debye, respectively, all of which are higher than the dipole moment of the free L, which is 4.33 Debye. These results show the strong dipole–dipole interactions within the metal chelates [39], emphasizing their crucial influence on both the structural and electronic properties of the systems. The optimized ground-state geometries of the compounds, along with key atomic numbering, are depicted in Figure 5. DFT analysis reveals that in the L-Pd complex, the metal ion coordinates with the O1, O2, N1, and N2 atoms, forming a square-planar structure. In contrast, the L-Mn and L-Cr complexes coordinate with the N1, N2, N3, and N4 atoms, as well as two oxygen atoms (O1 and O2), leading to the formation of an octahedral geometry. The Mulliken charge distribution offers important insights into the potential coordination sites of the ligand (L), helping to identify its chelating centers. By examining the gas-phase optimized geometry, the Mulliken charges of key atoms in L were determined. The atomic charges for N1 (−0.315), N2 (−0.451), and O1 (−0.348) were found to be negative, suggesting a strong tendency for electrophilic interactions. These negative charges indicate that these atoms are ideal coordination sites, facilitating the chelation of metal ions such as Pd^{2+}, Mn^{2+}, and Cr^{3+}.

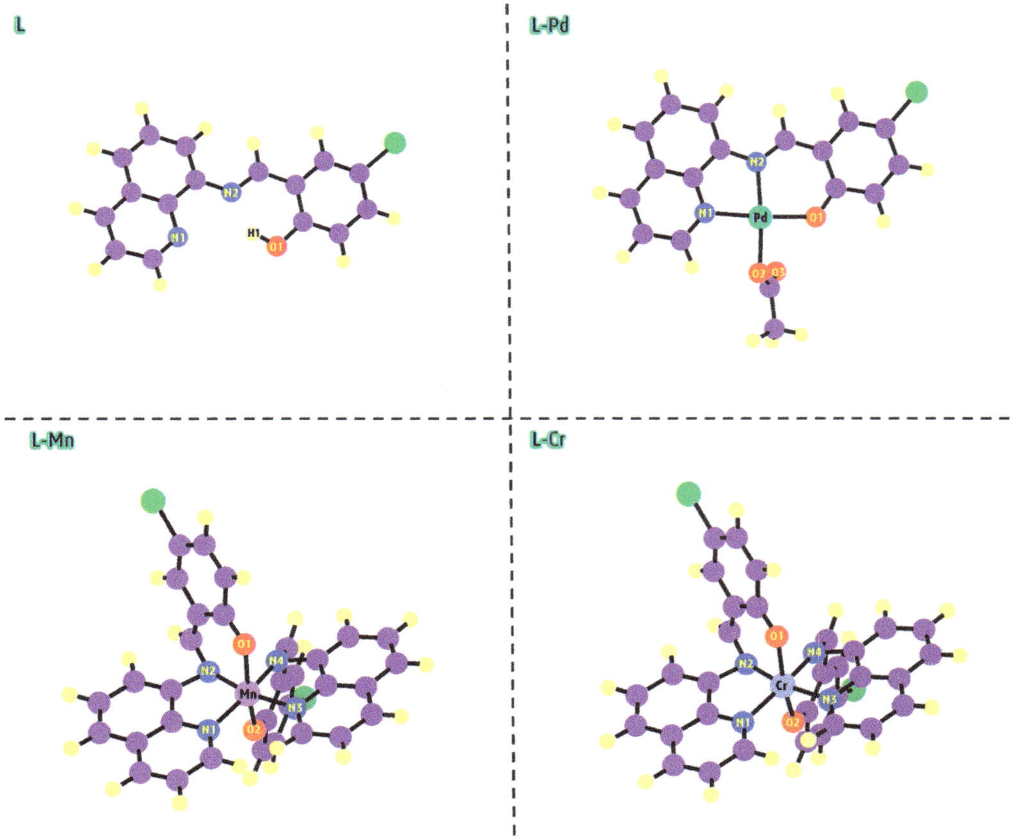

Figure 5. The proposed geometry of L, L-Pd, L-Mn, and L-Cr compounds by computational approach.

2.12.2. Electrophilic and Nucleophilic Reaction Sites

The MEP, known as the molecular potential surface map, is a valuable tool for analyzing charge-dependent interactions such as electrophilic and nucleophilic reactions, hydrogen bonding, and overall charge distribution within molecules. It also serves as an effective method for identifying reactive sites within a molecule [40]. The MEP maps of L, L-Pd, L-Mn, and L-Cr are illustrated in Figure 6, with color-coded representations indicating varying electrostatic potential levels. Typically, the electrostatic potential follows an increasing trend in the order: red < orange < yellow < green < blue [41]. In the visualization, green represents regions of neutral potential, yellow indicates areas with a slight electron density, and blue highlights nucleophilic zones with a higher positive charge distribution [42]. The MEP map of the L compound reveals a significant negative potential concentrated in the central region, specifically between the N1, N2, and O1 atoms. This area, highlighted in red, indicates a strong electrophilic reaction site, suggesting its potential role in interactions with positively charged species [43–45] such as Pd^{2+}, Mn^{2+}, and Cr^{3+}. These results align with the Mulliken charge distribution analysis, further confirming the active sites of the ligand (L) responsible for chelation with metal ions. In the L-Pd complex, the negative potential is predominantly concentrated around the O3 atoms, indicating that this region is highly favorable for electrophilic reactions and hydrogen bonding interactions. The MEP maps of the L-Mn and L-Cr complexes exhibit a comparable distribution, indicating similar regions of electrophilic and nucleophilic reactivity. The potential energy range for L, L-Pd, L-Mn, and L-Cr is −0.59 to +0.59 a.u., −1.06 to +1.06 a.u., −0.65 to +0.65 a.u., and −0.64 to +0.64 a.u., respectively.

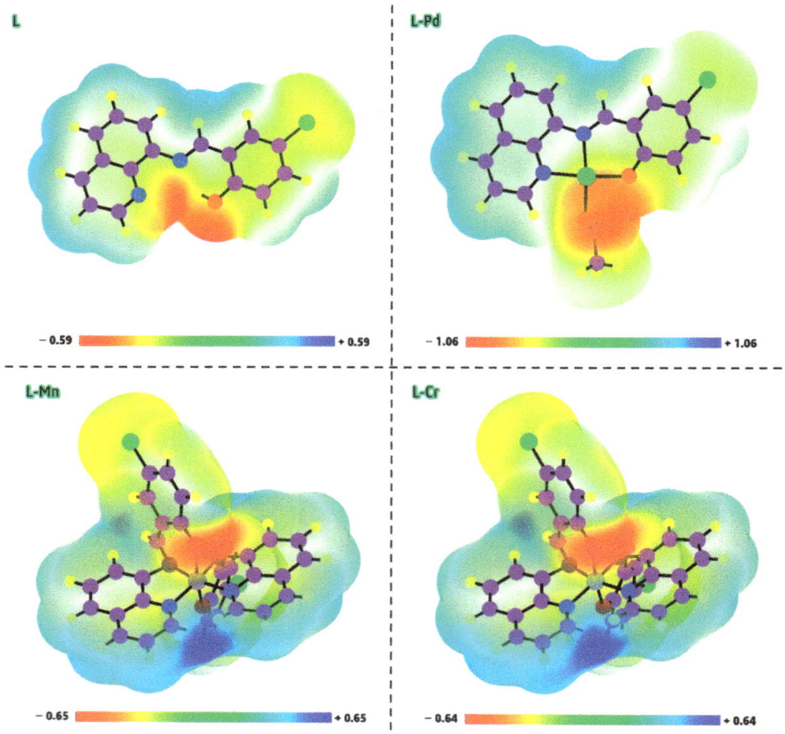

Figure 6. The MEP demonstration of L, L-Pd, L-Mn, and L-Cr compounds by computational approach.

2.12.3. Molecular Orbital Analysis

The HOMO and LUMO are key frontier orbitals that define a molecule's electronic properties and reactivity. These orbitals, representing the outermost boundaries of a molecule's electronic structure, play a fundamental role in various chemical reactions, particularly those involving electron transfer. HOMO is critical in electron-donating and nucleophilic reactions, as molecules with an intermediate HOMO energy serve as effective electron donors, readily participating in nucleophilic or oxidative processes. Conversely, LUMO represents the lowest accessible energy level for incoming electrons (Figure 7). Molecules with a low LUMO energy are excellent electron acceptors, making them highly reactive in electrophilic interactions. A crucial factor in molecular chemistry is the HOMO–LUMO energy gap. A smaller gap indicates higher reactivity, as the molecule requires less energy for excitation, enhancing its ability to engage in chemical transformations. The calculated energy values for L, L-Pd, L-Mn, and L-Cr are 3.72, 1.34, 2.71, and 2.66 eV, respectively. These findings suggest that the L-Pd complex exhibits the highest biological activity and chemical reactivity among the studied compounds. The observed trend in biological activity and reactivity follows the order of L-Pd > L-Cr > L-Mn > L, while the stability trend is reversed, following the sequence of L > L-Mn > L-Cr > L-Pd. Furthermore, the smaller energy gap of the Pd(II) complex enhances the efficiency of intramolecular charge transfer (ICT) along conjugated pathways. This facilitates electron transfer from electron-donating groups to electron-accepting groups within the molecule [46], thereby improving its electronic properties and reactivity. On the other hand, a smaller HOMO–LUMO gap typically indicates that the complex requires less energy for electronic transitions, thereby facilitating redox processes. The L-Pd complex, with its reduced HOMO–LUMO gap, is likely to possess a higher redox potential, making it more susceptible to electron exchange. This energy gap suggests that the L-Pd complex could exhibit enhanced redox activity, which may have important implications for its catalytic, electrochemical, or biological properties.

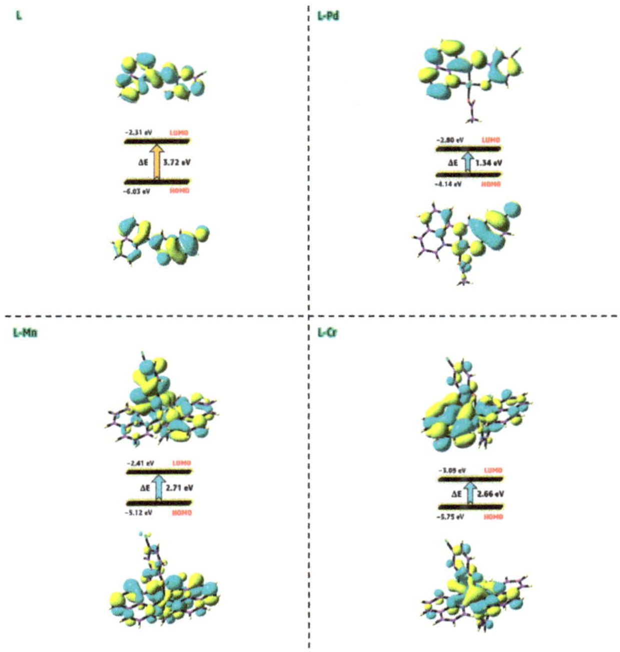

Figure 7. The predicted HOMO and LUMO contours of L, L-Pd, L-Mn, and L-Cr compounds.

2.12.4. Physicochemical Parameters

The energy values of the HOMO and LUMO in the compounds are essential for calculating several quantum indices, including chemical potential (P_i), maximum electronic charge transfer (ΔN_{max}), global electrophilicity (ω), absolute electronegativity (χ), absolute hardness (η), and absolute softness (σ). These parameters were derived using a series of mathematical equations, as outlined in Equations (1)–(6) [47].

$$Pi = -\chi \quad (1)$$

$$\chi = \frac{-(E_{HOMO} + E_{LUMO})}{2} \quad (2)$$

$$\eta = \frac{E_{LUMO} - E_{HOMO}}{2} \quad (3)$$

$$\sigma = \frac{1}{\eta} \quad (4)$$

$$\omega = \frac{Pi^2}{2\eta} \quad (5)$$

$$\Delta N_{max} = \frac{-Pi}{\eta} \quad (6)$$

Table 4 shows the values of the physicochemical parameters for the L, L-Pd, L-Mn, and L-Cr compounds, derived from the E_{HOMO} and E_{LUMO} values. A comprehensive analysis of the data provides valuable insights, highlighting key trends and characteristics of the L, L-Pd, L-Mn, and L-Cr compounds as follows:

(i) The degree of electron transfer within a compound can be assessed using the additional electronic charge parameter (ΔN_{max}), which measures a molecule's tendency to accept electrons from another species. Based on this parameter, the results suggest that the L-Pd, L-Mn, and L-Cr complexes possess enhanced electron transfer capabilities compared to the free ligand (L). Among these, the Pd(II) complex exhibits the highest electron-accepting ability, highlighting its superior electronic properties.

(ii) Balancing a compound's chemical reactivity with its hardness or softness is essential in determining its interaction potential. The Hard-Soft Acid-Base (HSAB) principle offers valuable insights into molecular reactivity, suggesting that hard acids preferentially bind to hard bases, while soft acids form more stable interactions with soft bases. In biological systems, key components such as cells and proteins are classified as soft molecules, making them more likely to interact with other soft molecules rather than hard ones. This explains why softer chemical environments generally enhance biological activity, whereas harder environments tend to suppress it [48]. Based on chemical hardness and softness parameters, the predicted reactivity trend for the studied compounds follows the order of L-Pd > L-Cr > L-Mn > L (Table 4), indicating that L-Pd exhibits the highest reactivity among them.

(iii) The global electrophilicity index (ω) quantifies a molecule's ability to accept electrons, classifying it as strong ($\omega > 1.5$ eV), moderate (0.9 eV $< \omega < 1.4$ eV), or marginal ($\omega < 0.8$ eV). The electrophilicity index values for the studied metal complexes range from 5.23 to 8.98 eV, confirming their strong electrophilic nature. This result suggests that these complexes possess significant reactivity, which may contribute to their potential biological activity [49].

Table 4. Physicochemical parameters for the L, L-Pd, L-Mn, and L-Cr compounds.

Parameter	L	L-Pd	L-Mn	L-Cr
E_{HOMO}	−6.03	−4.14	−5.12	−5.75
E_{LUMO}	−2.31	−2.80	−2.41	−3.09
$\Delta E_{(LUMO\text{–}HOMO)}$	3.72	1.34	2.71	2.66
χ	4.17	3.47	3.76	4.42
η	1.86	0.67	1.35	1.33
σ	0.53	1.49	0.74	0.75
P_i	−4.17	−3.47	−3.76	−4.42
ω	4.67	8.98	5.23	7.34
ΔN_{max}	2.24	5.18	2.78	3.32

Analyzing these parameters provides valuable insights into the reactivity, stability, and intermolecular interactions of molecules, along with their electrical and optical properties. This deeper understanding enables the development of novel materials with enhanced characteristics while also optimizing existing ones for a wide range of applications.

The DFT calculations presented in this study play a critical role in elucidating the electronic and structural features that govern the reactivity and biological behavior of the synthesized compounds. Key parameters such as HOMO–LUMO energy gaps, Mulliken charge distribution, molecular electrostatic potential (MEP), and global reactivity descriptors (electrophilicity, chemical hardness, and charge transfer capacity) offer detailed insights into the stability, coordination behavior, and potential interaction sites of the ligand and its metal complexes. Notably, the enhanced reactivity and biological affinity of the L-Pd complex can be rationalized by its low energy gap, high electrophilicity, and significant softness—attributes that align closely with its superior docking performance and experimental activity. These computational results, therefore, provide a predictive and mechanistic foundation that complements and substantiates our experimental findings.

2.13. Molecular Modeling

Molecular modeling is essential for accurately visualizing and predicting ligand–target interactions, making it a powerful tool for validating experimental results. In this study, molecular docking simulations were performed to reinforce the observed anticancer, antimicrobial, and antifungal properties of the synthesized compounds. The key docking representations highlighting their biological activity are summarized below.

To evaluate the antimicrobial potential of the compounds, the optimized structures were docked into the active site of 4EWP for *Micrococcus luteus* bacteria. Figure 8 provides a detailed visualization of the docking poses within the active site of 4EWP, along with the corresponding 3D interaction diagrams. The analysis revealed significant interactions with key amino acids, highlighting their potential effectiveness against microbial targets. According to the docking results, the free binding energies for L, L-Pd, L-Mn, and L-Cr were calculated as −6.67, −9.33, −8.63, and −8.68 kcal/mol, respectively. These findings indicate that the L-Pd complex exhibits the strongest binding affinity to the protein's active site compared to the other compounds, verifying the strong effect of this complex in the inhibition of *Micrococcus luteus*. The L-Pd complex establishes seven hydrophobic interactions between its aromatic rings and key amino acid residues of 4EWP, including ALA 338, LEU 236, MET 202, VAL 92, and VAL 156. Additionally, two hydrophobic interactions occur between a chlorine atom in the complex and the amino acids VAL 156 and PRO 95. Furthermore, a hydrogen bond is observed between the O1 atom and GLN 206, while a halogen bond is formed between the chlorine atom and THR 93.

To evaluate the anticancer potential of the compounds, molecular docking simulations were conducted using the 3HB5 breast cancer protein as the target. During the molecular docking process, L, L-Pd, L-Mn, and L-Cr were docked into the active site of 3HB5. The binding affinity values obtained were −6.78, −8.35, −6.98, and −7.03 kcal/mol, respectively. These results suggest that the L-Pd complex exhibits the strongest binding interaction with the target protein, indicating its potential for enhanced biological activity against breast cancer cells. A stronger interaction corresponds to a more negative binding energy. Accordingly, the binding strength follows the order of L-Pd > L-Cr > L-Mn > L. This trend aligns well with the experimental findings, further validating the computational results. The docking representation, along with the corresponding 3D interaction diagrams for the binding process of L, L-Pd, L-Mn, and L-Cr with 3HB5, is illustrated in Figure 9. The L-Pd complex forms five hydrophobic interactions between its aromatic rings and crucial amino acid residues of the breast cancer protein, including ILE 14, VAL 188, PHE 192, and PRO 187. Additionally, a chlorine atom in the complex engages in two hydrophobic interactions with LEU 149 and VAL 143. Moreover, two hydrogen bonds are established between the complex and the SER 142 and CYS 185 residues, further reinforcing its binding stability.

Our in vitro studies demonstrated the remarkable efficacy of these metal chelates in inhibiting fungal growth. To gain deeper insights into their mechanism of action, docking simulations were performed to analyze their molecular interactions in detail. The docking analysis specifically targeted the active site of *Candida albicans* (PDB code: 5V5Z), providing a comprehensive understanding of their antifungal properties. The binding free energy values of L, L-Pd, L-Mn, and L-Cr within the active site of *Candida albicans* were determined to be −7.16, −10.56, −8.26, and −10.35 kcal/mol, respectively. Notably, the L-Pd complex demonstrated the highest binding affinity, suggesting its potential for strong interactions and effective inhibition of *Candida albicans*. Figure 10 demonstrates the interaction diagrams between L, L-Pd, L-Mn, and L-Cr and the active site of *Candida albicans*. The L-Pd complex establishes three hydrogen bonds with the amino acid residues CYS 470, GLY 472, and ILE 471 through the O3 atom. Additionally, fifteen hydrophobic interactions were identified between the Pd(II) complex and the amino acid residues ILE 131, LEU 300, ILE 304, LEU 204, LEU 276, ALA 149, LEU 150, ALA 146, and LYS 143, demonstrating its strong binding affinity and potential stability within the biological system.

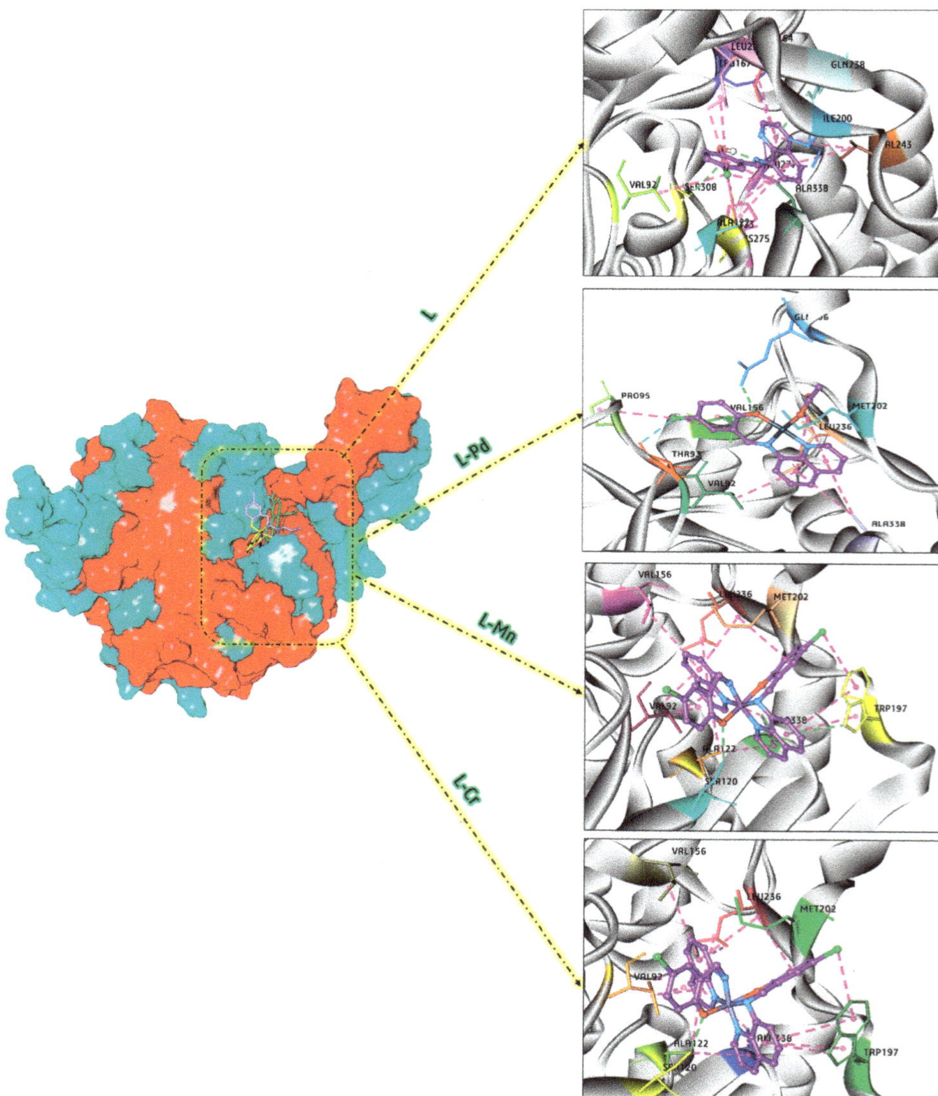

Figure 8. Docking analysis of L, L-Pd, L-Mn, and L-Cr in the interaction with the 4EWP target.

Theoretical calculations indicate that the studied chelates possess small energy gap values, suggesting their potential for efficient charge transfer to breast cancer protein,

Micrococcus luteus, and *Candida albicans*. Additionally, the high global electrophilicity index further confirms the enhanced reactivity of these compounds toward these biological targets. Consequently, all DFT parameters validate the superior biological activity of the L-Pd complex, as corroborated by both molecular docking and experimental findings.

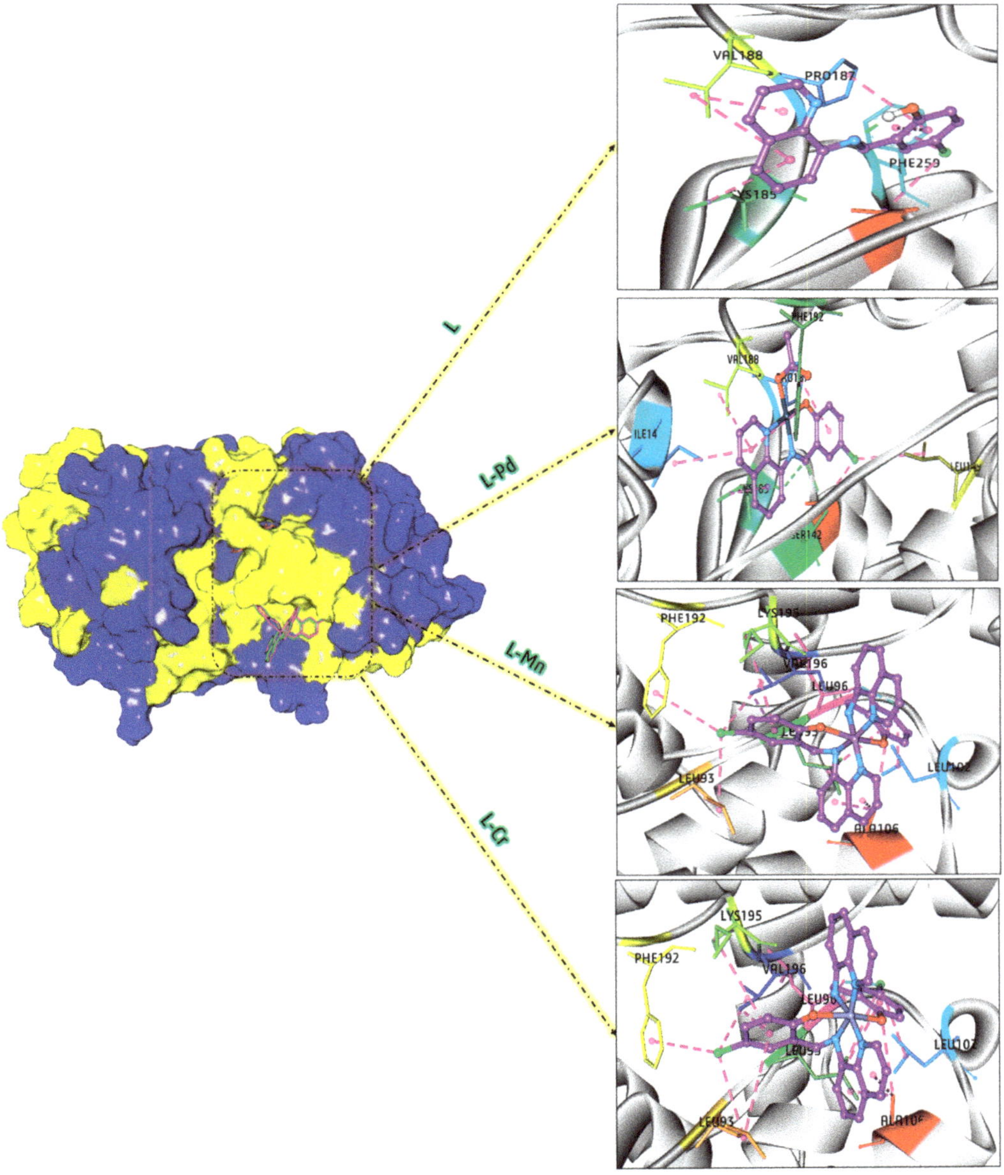

Figure 9. Docking analysis of L, L-Pd, L-Mn, and L-Cr in the interaction with the 3HB5 target.

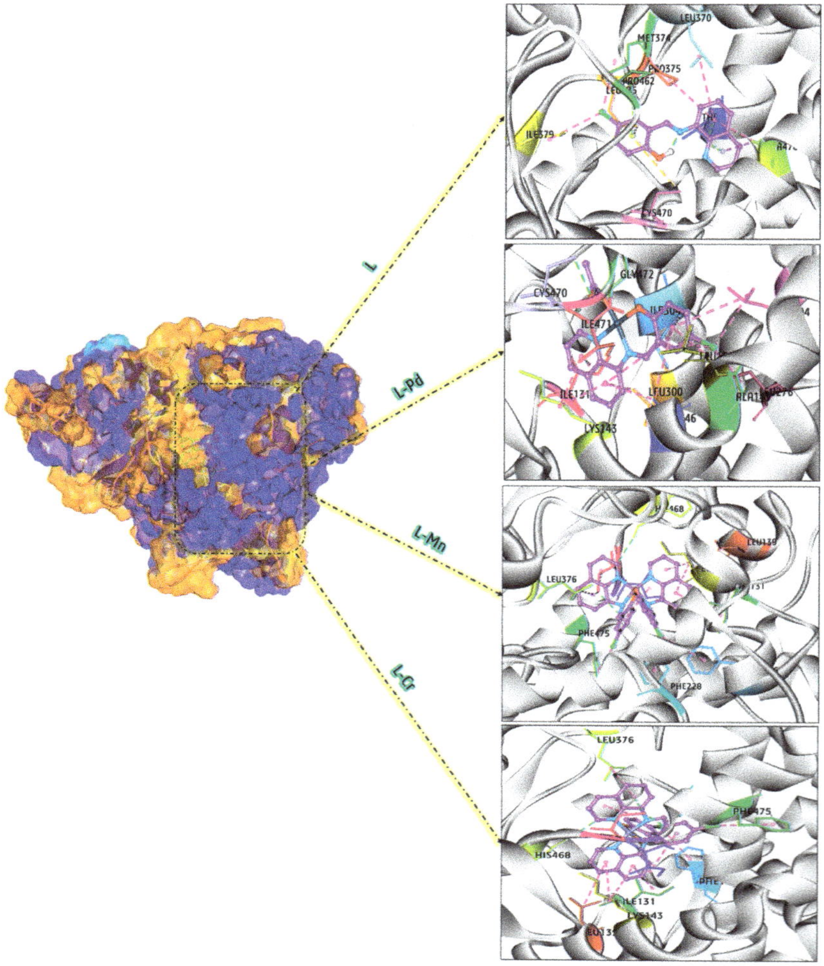

Figure 10. Docking analysis of L, L-Pd, L-Mn, and L-Cr in the interaction with the 5V5Z target.

2.14. Biological Activity

2.14.1. Antimicrobial Activity

The study evaluated the antimicrobial properties of a synthesized 4-chloro-2-(quinolin-8-yliminomethyl)-phenol ligand and its corresponding metal complexes against specific bacterial strains, including *S. marcescens*, *M. luteus*, and *E. coli*, as well as fungi such as (*A. flavus*, *C. candidum*, and *F. oxysporum*). Ofloxacin and fluconazole were employed as benchmarks for antibacterial and antifungal activities, respectively. The ligand–palladium complex (**L-Pd**) exhibited the most significant antimicrobial effect among the compounds examined, as indicated by the inhibition zone diameter measurements (mm/μg of sample). This enhanced activity in the metal complexes is likely attributed to chelation theory, which suggests that the formation of a chelate complex can increase the biological efficacy of the ligand by improving its ability to interact with biological targets [50,51]. The varying antimicrobial potencies observed may be a result of differences in the cellular structure, particularly the cell wall composition, of the microbial species under study [52]. The minimum concentration required to inhibit the growth of the microorganisms (MIC) for the synthesized compounds is presented in Table 5. Additionally, to affirm the accuracy of the antimicrobial assessment, an activity index calculation was performed, as outlined in Table 6. It is important to note that the concentration of dimethyl sulfoxide (DMSO) used in the preparation of these samples did not surpass 0.5%, ensuring that it would not interfere with the experimental outcomes [53]. The actual inhibition zones for both bacterial and fungal species are detailed in Tables S2 and S3 of the Supplementary Materials.

Table 5. The results of the minimum inhibition concentration (MIC) of the ligand and its metal complexes against the different strains of bacteria and fungi.

Compound	(MIC) Minimum Inhibition Concentration μg/mL					
	Bacteria			Fungi		
	S. marcescence	*E. coli*	*M. luteus*	*A. flavus*	*C. albicans*	*F. oxysporum*
L	7.5	8.25	6.5	8.75	6.25	7
L-Cr	2.75	3.5	2	3.75	2.25	3.25
L-Mn	3	3.75	2.5	4	3	3.75
L-Pd	2.25	3	1.75	3	2.25	2.75
Ofloxacin	2	2.75	1.5			
Fluconazole				2.25	1.75	2.5

Table 6. Antimicrobial activity index (%) of the ligand and its complexes.

Compound	Activity Index (%)					
	Bacteria			Fungi		
	S. marcescence	*E. coli*	*M. luteus*	*A. flavus*	*C. albicans*	*F. oxysporum*
L	44.05	38.48	37.7	39.41	34.96	33.65
L-Cr	94.82	91.58	95.06	92.87	94.09	91.23
L-Mn	89.94	86.97	93.22	88.26	91.26	86.28
L-Pd	97.71	98.2	98.74	96.23	95.37	95.85

2.14.2. Determination of Minimum Inhibition Concentration

The outcomes from the serial dilution assay (MIC) exhibit a strong congruence with the results obtained via the disk diffusion method, as depicted in Table 5 and Figures 11 and 12. In contrast to other complexes examined, the **L-Pd** complex has demonstrated remarkable bio-efficiency, with MIC values of 2.25, 3.00, and 1.75 M against *S. marcescens*, *E. coli*, and *M. luteus*, respectively. Amongst the bacterial species under study, *M. luteus* emerged as the most susceptible, while *E. coli* displayed the highest resilience. Notably, *G. candidum* was identified as the most sensitive fungus. This enhancement in antimicrobial potency can be rationalized in light of Overtone's principles [54–56], which suggest that chelation enhances the ligand's function as a more potent antibacterial agent by constraining bacterial proliferation. The significant increase in the inhibitory zones of metal chelates versus free ligands is indicative of the metal's ability to interact more effectively with biological targets due to the reduced polarity resulting from the shared positive charge with the ligand's hetero atoms and the delocalization of π electrons throughout the chelate framework. The bacterial cell wall and membrane, rich in lipids and polysaccharides, provide an optimal environment for metal ion interaction. The varying sensitivity of different bacteria to these complexes may be attributed to differences in their ribosomal structure or cellular permeability. Moreover, the metal complexes' effectiveness could be influenced by factors such as solubility, size, dipole moment, metal ion redox potential, bond length, complex geometry, and hydrophobicity. The steric hindrance may also affect their antimicrobial activity since limited lipid solubility can impede the metal ion's ability to reach the cell wall's active site, thereby hindering the complexes' antibacterial efficacy. The data suggest that the antibacterial performance of metal complexes is not solely determined by chelation but is instead a multifaceted phenomenon influenced by a confluence of factors. These include the complex's solubility, the metal–ligand bond length, the metal ion's redox state, its spatial arrangement within the complex, the molecule's overall shape, and the physicochemical properties that govern its interaction with bacterial membranes and subsequent cell penetration. Additionally, pharmacokinetic factors such as concentration and the molecule's tendency to partition between hydrophilic and hydrophobic environments contribute to the overall antimicrobial efficacy. Thus, the observed increase in antibacterial activity of metal complexes can be ascribed to a composite of various mechanisms beyond chelation, which may interact synergistically to inhibit microbial growth [57,58].

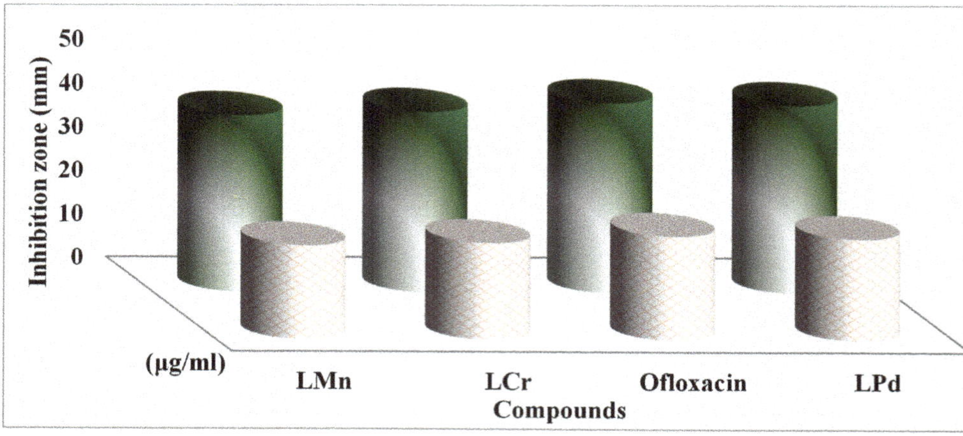

Figure 11. Antibacterial activity for the compounds under inspection against *M. luteus* bacteria.

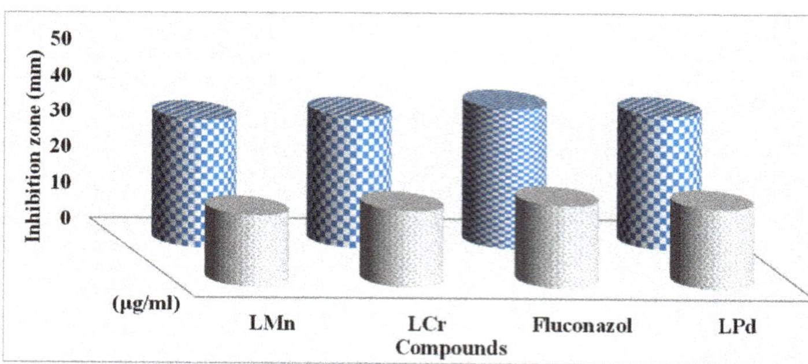

Figure 12. Antifungal activity for the compounds under inspection against *C. albicans* fungi.

2.15. Anti-Cancer

The study examined the antitumor properties of the 4-chloro-2-(quinolin-8-yliminomethyl)-phenol ligand and its associated complexes with Cr(III), Mn(II), and Pd(II) metals in HepG-2, HCT-116, and MCF-7 cell lines using IC_{50} values (Table S4 and Figure 13). It was found that both the ligand and its metal derivatives suppressed cell proliferation in a dose-dependent manner, although their efficiencies varied significantly (as depicted in Figure 13). Upon analysis of the IC_{50} values (Table S4), it was determined that all metal complexes exhibited stronger inhibitory effects than the ligand alone. Notably, the Pd(II) complex displayed the highest potency, with IC_{50} values of 3.82 and 4.17 µg/µL against MCF-7 and HepG-2 cells, respectively, which were comparable to the reference drug Cisplatin at 10.75 and 6.65 µg/µL. This suggests that the **L-Pd** complex holds promise as a potential treatment for hepatic tumors. Additionally, the **L-Cr** complex demonstrated substantial inhibitory activity against the MCF-7 cell line, with an IC_{50} value of 5.86 μg/μL. The relative cytotoxic potencies of the metal complexes can be arranged as follows: L-Pd > L-Cr > L-Mn > ligand. The sensitivity of the cancer cell lines to these substances was observed in the order of MCF-7 > HepG-2 > HCT-116. These results collectively suggest that the coordination of metal ions to the ligand enhances its anticancer capabilities. The primary factor contributing to the variation in biological activity among these compounds appears to be the alteration of coordination sites, which may influence the formation of more effective hydrogen bonds with the negatively charged DNA of cancer cells due to the metal ions' positive charge [59–62].

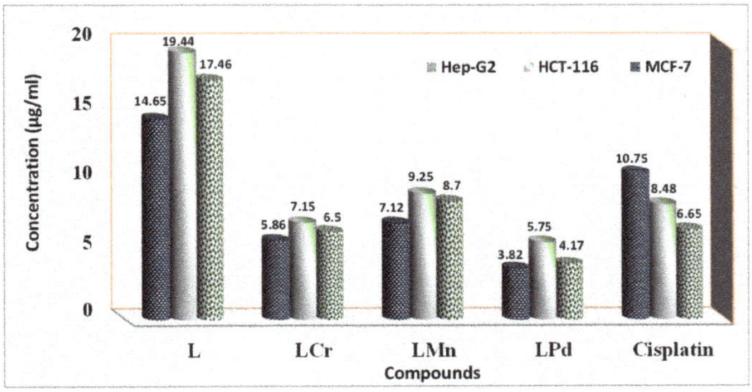

Figure 13. IC_{50} of the ligand, its metal chelates, and the medication Cisplatin versus Hep-G2, HCT-116, and MCF-7 cell lines.

The cytotoxicity assessment of the investigated compounds revealed significant differences in their biocompatibility toward normal human cells. The investigated imine ligand exhibited the highest IC_{50} values, averaging 356.63 and 415.42 μg/μL for human dermal fibroblasts and PBMCs, respectively, indicating excellent biocompatibility and minimal toxicity. In contrast, PdL nanoparticles showed lower cytotoxicity, with IC_{50} values ranging from 218.6 to 263.75 μg/μL, likely due to oxidative stress induced by Pd^{2+} ion release.

When compared with the standard anticancer agent cisplatin, which exhibited IC_{50} values of 192.85 μg/μL for dermal fibroblasts and 205.30 μg/μL for PBMCs, the investigated compounds demonstrated substantially lower toxicity profiles. This stark contrast highlights the potential of investigated imine complexes as safer alternatives for biomedical applications, especially in drug delivery or theranostics, where minimizing off-target toxicity is crucial.

Imine (Schiff base) metal complexes exhibit significant anticancer activity through multiple mechanisms [34,36] (Figure 14). They can interact with DNA by intercalation or covalent bonding, disrupting replication and transcription, ultimately leading to cell death. Additionally, these complexes induce reactive oxygen species (ROS) production, causing oxidative stress and cellular damage. They also inhibit key enzymes such as topoisomerases and proteasomes, preventing cancer cell proliferation. Furthermore, some imine complexes disrupt mitochondrial function, triggering apoptosis through intrinsic pathways. The metal center in these complexes can participate in redox reactions, enhancing cytotoxic effects. Overall, imine metal complexes exert anticancer effects by inducing DNA damage, oxidative stress, enzyme inhibition, and mitochondrial disruption, making them promising candidates for chemotherapy.

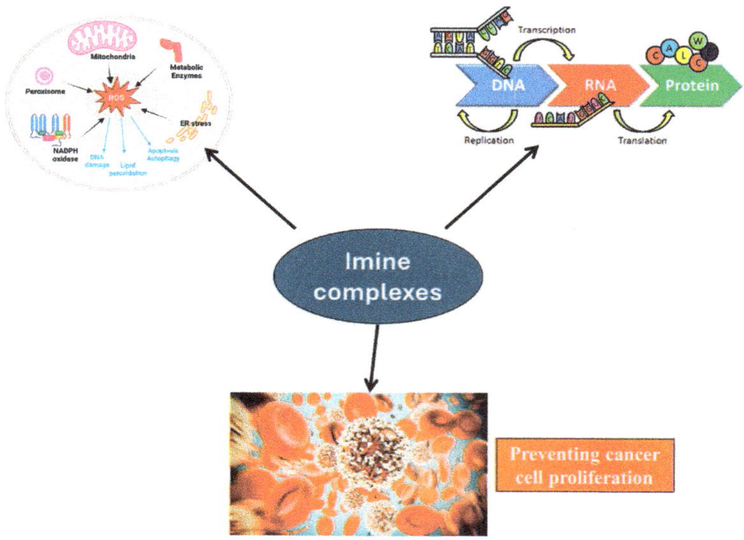

Figure 14. Pathway for anticancer mechanism of the investigated imine complexes. Figure generated using AI-assisted illustration tools for conceptual visualization purposes only.

2.16. Examination of DPPH Radicals Scavenging Efficiency

Oxidative processes mediated by free radicals are substantially implicated in the pathogenesis of various human ailments, including the aging process [63]. The novel synthesized compound, 4-chloro-2-(quinolin-8-yliminomethyl)-phenol, and its corresponding metal complexes were subjected to in vitro antioxidant assays to explore their therapeutic potential. The DPPH free radical scavenging method was employed for these evaluations due to its widespread use and practical advantages. The assay involved preparing solutions with varying concentrations of the ligand and its metal chelates (10, 25, 50, 100, and 150 μg ml^{-1}) to assess their antioxidant efficacy. Ascorbic acid served as a benchmark for comparison purposes. The findings reveal that the metal-bound forms of the ligand generally displayed greater antioxidant capacity than the free ligand itself. Specifically, it was observed that the antioxidant potency of all the complexes escalated in tandem with concentration augmentation. Further analysis demonstrated that the **L-Pd** complex presented the most proficient antioxidant activity, characterized by the lowest IC_{50} value of 15.35 μg/mL, as depicted in Figure 15 and Table S5. Conversely, the **L-Mn** complex exhibited the least antioxidant efficacy among the examined complexes, with an IC_{50} value of 25.60 μg ml^{-1}. This suggests that metal chelation can enhance the antioxidant properties of the ligand, potentially offering a promising avenue for the development of new antioxidant therapies.

The correlation between the prepared imine complexes and their biological activity is influenced by several factors. Firstly, the nanosized form provides a significantly higher surface-area-to-volume ratio, enhancing interactions with biological targets such as enzymes, DNA, and cellular membranes. This leads to improved solubility y and bioavailability, making the complexes more effective in biological systems. Additionally, their reduced size facilitates cellular uptake through mechanisms like endocytosis or passive diffusion, allowing for better penetration and interaction with intracellular targets, which can enhance antibacterial, antifungal, or anticancer activity.

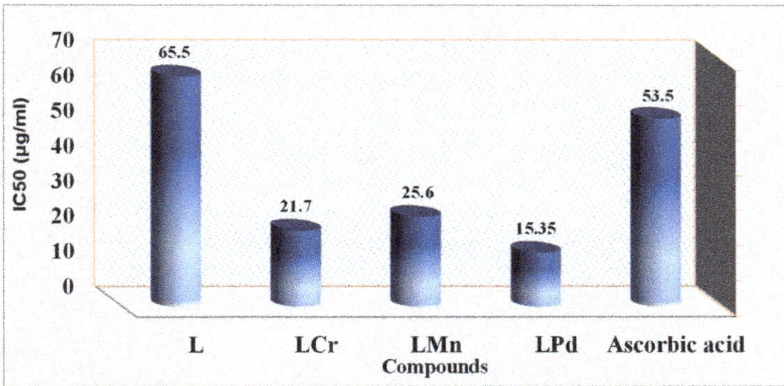

Figure 15. The ligand and its metal complexes inhibit the DPPH radical.

The nanoscale morphology of the solid complexes may influence parameters such as dissolution rate and bioavailability; however, upon dissolution in DMSO, the biologically active species are expected to be discrete metal–ligand molecules. Therefore, docking studies and mechanistic interpretations are based on the molecular form rather than nanoparticle aggregates. Their enhanced chemical and thermal stability prevent premature degradation, allowing for a sustained release of the active compound, which in turn improves therapeutic efficacy while minimizing toxicity. Many of these complexes incorporate metal ions such as Cr, Mn, or Pd, which contribute to biological activity through mechanisms like reactive oxygen species (ROS) generation, enzyme inhibition, or DNA cleavage, leading to potentiated antimicrobial and anticancer effects.

Another key advantage of nanosized imine complexes is their potential for selective targeting. Their surface can be functionalized to improve specificity toward diseased cells, such as cancerous cells, while reducing interactions with healthy cells, thereby minimizing off-target toxicity. This selective targeting, combined with their enhanced bioavailability and stability, makes nanosized imine complexes promising candidates for biomedical applications. Overall, their nanoscale properties significantly enhance their biological activity, making them more efficient and effective compared to their bulk counterparts.

3. Experimental

3.1. Reagents

All chemicals and solvents employed in the research were of analytical purity and utilized directly as obtained, without further purification. These comprised the precursors for the synthesis of imine ligands, namely 8-aminoquinoline and 5-chloro-2-hydroxybenzaldehyde (98%), as well as ethanol (99.9%), acetone (99%), N, N′-dimethylformamide (DMF) (98.5%), and dimethylsulfide (DMSO) (99%), obtained from Sigma-Aldrich, Schnelldorf, Germany. The study involved the use of these solvents without prior distillation. For the preparation of imine metal complexes, the following transition metal salts purchased from Sigma-Aldrich were utilized: $Mn(NO_3)_2{\cdot}4H_2O$ (99%), $Pd(OAc)_2$ (98.5%), and $Cr(NO_3)_3{\cdot}9H_2O$ (98%). These salts were selected for their established role in facilitating the formation of complex compounds and were essential to the experimental procedures conducted.

3.2. Instrumentation

All instrumentation and methods which used in the current investigation are supplied in the Supplementary Materials.

3.3. Synthesis of Tri-Dentate L Imine Ligand

The compound 4-chloro-2-(quinolin-8-yliminomethyl)-phenol (L) was synthesized using a condensation reaction (cf. Scheme 1). The process involved the combination of 8-aminoquinoline (2.88 g, 20 mmol) and 5-chloro-2-hydroxybenzaldehyde (3.13 g, 20 mmol) in a solvent of ethanol, EtOH (25 mL). To facilitate the reaction, the solution was heated to reflux for a duration of three hours. Upon completion of the heating period, the reaction mixture was allowed to cool to room temperature, which promoted the precipitation of a brown solid. The precipitate was isolated through the process of filtration to separate it from the remaining liquid components. Following the separation, the solid was further purified by recrystallization in ethanol, and dried in a desiccator to remove any adsorbed solvent molecules.

L ^1H NMR (400 MHz, DMSO-d_6): d (ppm) 13.98 (s, 1H, OH), 9.26 (s, 1H, CH=N), 8.83–8.80 (d, 1H, CH_{arm} adjacent nitrogen of quinoline moiety), 8.61–8.07 (d, J = 6.9 Hz 3H, CH_{arm} of pyridine ring in quinoline moiety), 7.90–6.83 (d, J = 7.1 Hz 3H, CH_{arm}), 7.42–6.81 (d, J = 7.1 2H, CH_{arm} of quinoline moiety). ^{13}C NMR (δ, ppm) in DMSO-d6: 167.30 (CH=N), 163.60 (C-OH), 152.20 (C=N of pyridine ring in quinoline moiety), 140.50 (right carbon adjacent nitrogen tom in quinoline moiety), 138.20 (left carbon adjacent nitrogen tom in quinoline moiety), 134.50 (carbon in para position for nitrogen tom in quinoline moiety), 132.10 (right carbon adjacent chloride atom in phenyl ring), 128.70 (left carbon adjacent chloride atom in phenyl ring), 127.10, 125.50, 124.70, 123.10, 121.10, 120.00, 115.20, 109.50.

3.4. Sono-Chemical Synthesis of Pd(II), Mn(II), and Cr(III) Complexes with 4-Chloro-2-(Quinolin-8-Yliminomethyl)-Phenol Ligand

Sono-chemical synthesis refers to the use of ultrasonic waves to promote chemical reactions. This method is particularly useful in synthesizing metal complexes, including **Salen metal complexes**, as it enhances reaction rates, improves yields, and leads to better product purity.

A methanolic solution (20 mL) of $Cr(NO_3)_3.9H_2O$ or $Mn(NO_3)_2.4H_2O$ or $Pd(OAc)_2$ (2 mmol) is prepared and added dropwise to a methanolic solution of the ligand (2 mmol) under continuous **ultrasonic irradiation** at 40 kHz. The reaction mixture is subjected to sonication at room temperature for 1–2 h, leading to the formation of the investigated complexes (cf. Scheme 1). The obtained precipitate is filtered, washed with ethanol and ether, and dried under vacuum.

3.5. Estimating the Stoichiometry of Chelates Using Job's and Molar Ratio Methods

The solubility and molar masses of Cr(III), Mn(II), and Pd(II) complexes were scrutinized in this study. Employing Job's continuous variation approach, the researchers examined the stability of these complexes in solution by systematically varying the molar ratios of the metal (M) and ligand (L) species. Following the preparation of these mixtures, the systems were meticulously equilibrated, and the digestibility rate for each specific combination was meticulously noted. The analysis was conducted by monitoring the absorbance of the individual solutions, with the ligand mole fraction ([L]/[L] + [M]) or the molar ratio ([L]/[M]) serving as key parameters to evaluate the stability and molecular weight of the formed complexes [64].

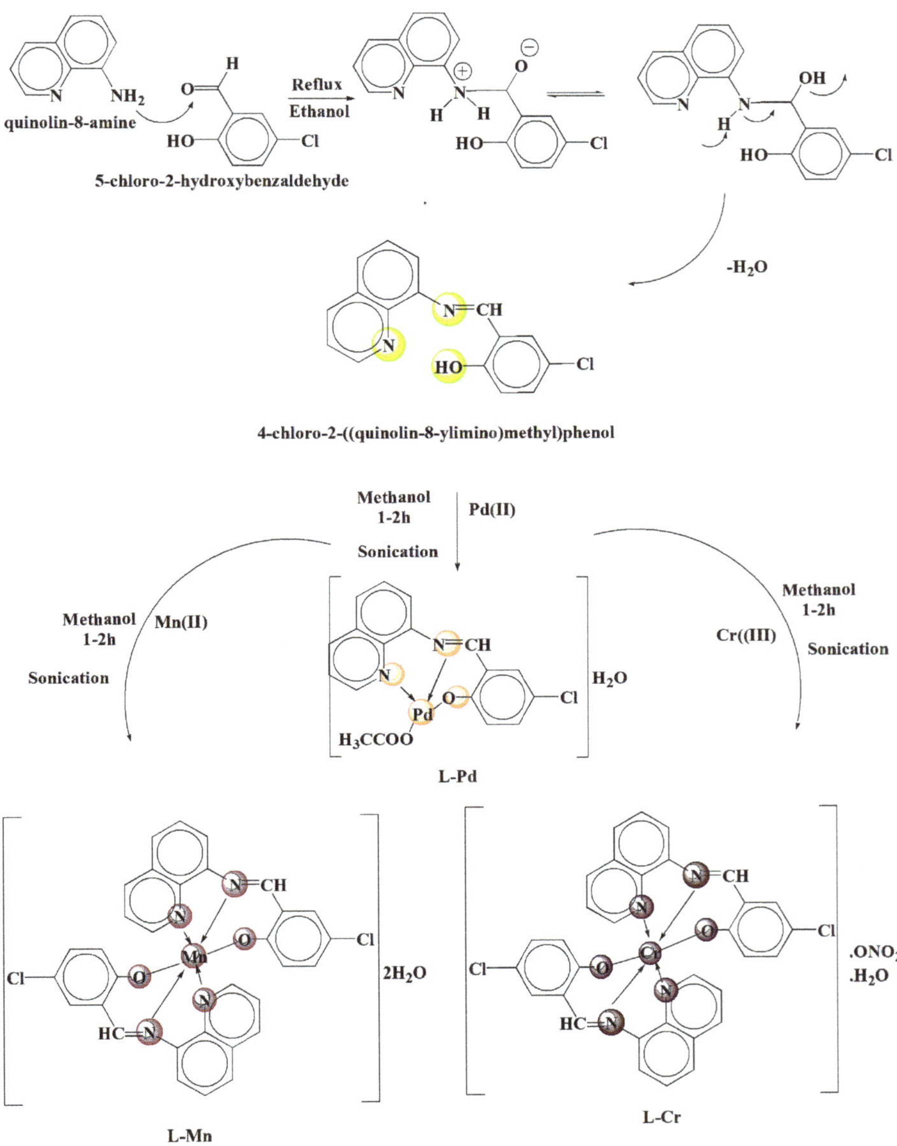

Scheme 1. Diagrammatic approach to the synthesis of the ligand and associated metal complexes.

3.6. Assessment of the Apparent Constant of Complexes

The formation constants (K_f) of the examined chelates have been derived for intensively studied systems, utilizing Job's method in conjunction with spectrophotometric analysis. This technique involves the following equations for different molar ratios:

For a 1:1 molar ratio [37,65],

$$k_f = \frac{\frac{A}{A_M}}{C\left(1 - A/A_m\right)^2} \tag{7}$$

Conversely, for a 1:2 molar ratio,

$$k_f = \frac{\frac{A}{A_M}}{4C^2\left(1 - A/A_m\right)^3} \tag{8}$$

Here, A represents the starting molar concentration of the metal ion, A_M is the maximum transmittance corresponding to the formation of the topmost chelate complex, and C denotes the transmittance at a specific wavelength within the transmittance range.

To evaluate the Gibbs free energy changes (ΔG) associated with these chelate formations, the following equation has been applied:

$$\Delta G = -RT \ln(K_f) \tag{9}$$

where R is the universal gas constant, T is the temperature in Kelvin, and K_f is the formation constant of the chelate. This relationship allows for the determination of the thermodynamic stability of the complexes formed.

3.7. Magnetic Moment Measurements

The assessment of magnetic susceptibility is a highly proficient strategy for examining the structural arrangement of complexes involving transition metals [66]. This process entails the determination of the molar susceptibility of the coordination complex, which is achieved by implementing diamagnetic corrections to account for the presence of other ions or molecules in the sample mixture. This methodology allows for the quantification of the susceptibility of a single paramagnetic metal atom within the material under investigation [67]. Several approaches are available for determining these diamagnetic adjustments; however, one prevalent technique relies on the premises established by Pascal. This approach incorporates specific relationships and equations that facilitate the precise calculation of magnetic susceptibility and the resultant effective magnetic moment.

$$\mu_{eff} = 2.83\sqrt{X_A.T} \tag{10}$$

$$X_A = X_M - (\text{diamagnetic correction}) \tag{11}$$

$$X_M = X_g.M.wt \tag{12}$$

The expression involves several variables: T represents absolute temperature in Kelvin, eff signifies the practical magnetic moment in Bohr Magnetons (B.M), X_g indicates the measured gm magnetic susceptibility, X_M denotes the molar magnetic permeability prior to correction, and X_A symbolizes the molar magnetic permeability following correction

3.8. Spectrophotometric Studies

For the purpose of examining ultraviolet–visible light wavelengths spanning from 200 to 700 nanometers, standard solutions of metal chelate with a concentration of 10^{-3} mol/L were concocted. This was accomplished by meticulously measuring and adding the required volumes of the metal compounds to dimethylformamide, serving as the solvent medium.

3.9. Thermogravimetric Analysis and Kinetic Studies

To evaluate the thermal stability of the complex, thermogravimetric analysis (TGA) was employed to determine the solvent's role within the coordination environment. Additionally, it is imperative to establish the metal content of the mixture. This heating process is governed by kinetics, which allows for the assessment of both kinetic and thermodynamic properties. For this purpose, the Coats–Redfern method was utilized, as detailed in the literature [68,69].

The kinetic equation used is given by

$$\ln = \left[\frac{g(\alpha)}{T^2}\right] = ln\left(\frac{AR}{\beta E}\right) - \frac{E}{RT} \quad (13)$$

R represents the gas constant, and g(α) is defined differently based on the reaction order, n. When n is not equal to 1, g(α) is expressed as $g(\alpha) = 1 - (1 - \alpha)^{1-n}/1 - n$. Conversely, for n = 1, g(α) simplifies to $g(\alpha) = -\ln(1 - \alpha)$. The correlation coefficient, denoted as r, was computed using the least squares method for a range of n values (n = 1.00, 0.66, 0.51, and n = 0.33). The relationship between $\ln [g(\alpha)/T^2]$ and 1/T was graphically presented, and the reaction order was selected as the n-value that yielded the most accurate fit ($r \approx 1$). The intercept provides the pre-exponential factor (A), and the slope offers the activation energy (Ea)/R. Thermodynamic parameters were also derived from this approach, analogous to the determination of A and E [70].

3.10. DFT and Docking Studies

3.10.1. DFT Calculations

Quantum mechanical atomistic simulations were conducted using DFT perspective to analyze various properties of L, L-Pd, L-Mn, and L-Cr compounds. These computations were carried out with the Gaussian 09W software [71], while molecular structures were visualized using GaussView 6.0. Optimization was achieved through the B3LYP hybrid functional, employing the 6-311g (d,p) basis set for lighter atoms and LANL2DZ for heavier ones. This approach facilitated the determination of molecular orbital structures and electrostatic potential surface maps. Additionally, the HOMO and LUMO energy levels, along with the energy gap (E_{gap}), were calculated for both the ligand and its metal complexes. Finally, the quantum reactivity of the studied compounds was assessed based on their molecular orbital energies.

3.10.2. Molecular Docking Approaches

The possible anticancer, antimicrobial, and antifungal effects of the L, L-Pd, L-Mn, and L-Cr compounds were investigated using an in silico docking approach to validate their experimentally observed biological activities. To acquire the initial crystallographic structures of 3HB5 [72]—a breast cancer protein (used for anticancer activity); 4EWP [73]—*Micrococcus luteus* (used for antimicrobial activity); and 5V5Z [74]—*Candida albicans* (used for antifungal activity), a thorough search was conducted in the Protein Data Bank (https://www.rcsb.org/, accessed on 2 June 2025). The ligand and its optimized complexes, obtained from DFT, were converted into PDB format through the GaussView 6.0 program and used as ligands for docking modeling. Molecular docking was performed using AutoDock Tools 1.5.6 in conjunction with Autogrid 4 and AutoDock 4.2 [75]. This process utilized the Lamarckian genetic algorithm (LGA) to identify the potential binding site and determine the optimal orientation of the ligand. For the initial preparation of the macromolecular structure, hydrogen atoms were added, and Kollman charges were appropriately assigned [76]. Furthermore, all water molecules, ligands, and non-essential components were removed from the protein structures [73]. Grid boxes were defined with dimensions of 80 × 90 × 90 Å for 3HB5, 96 × 86 × 96 Å for 4EWP, and 90 × 100 × 100 Å for 5V5Z, with a spacing of 0.375 Å. Ligand poses were extracted using the Chimera package [77] and subsequently analyzed for interaction diagrams using the Discovery Studio Visualizer 4.1 program.

3.11. Biological Evaluate

3.11.1. Antibacterial Activity

The in vitro biological properties of the ligand 4-chloro-2-(quinolin-8-yliminomethyl)-phenol and its complexes were scrutinized in a study to evaluate their efficacy against a range of bacterial strains, including *Micrococcus luteus* (Gram-positive), *Escherichia coli* (Gram-negative), and *Serratia marcescens* (Gram-negative). The antimicrobial activity was assessed using the well diffusion technique on nutrient agar plates [78]. Additionally, the antifungal potency of these substances was explored against diverse fungal species such as *Fusarium oxysporum*, *Geotrichum candidum*, and *Aspergillus flavus*, employing potato dextrose agar as the growth medium. To prepare the compounds for testing, they were dissolved in dimethyl sulfoxide (DMSO) to achieve concentrations of 10 mg/mL and 20 mg/mL. The experimental setup involved creating wells on the agar plates that had been previously inoculated with the target microorganisms [79,80]. These wells were then filled with the test solutions using a micropipette. For control purposes, some wells contained only DMSO, which served as the negative control, while others contained established antimicrobial agents like Ofloxacin for bacteria and fluconazole for fungi as positive controls. The plates were then incubated at optimal temperatures for the respective microbial groups: 24 h at 37 °C for bacteria and 72 h at 35 °C for fungi. Following the incubation period, the zone of inhibition surrounding the wells was measured using a specialized zone reader (Hi Antibiotic zone scale) to determine the compounds' inhibitory capabilities against the test microbes. It is noteworthy that the solvent DMSO, when used by itself, did not exhibit any antimicrobial activity, ensuring that the observed effects were solely attributed to the ligand and its complexes. The experiment was meticulously conducted with three replicates for each substance to ensure the reliability of the results.

3.11.2. Anticancer Evaluation of Ligand and Its Metal Complexes

The study utilized an ELISA microplate reader (Model Σ960 from Meter Tech, Taipei, Taiwan), which operates with a wavelength of 564 nm to evaluate the effects of the compounds on Hep-G2 (hepatocellular carcinoma, human liver cancer cell line) was obtained from the American Type Culture Collection (ATCC, Manassas, VA, USA), MCF-7 (human breast adenocarcinoma cell line) was purchased from ATCC (Manassas, VA, USA), and HCT-116 (human colorectal carcinoma cell line) was also purchased from ATCC (Manassas, VA, USA). The cellular analysis commenced with the distribution of the cells at a density of 10,000 cells per well in a 96-well plate, allowing them to adhere for 24 h at 37 °C. Afterward, the cells were subjected to different concentrations of the compounds, dissolved in DMSO (0, 1, 2.5, 5, and 10 μM), for a 48 h incubation period within a 5% CO_2 controlled environment at the same temperature. Post-incubation, the plate was processed for fixation and staining with sulfo-rhodamine B. Unwanted stain residues were removed through acetic acid washes, and the plate was treated with Tris-EDTA buffer. The optical density of the stained cells, which correlates with cell viability, was measured by the ELISA reader. The IC_{50} value, indicating the compounds' efficacy, was determined following standard procedures outlined in the literature [81–84].

$$\text{Inhibition concentration (IC) \%} = (\text{Control O.D} - \text{Ligand O.D}) \times 100/\text{Control O.D} \tag{14}$$

3.11.3. Assessment of Antioxidant Activity Using DPPH Radical Scavenging Assay

The investigation into the antioxidant characteristics of the molecules in question was performed via an assessment of their capacity to scavenge DPPH (1,1-diphenyl-2-picrylhydrazyl) free radicals, adhering to the methodological guidelines established [85]. This methodology is a prevalent technique for evaluating the antioxidative potential of substances. It is predicated on the interaction between the antioxidants and the stable DPPH free radical, which results in a decrease in optical density and a transformation in color from purple to yellow. This change occurs as a consequence of the antioxidants' capability to neutralize the free radicals. The experimental procedure entailed combining 2.0 mL of the chelates under study with 6.0 mL of the DPPH solution. Notably, the chelates were omitted from the control and blank samples. These solutions were then placed in a darkened setting and allowed to incubate for a duration of 0.5 h at a constant temperature of 25 degrees Celsius. Upon completion of the incubation period, spectrophotometric analysis of the transmittance was conducted at a wavelength of 519 nm, which is particularly relevant for monitoring the DPPH reaction's progress. The absorbance of the samples was determined by comparing them to both blank and reference solutions. The calculation of the IC_{50}, which represents the concentration at which the chelates exerted a 50% inhibition of DPPH free radicals, was executed using the appropriate formula:

$$\text{Scavenging activity}(\%) = {A_0 - A_t}\big/{A_0} \times 100 \tag{15}$$

where A_0 represents the initial absorbance of the blank (standard) control at time t = 0, and A_t is the absorbance at the end of the reaction. This approach is effective in quantifying the antioxidant effectiveness of the molecules. The assay involved a serial dilution to expose each sample to varying concentrations, and all experiments were triplicated to ensure the reliability of the data [85].

4. Conclusions

Novel nanosized Cr(III), Mn(II), and Pd(II) complexes incorporating the chloro-2-(quinolin-8-yliminomethyl)-phenol imine ligand were synthesized and characterized via different physicochemical and spectroscopic tools. The analytical outcomes indicate that complexes formed with Cr(III) and Mn(II) adhere to a 1:2 metal-to-ligand ratio, whereas the Pd(II) complex has a 1:1 stoichiometry. Conductivity studies revealed that the **L-manganese** and **L-palladium** systems behave as non-electrolytes, whereas the **L-chromium** complex exhibits mono-electrolyte properties due to its elevated conductivity. The ligand's interaction with metal ions is established through the nitrogen atom of the quinoline ring, the oxygen atom of the hydroxyl group, and the nitrogen atom in the azomethine bond. Thermal analysis, including thermogravimetric data, elucidated the degradation patterns of the synthesized complexes, which assisted in their characterization. The complexes' distinct geometrical structures were inferred from the analytical and spectral data: Cr(III) and Mn(II) complexes adopt a high-spin octahedral geometry, while the Pd(II) complex assumes a square-planar conformation. These structural predictions were substantiated by theoretical computations, reinforcing the study's foundational understanding. Biological assessments unveiled significant antifungal and antibacterial properties for the palladium-integrated complex, **L-Pd**, which showcased substantial inhibitory action across a diverse range of microbial strains. The antioxidant potential of the synthesized compounds was evaluated by DPPH assays, showcasing moderate-to-excellent activities in comparison to ascorbic acid, with particular effectiveness against DPPH radicals. The biomedical evaluation of these complexes highlighted their anticancer capabilities, with the palladium complex, **L-Pd**, displaying noteworthy cytotoxicity against three separate cancer cell lines: HCT-116 (colorectal), MCF-7 (breast), and HepG-2 (liver). The most pronounced effect was observed against MCF-7 cells, hinting at a potential breakthrough in the development of antineoplastic agents. This enhanced activity of the L-Pd complex may be attributed to the square-planar geometry of Pd(II), which facilitates stronger interactions with biological targets such as DNA and microbial enzymes. Additionally, the Pd(II) center may form more stable complexes with biomolecules, improving cellular uptake and bioavailability. These factors likely contribute to the observed superior antifungal, antibacterial, and anticancer performance of the L–Pd complex.

Supplementary Materials: The following supporting information can be downloaded at: https://www.mdpi.com/article/10.3390/inorganics13080271/s1, Figure S1: IR spectra of ligand(L) and its metal complexes; Figure S2: ^{1}H-NMR spectra of ligand (L); Figure S3: ^{13}C-NMR spectra of ligand (L); Figure S4: TGA -DTGA curves for the prepared complexes; Figure S5: Molar ratio plots for the prepared complexes in DMF; Table S1: Electronic Spectra of the ligand (L) and its metal complexes; Table S2: Antibacterial activity inhibition zone of the prepared compounds against the selected strains of bacteria; Table S3: Antifungal activity inhibition zone of the prepared compounds against the selected strains of fungi; Table S4: Cytotoxic activity (IC_{50}) of the prepared compounds against Colon carcinoma cells, (HCT-116 cell line), hepatic cellular carcinoma cells, (HepG-2), breast carcinoma cells (MCF-7) and norma cells (human dermal fibroblasts and PBMCs) for (L) ligand and its metal chelates; Table S5: Antioxidant activity of the investigated compounds.

Funding: This research was funded by Taif University, Saudi Arabia, through project number TU-DSPP-2024-216.

Data Availability Statement: The data that support the findings of this study are available from the corresponding author upon reasonable request.

Acknowledgments: The authors extend their appreciation to Taif University, Saudi Arabia, for supporting this work through project number TU-DSPP-2024-216.

Conflicts of Interest: The author declares that they have no known competing financial interests or personal relationships that could have appeared to influence the work reported in this paper.

References

1. Gogoi, H.P.; Barman, P. Salophen type ONNO donor Schiff base complexes: Synthesis, characterization, bioactivity, computational, and molecular docking investigation. *Inorg. Chim. Acta* **2023**, *556*, 121668. [CrossRef]
2. Al-Farraj, E.S.; Alharbi, S.K.; Feizi-Dehnayebi, M.; Asghar, B.H.; Alahmadi, N.; Eskander, T.N.A.; Alghamdi, M.A.; Abu-Dief, A.M. Molecular, Stochiometric, Stability and Biological Investigations of Novel Multifunctional Salen Metal Chelates: From Synthesis to Therapeutic Potential Supported by Theoretical Approaches. *Appl. Organomet. Chem.* **2025**, *39*, e70273. [CrossRef]
3. El-Kasaby, R.A.; Al-Farraj, E.S.; Abdou, A.; Abu-Dief, A.M. Synthesis, spectral analysis, physicochemical investigation and biomedical potential of some novel Cu(II), Ru(III) and VO(II) complexes with anthraquinone-based Schiff base supported by DFT and molecular docking insights. *J. Mol. Struct.* **2025**, *1345*, 143010. [CrossRef]
4. Abdel-Rahman, L.H.; Abu-Dief, A.M.; Abdel-Mawgoud Azza, A.H. Novel Di-and Tri-azomethine compounds as chemo sensors for the detection of various metal ions. *Int. J. Nano Chem.* **2019**, *5*, 1–17.
5. Vernekar, B.K.; Sawant, P.S. Interaction of metal ions with Schiff bases having N_2O_2 donor sites: Perspectives on synthesis, structural features, and applications. *Results Chem.* **2023**, *6*, 101039. [CrossRef]
6. Bulatov, E.; Sayarova, R.; Mingaleeva, R.; Miftakhova, R.; Gomzikova, M.; Ignatyev, Y.; Petukhov, A.; Davidovich, P.; Rizvanov, A.; Barlev, N.A. Isatin-Schiff base-copper (II) complex induces cell death in p53-positive tumors. *Cell Death Discov.* **2018**, *4*, 103. [CrossRef]
7. Hosny, S.; Shehata, M.R.; Aly, S.A.; Alsehli, A.H.; Salaheldeen, M.; Abu-Dief, A.M.; Abu-El-Wafa, S.M. Synergistic broad-spectrum bioactivity of some multifunctional novel Anil metal chelates: Design, synthesise, nonlinear optical properties, and biomedical applications supported by DFT and molecular docking insights. *J. Mol. Struct.* **2025**, *1339*, 142390. [CrossRef]
8. Hosny, S.; Shehata, M.R.; Aly, S.A.; Alsehli, A.H.; Salaheldeen, M.; Abu-Dief, A.M.; Abu-El-Wafa, S.M. Designing of novel nano-sized coordination compounds based on Spinacia oleracea extract: Synthesis, structural characterization, molecular docking, computational calculations, and biomedical applications. *Inorg. Chem. Commun.* **2024**, *160*, 111994. [CrossRef]
9. El-Lateef, H.M.A.; Khalaf, M.M.; Shehata, M.R.; Abu-Dief, A.M. Fabrication, DFT Calculation, and Molecular Docking of Two Fe(III) Imine Chelates as Anti-COVID-19 and Pharmaceutical Drug Candidate. *Int. J. Mol. Sci.* **2022**, *23*, 3994. [CrossRef]
10. Abu-Dief, A.M.; Said, M.A.; Elhady, O.; Al-Abdulkarim, H.A.; Alzahrani, S.; Eskander, T.N.A.; El-Remaily, M.A.E.A.A.A. Innovation of Fe(III), Ni(II), and Pd(II) complexes derived from benzothiazole imidazolidin-4-ol ligand: Geometrical elucidation, theoretical calculation, and pharmaceutical studies. *Appl. Organomet. Chem.* **2023**, *37*, e7162. [CrossRef]
11. Gupta, K.C.; Sutar, A.K. Catalytic activities of Schiff base transition metal complexes. *Coord. Chem. Rev.* **2008**, *252*, 1420–1450. [CrossRef]

12. Shaaban, S.; Yousef, T.A.; Al-Janabi, A.S.; Alammar, T.; Alaasar, M.; Shalabi, K.; Al-Karmalawy, A.A.; Ferjani, H.; Al-Dakhil, A.; Abu-Dief, A.M. Zn(II), Cu(II), and Fe(III) complexes of 2-(((4-(Methylselanyl) phenyl)imino)methyl)phenol: Synthesis, characterization, and multidisciplinary investigations. *Polyhedron* **2025**, *279*, 117652. [CrossRef]
13. Al-Ghamdi, K.; Alharas, M.M.; Abdel-Latif, S.A.; Alhashmialameer, D.; Al-Farraj, E.S.; Almalki, M.A.; El-Khatib, R.M.; Abu-Dief, A.M. Selective Novel Metal-Coordinated Biomedical Agents Encompassing Tetradentate Salen Ligand: Structural Elucidation, DFT Calculation, Cytotoxic, and Antioxidant Activities Supported by Molecular Docking Approach. *Appl. Organomet. Chem.* **2025**, *39*, e7991. [CrossRef]
14. Al-Abdulkarim, H.A.; El-khatib, R.M.; Aljohani, F.S.; Mahran, A.; Alharbi, A.; Mersal, G.A.; El-Metwaly, N.M.; Abu-Dief, A.M. Optimization for synthesized quinoline-based Cr^{3+}, VO^{2+}, Zn^{2+} and Pd^{2+} complexes: DNA interaction, biological assay and in-silico treatments for verification. *J. Mol. Liq.* **2021**, *339*, 116797. [CrossRef]
15. Abu-Dief, A.M.; Shehata, M.R.; Hassan, A.E.; Altayeb, B.M.; Aljohani, F.S.; Barnawi, I.O.; Abo-Dief, H.M.; Ragab, M.S. Tailoring of novel water soluble Pd(II), Cu(II), Fe(III) and VO(II) chelates based on 4-[(5-bromo-2-hydroxy-benzylidene)-amino]-benzenesulfonate ligand: Synthesis, spectral investigations, DNA interaction and pharmaceutical applications supported by molecular docking approach. *J. Mol. Struct.* **2025**, *1334*, 141780.
16. Al-Fakeh, M.S.; Alsikhan, M.A.; Alnawmasi, J.S. Physico-chemical study of Mn(II), Co(II), Cu(II), Cr(III), and Pd(II) complexes with schiff-base and aminopyrimidyl derivatives and anti-cancer, antioxidant, antimicrobial applications. *Molecules* **2023**, *28*, 2555. [CrossRef]
17. Suslick, K.S. Sonochemistry. *Science* **1990**, *247*, 1439–1445. [CrossRef] [PubMed]
18. Mason, T.J.; Peters, D. *Practical Sonochemistry: Power Ultrasound Uses and Applications*; Woodhead Publishing: Cambridge, UK, 2002.
19. Abu-Dief, A.M.; Said, M.A.; Elhady, O.; Alahmadi, N.; Alzahrani, S.; Eskander, T.N.A.; Ali, M.A.E.A.A. Designing of some novel Pd(II), Ni(II) and Fe(III) complexes: Synthesis, structural elucidation, biomedical applications, DFT and docking approaches against COVID-19. *Inorg. Chem. Commun.* **2023**, *155*, 110955. [CrossRef]
20. Gozdas, S.; Kose, M.; Mckee, V.; Elmastas, M.; Demirtas, I.; Kurtoglu, M. Crystal structures, electronic spectra and anticancer properties of new azo-azomethines and their nickel(II) and copper(II) chelates. *J. Mol. Struct.* **2024**, *1304*, 137691. [CrossRef]
21. Shaaban, S.; Abdullah, K.T.; Shalabi, K.; Yousef, T.A.; Al Duaij, O.K.; Alsulaim, G.M.; Althikrallah, H.A.; Alaasar, M.; Al-Janabi, A.S.; Abu-Dief, A.M. Synthesis, structural characterization, anticancer, antimicrobial, antioxidant, and computational assessments of zinc(II), iron(II), and copper(II) chelates derived from selenated Schiff base. *Appl. Organomet. Chem.* **2024**, *38*, e7712. [CrossRef]
22. Khalaf, M.M.; Abd El-Lateef, H.M.; Taha, A.A.; Abdou, A. Binuclear Fe(II) and Cu(II) complexes with 2-(pyridin-2-yl)-1H-benzimidazole and 4,4′-bipyridine: Synthesis, characterization, DFT insights, broad-spectrum bioactivity and docking study. *Inorganica Chim. Acta* **2025**, *577*, 122505. [CrossRef]
23. Abdel-Rhman, M.H.; Motawea, R.; Belal, A.; Hosny, N.M. Spectral, structural and cytotoxicity studies on the newly synthesized n'^1, n'^3-diisonicotinoylmalonohydrazide and some of its bivalent metal complexes. *J. Mol. Struct.* **2022**, *1251*, 131960. [CrossRef]
24. Hosny, N.M.; Ibrahim, O.A.; Belal, A.; Hussien, M.A.; Abdel-Rhman, M.H. Synthesis, characterization, DFT, cytotoxicity evaluation and molecular docking of a new carbothioamide ligand and its coordination compounds. *Results Chem.* **2023**, *5*, 100776. [CrossRef]
25. Ali, I.; Wani, W.A.; Saleem, K. Empirical formulae to molecular structures of metal complexes by molar conductance. *Synth. React. Inorg. Met.-Org. Nano-Met. Chem.* **2013**, *43*, 1162–1170. [CrossRef]
26. Khalaf, M.M.; Abd El-Lateef, H.M.; Gouda, M.; Sayed, F.N.; Mohamed, G.G.; Abu-Dief, A.M. Design, structural inspection and bio-medicinal applications of some novel imine metal complexes based on acetylferrocene. *Materials* **2022**, *15*, 4842. [CrossRef] [PubMed]
27. Ali, O.A. Characterization, thermal and fluorescence study of Mn(II) and Pd(II) Schiff base complexes. *J. Therm. Anal. Calorim.* **2017**, *128*, 1579–1590. [CrossRef]
28. Soliman, M.H.; Mohamed, G.G. Cr(III), Mn(II), Fe(III), Co(II), Ni(II), Cu(II) and Zn(II) new complexes of 5-aminosalicylic acid: Spectroscopic, thermal characterization and biological activity studies. *Spectrochim. Acta Part A Mol. Biomol. Spectrosc.* **2013**, *107*, 8–15. [CrossRef]
29. Zare, N.; Zabardasti, A. A new nano-sized mononuclear Cu (II) complex with N,N-donor Schiff base ligands: Sonochemical synthesis, characterization, molecular modeling and biological activity. *Appl. Organomet. Chem.* **2019**, *33*, e4687. [CrossRef]
30. Al-Hazmi, G.A.; Abou-Melha, K.S.; Althagafi, I.; El-Metwaly, N.; Shaaban, F.; Abdul Galil, M.S.; El-Bindary, A.A. Synthesis and structural characterization of oxovanadium(IV) complexes of dimedone derivatives. *Appl. Organomet. Chem.* **2020**, *34*, e5672. [CrossRef]
31. Refat, M.S.; Altalhi, T.; Bakare, S.B.; Al-Hazmi, G.H.; Alam, K. New Cr(III), Mn(II), Fe(III), Co(II), Ni(II), Zn(II), Cd(II), and Hg(II) gibberellate complexes: Synthesis, structure, and inhibitory activity against COVID-19 protease. *Russ. J. Gen. Chem.* **2021**, *91*, 890–896. [CrossRef] [PubMed]

32. El-Remaily, M.A.E.A.A.A.; Eskander, T.N.A.; Elhady, O.; Alhashmialameer, D.; Alsehli, M.; Kamel, M.S.; Feizi-Dehnayebi, M.; Abu-Dief, A.M. A comparative study for the efficiency of Pd(II) and Fe(III) complexes as efficient catalysts for synthesis of dihydro-7H-5-thia-hexaaza-s-indacen-6-one derivatives supported with DFT approach. *Appl. Organomet. Chem.* **2024**, *38*, e7653. [CrossRef]
33. Abdel-Rahman, L.H.; Abu-Dief, A.M.; Aboelez, M.O.; Abdel-Mawgoud, A.A.H. DNA interaction, antimicrobial, anticancer activities and molecular docking study of some new VO(II), Cr(III), Mn(II) and Ni(II) mononuclear chelates encompassing quaridentate imine ligand. *J. Photochem. Photobiol. B Biol.* **2017**, *170*, 271–285. [CrossRef] [PubMed]
34. Yousef, T.A.; El-Reash, G.A.; El-Gammal, O.A.; Bedier, R.A. Co(II), Cu(II), Cd(II), Fe(III) and U(VI) complexes containing a NSNO donor ligand: Synthesis, characterization, optical band gap, in vitro antimicrobial and DNA cleavage studies. *J. Mol. Struct.* **2012**, *1029*, 149–160. [CrossRef]
35. Abdel-Rahman, L.H.; Abu-Dief, A.M.; Moustafa, H.; Hamdan, S.K. Ni(II) and Cu(II) complexes with ONNO asymmetric tetradentate Schiff base ligand: Synthesis, spectroscopic characterization, theoretical calculations, DNA interaction and antimicrobial studies. *Appl. Organomet. Chem.* **2017**, *31*, e3555. [CrossRef]
36. Aljohani, E.T.; Shehata, M.R.; Alkhatib, F.; Alzahrani, S.O.; Abu-Dief, A.M. Development and structure elucidation of new VO^{2+}, Mn^{2+}, Zn^{2+}, and Pd^{2+} complexes based on azomethine ferrocenyl ligand: DNA interaction, antimicrobial, antioxidant, anticancer activities, and molecular docking. *Appl. Organomet. Chem.* **2021**, *35*, e6154. [CrossRef]
37. Munde, A.S.; Jagdale, A.N.; Jadhav, S.M.; Chondhekar, T.K. Synthesis, characterization and thermal study of some transition metal complexes of an asymmetrical tetradentate Schiff base ligand. *J. Serbian Chem. Soc.* **2010**, *75*, 349–359. [CrossRef]
38. El-Remaily, M.A.E.A.A.A.; Elhady, O.; Eskander, T.N.A.; Mohamed, S.K.; Abu-Dief, A.M. Development of novel guanidine iron (III) complexes as a powerful catalyst for the synthesis of tetrazolo [1,5-a] pyrimidine by green protocol. *Sohag J. Sci.* **2024**, *9*, 7–15. [CrossRef]
39. Abu-Dief, A.M.; Said, M.A.; Elhady, O.; Alzahrani, S.; Aljohani, F.S.; Eskander, T.N.A.; Ali, M.A.E.A.A. Design, structural inspection of some new metal chelates based on benzothiazol-pyrimidin-2-ylidene ligand: Biomedical studies and molecular docking approach. *Inorg. Chem. Commun.* **2023**, *158*, 111587. [CrossRef]
40. Abu-Dief, A.M.; Feizi-Dehnayebi, M.; Nafady, A.; Almalki, M.A.; Kassem, A.M.; Abu Al-Ola, K.A.; Altayeb, B.M.; Abdel-Rahman, L.H. Fabrication, structural inspection, stability studies in solution and DFT calculations of some novel complexes drived from 4-(Benzothiazol-2-yliminomethyl)-phenol ligand: Pharmaceutical applications supported by molecular docking approach. *J. Mol. Struct.* **2025**, *1328*, 141284. [CrossRef]
41. Kazachenko, A.S.; Issaoui, N.; Holikulov, U.; Al-Dossary, O.M.; Ponomarev, I.S.; Kazachenko, A.S.; Akman, F.; Bousiakou, L.G. Noncovalent interactions in *N*-methylurea crystalline hydrates. *Z. Phys. Chem.* **2024**, *238*, 89–114. [CrossRef]
42. Akman, F.; Demirpolat, A.; Kazachenko, A.S.; Kazachenko, A.S.; Issaoui, N.; Al-Dossary, O. Molecular structure, electronic properties, reactivity (ELF, LOL, and Fukui), and NCI-RDG studies of the binary mixture of water and essential oil of Phlomis bruguieri. *Molecules* **2023**, *28*, 2684. [CrossRef]
43. Dehghani, N.; Ziarani, G.M.; Feizi-Dehnayebi, M.; Mirhosseyni, M.; Badiei, A. Synthesis and theoretical investigation of a new Hg^{2+} chemosensor based nanomagnetic particle (Fe_3O_4@SiO_2@ Pr-NCIM). *Mater. Res. Bull.* **2025**, *183*, 113200. [CrossRef]
44. Mohammadi Ziarani, G.; Rezakhani, M.; Feizi-Dehnayebi, M.; Nikolova, S. Fumed-Si-Pr-Ald-Barb as a fluorescent chemosensor for the Hg^{2+} detection and $Cr_2O_7{}^{2-}$ ions: A combined experimental and computational perspective. *Molecules* **2024**, *29*, 4825. [CrossRef]
45. Mohammadi Ziarani, G.; Ebrahimi, D.; Feizi-Dehnayebi, M.; Badiei, A.; Abu-Dief, A.M. Tailored Silica-Based Sensors (SBA-Pr-Ald-MA) for Efficient Detection of Iron (III) Ions: A Comprehensive Theoretical and Experimental Viewpoint. *Appl. Organomet. Chem.* **2025**, *39*, e7917. [CrossRef]
46. Zamiran, F.; Mohammadi Ziarani, G.; Feizi-Dehnayebi, M.; Mirhosseyni, M.; Badiei, A.; Abu-Dief, A.M. Design, Preparation, Characterization, Density Functional Theory, and HOMO-LUMO Perspective of Fe_3O_4@SiO_2-Pr-NH-IC as a New Nanomagnetic Chemosensor. *Appl. Organomet. Chem.* **2025**, *39*, e7998. [CrossRef]
47. Gungor, O.; Gul, A.; Comertpay, S.; McKee, V.; Yalcinkaya, O.B.; Kose, M. Copper(II) and Zinc(II) complexes of new water-soluble Schiff base ligands and their antiproliferative properties towards mesothelioma cell line. *J. Photochem. Photobiol. A Chem.* **2025**, *459*, 116049. [CrossRef]
48. Canakdag, M.; Feizi-Dehnayebi, M.; Kundu, S.; Sahin, D.; İlhan, İ.Ö.; Alhag, S.K.; Al-Shuraym, L.A.; Akkoc, S. Comprehensive evaluation of purine analogues: Cytotoxic and antioxidant activities, enzyme inhibition, DFT insights, and molecular docking analysis. *J. Mol. Struct.* **2025**, *1323*, 140798. [CrossRef]
49. El-Remaily, M.A.E.A.A.A.; El-Dabea, T.; El-Khatib, R.M.; Abdou, A.; El Hamd, M.A.; Abu-Dief, A.M. Efficiency and development of guanidine chelate catalysts for rapid and green synthesis of 7-amino-4,5-dihydro-tetrazolo [1,5-a] pyrimidine-6-carbonitrile derivatives supported by density functional theory (DFT) studies. *Appl. Organomet. Chem.* **2023**, *37*, e7262. [CrossRef]
50. Pérez, P.; Domingo, L.R.; Aizman, A.; Contreras, R. The electrophilicity index in organic chemistry. *Theor. Comput. Chem.* **2007**, *19*, 139–201.

51. Raman, N.; Raja, J.D. Synthesis, structural characterization and antibacterial studies of some biosensitive mixed ligand copper (II) complexes. *Indian J. Chem. Sect. A* **2007**, *46*, 1611–1614.
52. Abdel-Rahman, L.H.; Abdelhamid, A.A.; Abu-Dief, A.M.; Shehata, M.R.; Bakheet, M.A. Facile synthesis, X-ray structure of new multi-substituted aryl imidazole ligand, biological screening and DNA binding of its Cr(III), Fe(III) and Cu(II) coordination compounds as potential antibiotic and anticancer drugs. *J. Mol. Struct.* **2020**, *1200*, 127034. [CrossRef]
53. Epand, R.M.; Walker, C.; Epand, R.F.; Magarvey, N.A. Molecular mechanisms of membrane targeting antibiotics. *Biochim. Biophys. Acta (BBA)-Biomembr.* **2016**, *1858*, 980–987. [CrossRef]
54. Ansel, H.C.; Norred, W.P.; Roth, I.L. Antimicrobial activity of dimethyl sulfoxide against *Escherichia coli*, *Pseudomonas aeruginosa*, and *Bacillus megaterium*. *J. Pharm. Sci.* **1969**, *58*, 836–839. [CrossRef] [PubMed]
55. Tweedy, B.G. Possible mechanism for reduction of elemental sulfur by *Monilinia fructicola*. *Phytopathology* **1964**, *54*, 910.
56. Abu-Dief, A.M.; El-Metwaly, N.M.; Alzahrani, S.O.; Alkhatib, F.; Abualnaja, M.M.; El-Dabea, T.; Ali, M.A.E.A.A. Synthesis and characterization of Fe(III), Pd(II) and Cu(II)-thiazole complexes; DFT, pharmacophore modeling, in-vitro assay and DNA binding studies. *J. Mol. Liq.* **2021**, *326*, 115277. [CrossRef]
57. Sindhu, Y.; Athira, C.J.; Sujamol, M.S.; Joseyphus, R.S.; Mohanan, K. Synthesis, characterization, DNA cleavage, and antimicrobial studies of some transition metal complexes with a novel Schiff base derived from 2-aminopyrimidine. *Synth. React. Inorg. Met.-Org. Nano-Met. Chem.* **2013**, *43*, 226–236. [CrossRef]
58. Mohamed, G.G.; Soliman, M.H. Synthesis, spectroscopic and thermal characterization of sulpiride complexes of iron, manganese, copper, cobalt, nickel, and zinc salts. Antibacterial and antifungal activity. *Spectrochim. Acta Part A Mol. Biomol. Spectrosc.* **2010**, *76*, 341–347. [CrossRef]
59. Aljohani, F.S.; Abu-Dief, A.M.; El-Khatib, R.M.; Al-Abdulkarim, H.A.; Alharbi, A.; Mahran, A.; Khalifa, M.E.; El-Metwaly, N.M. Structural inspection for novel Pd(II), VO(II), Zn(II) and Cr(III)-azomethine metal chelates: DNA interaction, biological screening and theoretical treatments. *J. Mol. Struct.* **2021**, *1246*, 131139. [CrossRef]
60. El-Wakiel, N.A. Synthesis and characterization of azo sulfaguanidine complexes and their application for corrosion inhibition of silicate glass. *Appl. Organomet. Chem.* **2016**, *30*, 664–673. [CrossRef]
61. Abdel-Rahman, L.H.; Adam, M.S.S.; Abu-Dief, A.M.; Moustafa, H.; Basha, M.T.; Aboraia, A.S.; Al-Farhan, B.S.; Ahmed, H.E.S. Synthesis, theoretical investigations, biocidal screening, DNA binding, in vitro cytotoxicity and molecular docking of novel Cu(II), Pd(II) and Ag(I) complexes of chlorobenzylidene Schiff base: Promising antibiotic and anticancer agents. *Appl. Organomet. Chem.* **2018**, *32*, e4527. [CrossRef]
62. Abdel-Rahman, L.H.; Abu-Dief, A.M.; Shehata, M.R.; Atlam, F.M.; Abdel-Mawgoud, A.A.H. Some new Ag(I), VO(II) and Pd(II) chelates incorporating tridentate imine ligand: Design, synthesis, structure elucidation, density functional theory calculations for DNA interaction, antimicrobial and anticancer activities and molecular docking studies. *Appl. Organomet. Chem.* **2019**, *33*, e4699. [CrossRef]
63. Gaetke, L.M.; Chow, C.K. Copper toxicity, oxidative stress, and antioxidant nutrients. *Toxicology* **2003**, *189*, 147–163. [CrossRef]
64. El-Remaily, M.A.E.A.A.A.; Elhady, O.; Alzubi, M.S.H.; Eskander, T.N.A.; El Hamd, M.A.; Al-Ghamdi, K.; Abu-Dief, A.M. Development of new thiazole-guanidine complexes as rapid and recoverable catalysts for the synthesis of 6-piperidin-dihydro-thia-hexaaza-s-indacene derivatives supported by DFT studies. *Appl. Organomet. Chem.* **2024**, *38*, e7454. [CrossRef]
65. El-Remaily, M.A.E.A.A.A.; Eskander, T.N.A.; Alzahrani, A.Y.A.; Alsehli, M.; Al-Ghamdi, K.; ALMughram, M.H.; El-Hady, O.M.; Feizi-Dehnayebi, M.; Abu-Dief, A.M. Tailoring of novel Ru(III) and Cr(III) Salen complexes as catalysts for a sustainable and green synthesis of dihydro-tetrazolo [1,5-a] Thiazolo [4,5-d] pyrimidin-6-yl morpholine: Experimental and theoretical approaches. *Appl. Organomet. Chem.* **2025**, *39*, e7879. [CrossRef]
66. Ismael, M.; Abdel-Mawgoud, A.M.M.; Rabia, M.K.; Abdou, A. Design and synthesis of three Fe(III) mixed-ligand complexes: Exploration of their biological and phenoxazinone synthase-like activities. *Inorganica Chim. Acta* **2020**, *505*, 119443. [CrossRef]
67. Al-Shamry, A.A.; Khalaf, M.M.; El-Lateef, H.M.A.; Yousef, T.A.; Mohamed, G.G.; El-Deen, K.M.K.; Gouda, M.; Abu-Dief, A.M. Development of new azomethine metal chelates derived from isatin: DFT and pharmaceutical studies. *Materials* **2022**, *16*, 83. [CrossRef]
68. Alkhatib, F.; Hameed, A.; Sayqal, A.; Bayazeed, A.A.; Alzahrani, S.; Al-Ahmed, Z.A.; Zaky, R.; El-Metwaly, N.M. Green-synthesis and characterization for new Schiff-base complexes; spectroscopy, conductometry, Hirshfeld properties and biological assay enhanced by in-silico study. *Arab. J. Chem.* **2020**, *13*, 6327–6340. [CrossRef]
69. Al-Farraj, E.S.; Qasem, H.A.; Aouad, M.R.; Al-Abdulkarim, H.A.; Alsaedi, W.H.; Khushaim, M.S.; Feizi-Dehnayebi, M.; Al-Ghamdi, K.; Abu-Dief, A.M. Synthesis, Structural Determination, DFT Calculation and Biological Evaluation Supported by Molecular Docking Approach of Some New Complexes Incorporating (E)-N′-(3,5-di-Tert-Butyl-2-Hydroxybenzylidene) Isonicotino Hydrazide Ligand. *Appl. Organomet. Chem.* **2025**, *39*, e7768. [CrossRef]
70. Ali, M.A.E.A.A.; Elhady, O.; Abdou, A.; Alhashmialameer, D.; Eskander, T.N.A.; Abu-Dief, A.M. Development of new 2-(Benzothiazol-2-ylimino)-2,3-dihydro-1H-imidazol-4-ol complexes as a robust catalysts for synthesis of thiazole 6-carbonitrile derivatives supported by DFT studies. *J. Mol. Struct.* **2023**, *1292*, 136188.

71. Frisch, M.; Trucks, G.; Schlegel, H.B.; Scuseria, G.E.; Robb, M.A.; Cheeseman, J.R.; Scalmani, G.; Barone, V.; Mennucci, B.; Petersson, G. *Gaussian 09, Revision D.01*; Gaussian, Inc.: Wallingford, CT, USA, 2009; 201p.
72. Mazumdar, M.; Fournier, D.; Zhu, D.W.; Cadot, C.; Poirier, D.; Lin, S.X. Binary and ternary crystal structure analyses of a novel inhibitor with 17β-HSD type 1: A lead compound for breast cancer therapy. *Biochem. J.* **2009**, *424*, 357–366. [CrossRef] [PubMed]
73. Pereira, J.H.; Goh, E.B.; Keasling, J.D.; Beller, H.R.; Adams, P.D. Structure of FabH and factors affecting the distribution of branched fatty acids in Micrococcus luteus. *Biol. Crystallogr.* **2012**, *68*, 1320–1328. [CrossRef]
74. Keniya, M.V.; Sabherwal, M.; Wilson, R.K.; Woods, M.A.; Sagatova, A.A.; Tyndall, J.D.; Monk, B.C. Crystal structures of full-length lanosterol 14α-demethylases of prominent fungal pathogens *Candida albicans* and *Candida glabrata* provide tools for antifungal discovery. *Antimicrob. Agents Chemother.* **2018**, *62*, 10–1128. [CrossRef] [PubMed]
75. Morris, G.M.; Huey, R.; Lindstrom, W.; Sanner, M.F.; Belew, R.K.; Goodsell, D.S.; Olson, A.J. AutoDock4 and AutoDockTools4: Automated docking with selective receptor flexibility. *J. Comput. Chem.* **2009**, *30*, 2785–2791. [CrossRef]
76. Feizi-Dehnayebi, M.; Ziarani, G.M.; Lohith, T.N.; Ghareghomi, S.; Panahande, Z.; Farsadrooh, M.; Majidinia, M. Probing the biological activity of isatin derivatives against human lung cancer A549 cells: Cytotoxicity, CT-DNA/BSA binding, DFT/TD-DFT, topology, ADME-Tox, docking and dynamic simulations. *J. Mol. Liq.* **2025**, *428*, 127475. [CrossRef]
77. Pettersen, E.F.; Goddard, T.D.; Huang, C.C.; Couch, G.S.; Greenblatt, D.M.; Meng, E.C.; Ferrin, T.E. UCSF Chimera—A visualization system for exploratory research and analysis. *J. Comput. Chem.* **2004**, *25*, 1605–1612. [CrossRef] [PubMed]
78. Todorova, M.; Bakalska, R.; Feizi-Dehnayebi, M.; Ziarani, G.M.; Pencheva, M.; Stojnova, K.; Milusheva, M.; Nedialkov, P.; Cherneva, E.; Kolev, T.; et al. Synthesis, Anti-Inflammatory Activity, and Docking Simulation of a Novel Styryl Quinolinium Derivative. *Appl. Sci.* **2024**, *15*, 284. [CrossRef]
79. Abdel-Rahman, L.H.; El-Khatib, R.M.; Nassr, L.A.; Abu-Dief, A.M.; Ismael, M.; Seleem, A.A. Metal based pharmacologically active agents: Synthesis, structural characterization, molecular modeling, CT-DNA binding studies and in vitro antimicrobial screening of iron (II) bromosalicylidene amino acid chelates. *Spectrochim. Acta Part A Mol. Biomol. Spectrosc.* **2014**, *117*, 366–378. [CrossRef]
80. Abu-Dief, A.M.; Nassr, L.A. Tailoring, physicochemical characterization, antibacterial and DNA binding mode studies of Cu(II) Schiff bases amino acid bioactive agents incorporating 5-bromo-2-hydroxybenzaldehyde. *J. Iran. Chem. Soc.* **2015**, *12*, 943–955. [CrossRef]
81. Abdel-Rahman, L.H.; Abu-Dief, A.M.; El-Khatib, R.M.; Abdel-Fatah, S.M. Sonochemical synthesis, DNA binding, antimicrobial evaluation and in vitro anticancer activity of three new nano-sized Cu(II), Co(II) and Ni(II) chelates based on tri-dentate NOO imine ligands as precursors for metal oxides. *J. Photochem. Photobiol. B Biol.* **2016**, *162*, 298–308. [CrossRef]
82. Abdel-Rahman, L.H.; Abu-Dief, A.M.; Newair, E.F.; Hamdan, S.K. Some new nano-sized Cr(III), Fe(II), Co(II), and Ni(II) complexes incorporating 2-((E)-(pyridine-2-ylimino) methyl) napthalen-1-ol ligand: Structural characterization, electrochemical, antioxidant, antimicrobial, antiviral assessment and DNA interaction. *J. Photochem. Photobiol. B Biol.* **2016**, *160*, 18–31. [CrossRef]
83. Abdel-Rahman, L.H.; Ismail, N.M.; Ismael, M.; Abu-Dief, A.M.; Ahmed, E.A.H. Synthesis, characterization, DFT calculations and biological studies of Mn(II), Fe(II), Co(II) and Cd(II) complexes based on a tetradentate ONNO donor Schiff base ligand. *J. Mol. Struct.* **2017**, *1134*, 851–862. [CrossRef]
84. Abu-Dief, A.M.; Nassar, I.F.; Elsayed, W.H. Magnetic $NiFe_2O_4$ nanoparticles: Efficient, heterogeneous and reusable catalyst for synthesis of acetylferrocene chalcones and their anti-tumour activity. *Appl. Organomet. Chem.* **2016**, *30*, 917–923. [CrossRef]
85. Ferrari, E.; Asti, M.; Benassi, R.; Pignedoli, F.; Saladini, M. Metal binding ability of curcumin derivatives: A theoretical vs. experimental approach. *Dalton Trans.* **2013**, *42*, 5304–5313. [CrossRef] [PubMed]

inorganics MDPI

Article

Manganese(II) Complexes with 3,5–Dibromosalicylaldehyde: Characterization and Interaction Studies with DNA and Albumins

Vasia Theodoulou, Ariadni Zianna, Antonios G. Hatzidimitriou and George Psomas *

Laboratory of Inorganic Chemistry, Department of Chemistry, Aristotle University of Thessaloniki, GR–54124 Thessaloniki, Greece
* Correspondence: gepsomas@chem.auth.gr

Abstract

The interaction of manganese(II) with deprotonated 3,5–dibromo–salicylaldehyde (3,5–diBr–saloH) in the absence or the presence of the *N,N′*-donors 2,2′–bipyridylamine (bipyam), 2,2′–bipyridine (bipy), 1,10–phenanthroline (phen), and 2,9–dimethyl–1,10–phenanthroline (neoc) as co-ligands yielded five neutral mononuclear complexes, namely Mn(3,5-diBr-salo)$_2$(CH$_3$OH)$_2$] (complex **1**), [Mn(3,5-diBr-salo)$_2$(bipyam)] (complex **2**), [Mn(3,5-diBr-salo)$_2$(bipy)] (complex **3**), [Mn(3,5-diBr-salo)$_2$(phen)] (complex **4**), and [Mn(3,5-diBr-salo)$_2$(neoc)] (complex **5**), respectively. The resultant complexes were characterized with physicochemical and spectroscopic techniques, and single-crystal X-ray crystallography was applied to determine the crystal structure of complex **2**. The evaluation of the potential biological profile of the complexes focused on the interaction with linear calf-thymus (CT) DNA, and bovine (BSA) and human (HSA) serum albumin. According to the data derived, the complexes interact intercalatively and strongly with CT DNA and associate tightly and reversibly with both albumins studied.

Keywords: manganese(II) complexes; 3,5–dibromosalicylaldehyde; interaction with DNA; interaction with albumins

1. Introduction

Manganese is a significant biometal because of its presence in the active center of many vital enzymes (oxygen-evolving center, superoxide dismutase, and catalase), its involvement in various functions such as glucose metabolism, energy production, protein digestion, and synthesis of cholesterol and fatty acids [1–4], and its role as a cofactor in the synthesis and activation of important enzymes including transferases, isomerases, hydrolases [5], and glutamine synthetase [6]. In addition, manganese compounds are gaining an increasing role in the research of metallodrugs since there are examples of manganese complexes that have shown in vitro anticancer [7–10], antimicrobial [11–14], and antioxidant [15,16] potency, although the use of manganese chemotherapeutics is limited to SC–52608 and Teslascan as anticancer and MRI contrast agents, respectively [17].

Substituted 2–hydroxy–benzaldehydes or salicylaldehydes (X–saloH) are compounds presenting potency against bacteria and yeasts [18,19] as well as antioxidant activity [20]. The insertion of diverse groups on the benzene ring (such as alkyl, alkoxy, nitro, and halogen groups) may alter the biological efficacy of these compounds [21]. In particular, the halogenation is an important modification of the ring that results in an improved

 https://doi.org/10.3390/inorganics13080263

biological profile; halogenated compounds contribute to an increase in the permeability of the cell membrane and a decrease in the degradation of metabolism [22,23] and are among the most active drugs [21,24]. The reports of manganese complexes with substituted salicylaldehydes as ligands are rather limited till now [25,26].

3,5–dibromosalicylaldehyde (3,5–diBr–saloH, Figure 1) is a dibromo-substituted salicylaldehyde inhibiting the inositol-requiring enzyme 1 (IRE1), which is related to the treatment of ischemic stroke [27,28]. In the literature, diverse metal complexes with 3,5–dibromo–salicylaldehyde as a ligand, including mononuclear zinc(II) [29], copper(II) [30], nickel(II) [31], iron(III) [32], and palladium(II) complexes [33], as well as a tetranuclear nickel(II) complex [34], have been structurally characterized and biologically evaluated.

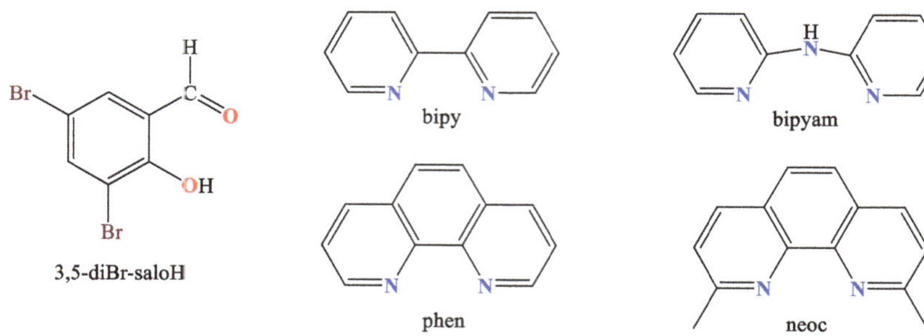

Figure 1. Syntax formula of 3,5–dibromo–salicylaldehyde (3,5–diBr–saloH), 2,2′–bipyridine (bipy), 2,2′–bipyridylamine (bipyam), 1,10–phenanthroline (phen), and 2,9–dimethyl–1,10–phenanthroline (neoc).

In continuation of our current research project regarding metal complexes of 3,5–dibromo–salicylaldehyde [29–33], five novel neutral mononuclear manganese(II) complexes of 3,5–diBr–saloH were synthesized in the absence or the presence of the *N,N′*-donors 2,2′–bipyridylamine (bipyam), 2,2′–bipyridine (bipy), 1,10–phenanthroline (phen), and neocuproine (2,9–dimethyl–1,10–phenanthroline, neoc) (Figure 1). The resultant complexes [Mn(3,5-diBr-salo)$_2$(CH$_3$OH)$_2$] (complex **1**), [Mn(3,5-diBr-salo)$_2$(bipyam)] (complex **2**), [Mn(3,5-diBr-salo)$_2$(bipy)] (complex **3**), [Mn(3,5-diBr-salo)$_2$(phen)] (complex **4**), and [Mn(3,5-diBr-salo)$_2$(neoc)] (complex **5**) were characterized with physicochemical and spectroscopic (FT–IR and UV–vis) techniques, while single-crystal X-ray crystallography was applied for complex **2**. The biological evaluation of compounds **1–5** focused on the interaction with calf-thymus (CT) DNA monitored with titration studies (measurements of DNA viscosity and UV–vis spectroscopy) and via the replacement ability of ethidium bromide (EB) from the EB–DNA adduct, and the affinity for bovine serum albumin (BSA) and human serum albumin (HSA) monitored with fluorescence emission spectroscopy. In these studies, the interaction mode was evaluated, and the corresponding binding constants of the complexes were also calculated.

2. Results and Discussion

2.1. Synthesis and Spectroscopic Characterization

Complexes **1–5** were synthesized in methanol in satisfactory yield via the aerobic reaction of MnCl$_2$·4H$_2$O with deprotonated (with CH$_3$ONa) 3,5-diBr-salo$^-$ in the absence (in a 1:2 Mn^{2+}:(3,5-diBr-salo)$^-$ ratio) or in the presence of the corresponding *N,N′*-donor (in a 1:2:1 Mn^{2+}:(3,5-diBr-salo)$^-$:(*N,N′*-donor) ratio). The resultant complexes were characterized with diverse physicochemical techniques (elemental analysis, molar conductivity measurements), room temperature (RT) magnetic measurements, IR and UV–vis spectroscopies, and single-crystal X-ray crystallography (for complex **2**).

Complexes **1–5** are neutral (molar conductivity values in DMSO solution were found to be <15 $S \cdot cm^2 \cdot mol^{-1}$) [35] and possess a 1:2 Mn(II):(3,5–diBr–salo) (for **1**) or a 1:2:1 Mn(II):(3,5–diBr–salo):(*N,N′*-donor) (for **2–5**) composition. The values of μ_{eff} derived for complexes **1–5** at RT (μ_{eff} = 5.85–5.96 BM) are close to the spin-only value (μ_{eff} = 5.92 BM) and are characteristic for isolated high-spin manganese(II) ions (i.e., mononuclear complexes) bearing a d^5 configuration (S = 5/2) [36].

The bands observed at ~3200 cm^{-1} and 1400 cm^{-1} in the spectrum of the free 3,5-diBr-saloH [29] attributable to the stretching and bending vibrations, respectively, of the phenolic –OH [37] disappeared in the IR spectra of the complexes (Figure S1), indicating the deprotonation of the phenolato group. In addition, the binding of the phenolato oxygen with Mn(II) was proved from the presence of the stretching vibration v(C–O $\rightarrow$ Mn) located at 1306–1328 cm^{-1} [37]. Furthermore, the coordination of the carbonyl oxygen was confirmed from the shift of v(C=O) from 1679 cm^{-1} in free 3,5-diBr-saloH towards lower wavenumbers (1625–1641 cm^{-1}) concluding therefore an overall bidentate coordination of deprotonated 3,5–diBr–salo⁻ ligands. The existence and the coordination of the *N,N′*-donor co-ligands were verified from the presence of the corresponding out-of-plane ρ(C–H) vibrations [37]: at 767 cm^{-1} for $\rho(C–H)_{bipyam}$ in complex **2**, 760 cm^{-1} for $\rho(C–H)_{bipy}$ in complex **3**, 729 cm^{-1} for $\rho(C–H)_{phen}$ in complex **4**, and 735 cm^{-1} for $\rho(C–H)_{neoc}$ in complex **5**.

The UV–vis spectra of the compounds (Figure S3) were recorded as nujol mull (as solid state), in DMSO solution, and in the presence of buffer solution (150 mM NaCl and 15 mM trisodium citrate at pH = 7) to investigate whether the complexes remain stable in the presence of buffer solution used in the biological experiments. The UV–vis spectra are similar (no significant changes, e.g., shift of the λ_{max} or new peaks, were observed), constituting proof of the integrity and the structure of the complexes in solution [25,26,29–32].

These features concerning the behavior of complexes **1–5** in solution (electronic spectra and molar conductivity measurements) are evidence of their stability in solution.

2.2. Structure of the Complexes

2.2.1. Crystal Structure of Complex 2

Among complexes **1–5**, single-crystals were obtained only for [Mn(3,5-diBr-salo)$_2$ (bipyam)] (complex **2**) in order to determine its molecular structure with single-crystal X-ray crystallography. Complex **2** crystallized in the monoclinic crystal system and the *Cc* space group (Table S1). The molecular structure is depicted in Figure 2, and selected bond lengths and angles are summarized in Table 1.

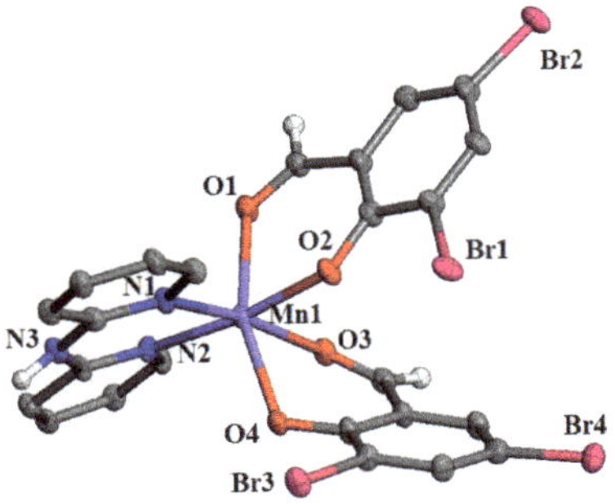

Figure 2. Molecular structure of complex [Mn(3,5-diBr-salo)$_2$(bipyam)]. Aromatic hydrogen atoms were omitted for clarity. Carbon and hydrogen atoms are grey and white, respectively.

Table 1. Selected bond lengths (Å) and angles (°) for [Mn(3,5-diBr-salo)$_2$(bipyam)] (complex **2**).

Bond	Length (Å)	Bond	Length (Å)
Mn1—O1	2.250(6)	Mn1—O4	2.091(6)
Mn1—O2	2.105(6)	Mn1—N1	2.137(7)
Mn1—O3	2.281(6)	Mn1—N2	2.157(7)
Bond	**Angle (°)**	**Bond**	**Angle (°)**
O1—Mn1—O2	84.0(2)	O2—Mn1—O3	89.3(2)
O1—Mn1—O3	82.8(2)	O2—Mn1—O4	87.9(2)
O1—Mn1—O4	157.1(2)	O2—Mn1—N1	172.7(3)
O1—Mn1—N1	90.4(2)	O2—Mn1—N2	92.1(2)
O1—Mn1—N2	105.5(3)	O4—Mn1—N1	99.0(2)
O3—Mn1—O4	75.7(2)	O4—Mn1—N2	96.1(2)
O3—Mn1—N1	94.7(2)	N1—Mn1—N2	84.8(3)
O3—Mn1—N2	171.6(3)		

It is a mononuclear manganese(II) complex, and the 3,5–diBr–salo$^-$ ligands are deprotonated and bound to Mn1 through the aldehyde oxygens O1 and O3 and the phenolato oxygen atoms O2 and O4. Two nitrogen atoms from the chelating bipyam ligand complete the coordination sphere of the six-coordinate Mn1 ion, which bears a distorted octahedral geometry. The Mn–O$_{phenolato}$ are the shortest bond distances (2.091(6)–2.105(6) Å), while the Mn–O$_{aldehydo}$ are the longest ones (2.250(6)–2.281(6) Å) in the coordination sphere. A similar arrangement of X–salo$^-$ and α–diimine ligands around the central metal was also reported in a series of Co(II) [38], Ni(II) [31], Zn(II) [29,39], and Cd(II) [40].

The crystal structure of complex **2** is further stabilized by an intermolecular hydrogen bond developed between the imino hydrogen H31 of the bipyam ligand with the non-coordinated oxygen O4^i (symmetry code: (i) x, $-y+1$, $z-1/2$) of an adjacent molecule (N3—H31 = 0.86 Å, H31···O4^i = 2.30 Å, N3···O4^i = 2.952(13) Å, N3—H31···O4^i = 133°).

2.2.2. Proposed Structures for Complexes 1 and 3–5

Despite all our efforts, single-crystals were not obtained for all complexes (i.e., **1** and **3–5**) under study. Thus, their structural characterization succeeded according to existing experimental data derived from IR and UV–vis spectroscopy, molar conductivity, and RT magnetic measurements and after a comparison with similar reported manganese(II)-(X-salo) complexes. On the basis of RT magnetic measurements (μ_{eff} = 5.85–5.96 BM), the Mn(II) complexes are mononuclear. Based on IR spectra, the 3,5-diBr-salo ligands bind to Mn(II) ions in a chelating bidentate fashion through the phenolato and the carbonyl oxygen atoms, and the presence of the *N,N′*-donors as co-ligands in complexes **3–5** was also confirmed.

In conclusion, complex **1** (Figure S2) is expected to have a similar structure to the reported Mn(II) complexes [Mn(5–NO$_2$–salo)$_2$(MeOH)$_2$] [26], [Mn(5–Cl–salo)$_2$(MeOH)$_2$], and [Mn(5–Br–salo)$_2$(MeOH)$_2$] [25] (5–NO$_2$–saloH = 5–nitro–salicylaldehyde, 5–Cl–saloH = 5–chloro–salicylaldehyde, and 5–Br–saloH = 5–bromo–salicylaldehyde). On the other hand, complexes **3–5** are expected to have a similar structure (Figure S2) to complex **2** and analogous reported Mn(II) complexes [Mn(5–NO$_2$–salo)$_2$(phen)] [26], [Mn(4–OMe–salo)$_2$(phen)], [Mn(4–OMe–salo)$_2$(bipy)], and [Mn(4–OMe–salo)$_2$(bipyam)] [25] (4–OMe–saloH = 4–methoxo–salicylaldehyde), as well as the nickel(II) complex [Ni(3,5–diBr–salo)$_2$(neoc)] [31].

2.3. *Interaction of the Complexes with CT DNA*

Biologically active compounds may interact with DNA in diverse modes depending on their stability, their structure, and the nature of their ligands [41]. Metal complexes interact with the double helix of DNA in a covalent way or in a noncovalent mode or via the cleavage of the DNA helix [42,43]. In order to study the affinity of complexes **1–5** for CT DNA, titration studies employing UV-vis spectroscopy, viscosity measurements, and EB-displacement monitored with fluorescence emission spectroscopy were performed.

UV–vis spectroscopy titration studies provide initial information regarding the interaction of complexes with CT DNA and aid the determination of the DNA-binding constant (K_b). The UV–vis spectra of the complexes were recorded in the presence of increasing amounts of CT DNA solution (Figures 3 and S4). The intraligand bands in the region 300–315 nm and at 425 nm presented upon addition of CT DNA a slight hyperchromism, while the positions of the corresponding λ_{max} remained stable (Table 2). Such changes reveal the interaction of the compounds with CT DNA [44] but cannot lead to a safe conclusion regarding the mode of this interaction, thus necessitating the employment of more techniques such as DNA-viscosity measurements and EB-displacement studies.

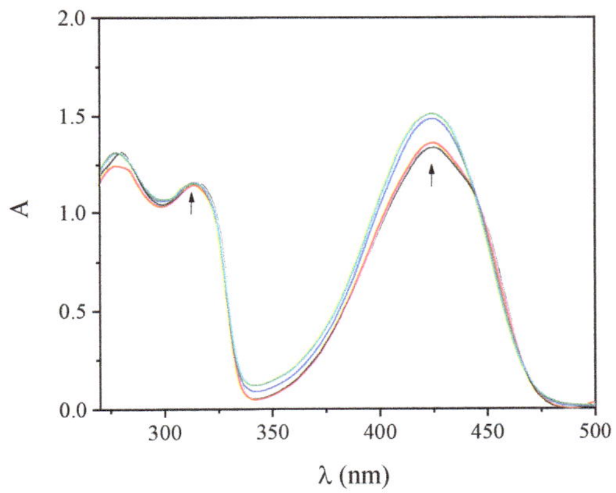

Figure 3. UV–vis spectra of a DMSO solution of complex **2** (100 μM) in the presence of increasing amounts of CT DNA. The arrows show the changes upon increasing amounts of CT DNA.

Table 2. UV–vis spectroscopic data concerning the interaction of complexes **1–5** with CT DNA: UV-band (λ_{max}, in nm) (percentage of the observed hyper/hypochromism ($\Delta A/A_0$, in %), blue-/red-shift of the λ_{max} ($\Delta\lambda$, in nm)); DNA-binding constant (K_b, in M^{-1}).

Compound	λ_{max}(nm) ($\Delta A/Ao$ (%)) [a], $\Delta\lambda$ (nm) [b])	K_b (M^{-1})
3,5-diBr-saloH [30]	337 (<−50, elimination); 427 (>+50, 0)	3.71(±0.14) × 10^5
[Mn(3,5-diBr-salo)$_2$(CH$_3$OH)$_2$], **1**	425 (+11, −1)	1.12(±0.32) × 10^5
[Mn(3,5-diBr-salo)$_2$(bipyam)], **2**	315 (+2, 0); 425 (+13, 0)	3.54(±0.42) × 10^5
[Mn(3,5-diBr-salo)$_2$(bipy)], **3**	425 (+18, −2)	6.58(±0.53) × 10^5
[Mn(3,5-diBr-salo)$_2$(phen)], **4**	300 (+2, 0); 425 (+11, −2)	5.25(±0.45) × 10^5
[Mn(3,5-diBr-salo)$_2$(neoc)], **5**	425 (+12, 0)	1.73(±0.42) × 10^5

[a] "+" denotes hyperchromism and "−" denotes hypochromism. [b] "+" denotes red-shift and "−" denotes blue-shift.

The K_b values of complexes **1–5** were calculated with the aid of the Wolfe–Shimer equation (Equation (S1)) [45] and the plots of [DNA]/(ε_A-ε_f) versus [DNA] (Figure S5). Most of the complexes present a higher K_b value than free 3,5-diBr-saloH (Table 2), revealing its stronger binding to CT DNA upon coordination, with complex **3** bearing the highest K_b (= 6.58(±0.53) × 10^5 M^{-1}) among the compounds studied herein. The DNA-binding constants of complexes **1–5** are in the range reported (= 2.83 × 10^5 – 3.37 × 10^6 M^{-1}) for a series of transition metal complexes with 3,5-diBr-salo ligands [29–33] and Mn(II) complexes with di- or mono-substituted salicylaldehydes ($K_b = 9.92 \times 10^4 - 3.08 \times 10^7\ M^{-1}$) [25,26]. The K_b values of the complexes are higher than that of EB ($K_b = 1.23 \times 10^5\ M^{-1}$) [46], which is a reference intercalator, revealing the potency of the compounds to displace EB from its intercalating site.

The relative viscosity of DNA is usually sensitive to DNA-length changes since they are related via the equation $L/L_0 = (\eta/\eta_0)^{1/3}$, where η/η_0 denotes the relative DNA viscosity and L/L_0 is the relative DNA length. Therefore, the measurement of the viscosity of a DNA solution in the presence of a compound provides significant information regarding changes in relative DNA length arising from specific DNA interaction modes [47]. In general, a classic intercalation will increase the separation distance between DNA bases to host the intercalating molecule in-between, thus resulting in an elongation of the relative DNA length in proportion to the relative DNA viscosity. An external interaction with DNA (an electrostatic or DNA-groove binding interaction) may bend slightly the DNA helix, leading to a slight shortening of the relative DNA helix, which may decrease relative DNA viscosity or leave it practically stable [47].

In the current study, the changes in viscosity of a CT DNA solution (0.1 mM) were monitored in the presence of increasing amounts of complexes **1–5** (up to the value of $r = 0.36$, Figure 4). Especially in the case of complexes **2** and **5**, the addition of the complexes initially resulted (up to $r \sim 0.1$, Figure 4) in a slight decrease in relative DNA viscosity, showing that an external interaction with DNA may probably take place, allowing the complexes to approach DNA. Further addition of the complexes into the DNA solution resulted in a gradual increase in the relative DNA viscosity, revealing an intercalative interaction [47].

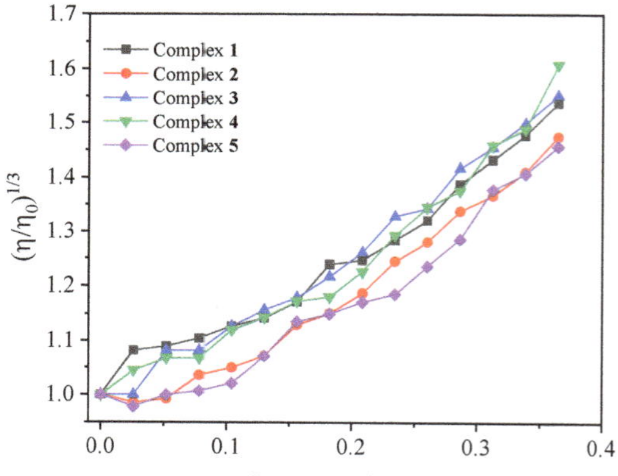

Figure 4. Relative viscosity $(\eta/\eta_0)^{1/3}$ of CT DNA (0.1 mM) in buffer solution (150 mM NaCl and 15 mM trisodium citrate at pH 7.0) in the presence of the compounds (initial concentration 0.1 mM) at increasing amounts (r = [compound]/[DNA] up to 0.36).

EB intercalates to DNA through its planar phenanthridine ring, which inserts in-between two adjacent DNA bases, resulting in an intense fluorescence emission band at $\lambda_{max} = 592–594$ nm upon excitation at 540 nm [48]. It is considered a DNA-intercalation marker, since the quenching of this band in the presence of intercalating compounds is an indication of competition for the same DNA-intercalation sites [48]. For this experiment, the fluorescence emission spectra of a buffer solution containing pretreated EB (40 μM) and CT DNA (40 μM) were recorded (with $\lambda_{excitation} = 540$ nm) in the presence of increasing amounts of complexes **1–5** (Figures 5A and S6). The quenching of the EB–DNA emission band at $\lambda_{max} = 594$ nm observed for all complexes (up to 54% of the initial fluorescence,

Figure 5B, Table 3) may be assigned to the displacement of the EB from EB-DNA induced by the complexes as a result of the competition for the same DNA sites. Such displacement of EB indirectly suggests intercalation as the most probable mode of interaction between CT DNA and complexes **1–5** [48].

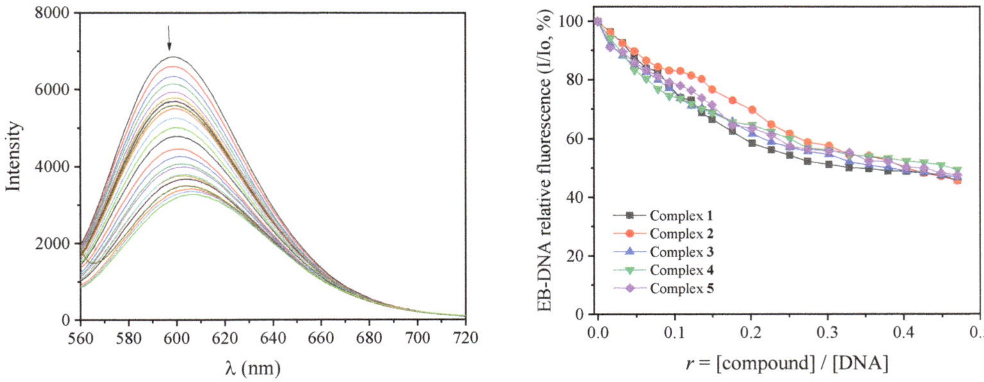

(A) **(B)**

Figure 5. (**A**) Fluorescence emission spectra ($\lambda_{excitation}$ = 540 nm) for EB–DNA ([EB] = 40 μM, [DNA] = 45 μM) in buffer solution (150 mM NaCl and 15 mM trisodium citrate at pH 7.0) in the absence and presence of increasing amounts of the complex **2**. The arrow shows the changes in intensity upon increasing amounts of the complex. (**B**) Plot of relative EB–DNA fluorescence emission intensity at $\lambda_{emission}$ = 594 nm (I/Io, %) versus *r* (*r* = [compound]/[DNA]) in the presence of complexes **1–5** (up to 45.7% of the initial EB–DNA fluorescence for complex **1**, 45.7% for complex **2**, 46.9% for complex **3**, 49.3% for complex **4**, and 47.5% for complex **5**).

Table 3. Fluorescence features of the EB-displacement studies for complexes **1–5**: percentage of EB–DNA fluorescence emission quenching ($\Delta I/I_0$, in %), Stern–Volmer (K_{SV}, in M^{-1}), and quenching constants (K_q, in $M^{-1}s^{-1}$).

Compound	ΔI/Io (%)	K_{sv} (M^{-1})	K_q ($M^{-1}s^{-1}$)
3,5-diBr-saloH [30]	56.3	$3.95(\pm0.10) \times 10^4$	$1.72(\pm0.04) \times 10^{12}$
$[Mn(3,5\text{-diBr-salo})_2(CH_3OH)_2]$, **1**	54.3	$1.09(\pm0.03) \times 10^5$	$4.72(\pm0.12) \times 10^{12}$
$[Mn(3,5\text{-diBr-salo})_2(\text{bipyam})]$, **2**	54.3	$4.15(\pm0.05) \times 10^4$	$1.80(\pm0.02) \times 10^{12}$
$[Mn(3,5\text{-diBr-salo})_2(\text{bipy})]$, **3**	53.1	$4.14(\pm0.01) \times 10^4$	$1.80(\pm0.04) \times 10^{12}$
$[Mn(3,5\text{-diBr-salo})_2(\text{phen})]$, **4**	50.7	$3.19(\pm0.09) \times 10^4$	$1.39(\pm0.04) \times 10^{12}$
$[Mn(3,5\text{-diBr-salo})_2(\text{neoc})]$, **5**	52.4	$3.83(\pm0.08) \times 10^4$	$1.67(\pm0.35) \times 10^{12}$

The evaluation of the EB-displacement was assessed through the Stern–Volmer constants (K_{SV}) calculated with the Stern–Volmer equation (Equation (S2)) [48] and the corresponding Stern–Volmer plots (Figure S7). Most of the complexes studied herein present higher K_{SV} values than free 3,5-diBr-saloH (Table 3), with $[Mn(3,5\text{-diBr-salo})_2(CH_3OH)_2]$ bearing the highest constant ($K_{SV} = 1.09(\pm0.03) \times 10^5$ M^{-1}). The quenching constants (K_q) for the complexes were calculated with Equation (S3), applying the value of 23 ns as the fluorescence lifetime for the EB–DNA system (τ_0) [49]. The K_q values of all complexes are found to be of the 10^{12} $M^{-1}s^{-1}$ order and are much higher than the value of 10^{10} $M^{-1}s^{-1}$; these values suggest the presence of a static quenching mechanism and confirm the formation of a new DNA–compound adduct as the result of the displacement of EB and subsequently indicate an intercalative mode of interaction [48].

2.4. Interaction of the Complexes with Albumins

Serum albumin (SA) is the most abundant protein in human blood plasma with important key roles such as the reversible binding and the transportation of drugs or other compounds and the maintenance of the osmotic pressure [50,51]. Within this context, the interaction of complexes **1–5** with the homologue albumins HSA and BSA was studied with fluorescence emission spectroscopy. In addition, the subtraction of the spectra was performed from the free compound for precise quantitative studies. Furthermore, the inner-filter effect was assessed with Equation (S4), and it was found negligible to affect the measurements [52].

The fluorescence emission spectra (excitation at 295 nm) of the albumins (3 μM) in buffer solution were recorded in the presence of increasing amounts of complexes **1–5** (Figures 6, S8 and S9). The intense fluorescence emission band observed at 345 nm (for BSA) or 340 nm (for HSA) exhibited a significant quenching (up to 92% of the initial emission intensity) in the presence of the complexes (Figure 7). Such quenching may result from the changes in the albumin's secondary structure and may be attributed to possible alterations of tryptophan residues of Sas and serves as an indication of the interaction of the complexes with the albumins [48].

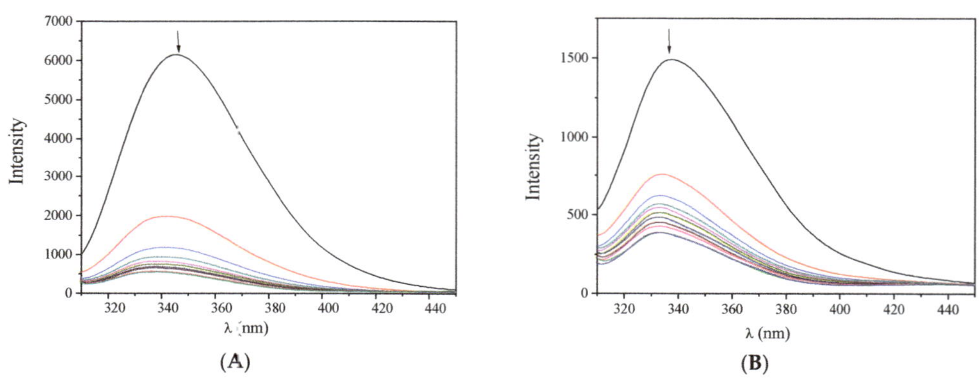

Figure 6. Fluorescence emission spectra ($\lambda_{excitation}$ = 295 nm) of (**A**) BSA (3 μM) and (**B**) HSA (3 μM) in buffer solution (150 mM NaCl and 15 mM trisodium citrate at pH 7.0) in the presence of increasing amounts of complex **1**. The arrows show the changes in intensity upon increasing amounts of the complex.

The interaction of the complexes with the albumins was assessed through the SA-quenching constants (K_q) and the SA-binding constants (K) calculated with the Stern–Volmer and Scatchard equations and the corresponding plots, respectively. The K_q constants were calculated with the Stern–Volmer equations (Equations (S2) and (S3)) and plots (Figures S10 and S11) applying the value of $\tau_0 = 10^{-8}$ s as the fluorescence lifetime of tryptophan in SAs [48]. The K_q values of the complexes for both albumins (Table 4) are of the order of 10^{12}–10^{13} $M^{-1}s^{-1}$ and are much higher than the value of 10^{10} $M^{-1}s^{-1}$, revealing the existence of a static quenching mechanism [48], which confirms the interaction of the compounds with the albumin.

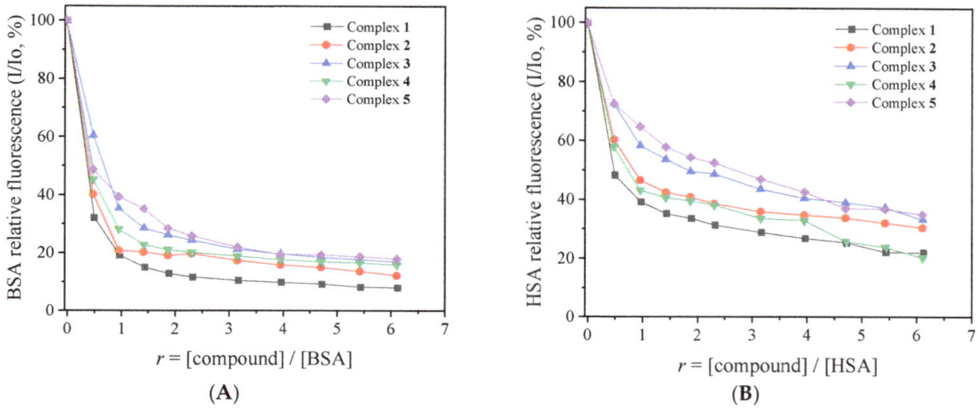

Figure 7. (**A**) Plot of % relative fluorescence intensity of BSA at λ_{em} = 345 nm (I/Io, %) versus r (r = [compound]/[BSA]) for its complexes **1–5** (up to 8.0% of the initial BSA fluorescence for complex **1**, 12.3% for complex **2**, 17.0% for complex **3**, 15.9% for complex **4**, and 18.0% for complex **5**). (**B**) Plot of % relative fluorescence intensity of HSA at λ_{em} = 340 nm (I/Io, %) versus r (r = [compound]/[HSA]) for complexes **1–5** (up to 21.9% of the initial HSA fluorescence for complex **1**, 30.4% for complex **2**, 33.3% for complex **3**, 20.3% for complex **4**, and 35.0% for complex **5**).

Table 4. The BSA/HSA-quenching constants (K_q, in $M^{-1}s^{-1}$) and the BSA/HSA-binding constants (K, in M^{-1}) calculated for complexes **1–5**.

Compound	$K_{q\,(BSA)}$ ($M^{-1}s^{-1}$)	$K_{(BSA)}$ (M^{-1})	$K_{q\,(HSA)}$ ($M^{-1}s^{-1}$)	$K_{(HSA)}$ (M^{-1})
3,5-diBr-saloH [30]	$3.65(\pm0.25) \times 10^{13}$	$2.97(\pm0.16) \times 10^{6}$	$1.72(\pm0.06) \times 10^{13}$	$4.04(\pm0.30) \times 10^{5}$
Complex **1**	$4.42(\pm0.26) \times 10^{13}$	$2.12(\pm0.06) \times 10^{6}$	$1.24(\pm0.04) \times 10^{13}$	$1.45(\pm0.07) \times 10^{6}$
Complex **2**	$3.18(\pm0.19) \times 10^{13}$	$1.36(\pm0.03) \times 10^{6}$	$6.29(\pm0.32) \times 10^{12}$	$8.18(\pm0.25) \times 10^{5}$
Complex **3**	$1.81(\pm0.10) \times 10^{13}$	$9.86(\pm0.28) \times 10^{5}$	$7.52(\pm0.30) \times 10^{12}$	$4.28(\pm0.09) \times 10^{5}$
Complex **4**	$1.24(\pm0.05) \times 10^{13}$	$1.71(\pm0.05) \times 10^{6}$	$1.93(\pm0.10) \times 10^{13}$	$9.05(\pm0.33) \times 10^{5}$
Complex **5**	$3.01(\pm0.12) \times 10^{13}$	$7.49(\pm0.23) \times 10^{5}$	$8.57(\pm0.21) \times 10^{12}$	$2.77(\pm0.05) \times 10^{5}$

The K values of the complexes for both SAs were calculated with the Scatchard equation (Equation (S5)) and corresponding plots (Figures S12 and S13). They are of 10^5–10^6 M^{-1} magnitude, showing a tight interaction of the complexes with the albumins. The K_q and K constants of complexes **1–5** (Table 4) are comparable with those recently reported (for BSA: $K_q = 6.83 \times 10^{12} - 2.11 \times 10^{14}$ $M^{-1}s^{-1}$, $K = 8.21 \times 10^4 - 2.11 \times 10^6$ M^{-1}; for HSA: $K_q = 6.00 \times 10^{12} - 4.22 \times 10^{13}$ $M^{-1}s^{-1}$, $K = 2.23 \times 10^4 - 1.99 \times 10^6$ M^{-1}) for metal complexes with 3,5-diBr-salo$^-$ ligands [29–33] and higher than most reported Mn(II) complexes with di- or mono-substituted salicylaldehydes ($K_{q(BSA)} = 3.32 \times 10^{12} - 3.14 \times 10^{13}$ $M^{-1}s^{-1}$, $K_{(BSA)} = 2.37 \times 10^4 - 4.79 \times 10^5$ M^{-1}) [25,26].

In addition, the SA-binding constants of all complexes under study are significantly lower than the value of 10^{15} M^{-1}, which is the highest binding constant found for the noncovalent interactions involving avidin. Based on this comparison, we may conclude that the herein reported complexes **1–5** can reversibly bind to the albumins and may be transferred and released at possible biological targets [53].

3. Experimental Section

3.1. Materials–Instrumentation–Methods

Information concerning materials, instrumentation, and physical measurements is cited in the Supplementary Materials (Section S1) [54] (abbreviations used for IR spectra: vs. = very strong; s = strong; sm = strong-to-medium; m = medium).

The procedure regarding the determination of crystal structure of complex **2** with single-crystal X-ray crystallography is described in the Supporting Information file (Section S2) [55–58]. Details of crystal data and structure refinement parameters are shown in Table S1 [59,60].

All the procedures and relevant equations used in the in vitro study of the biological activity (interaction with CT DNA, HSA, and BSA) of the compounds are described in the Supplementary Materials (Sections S3 and S4) [61].

3.2. Synthesis of the Complexes

3.2.1. Synthesis of [Mn(3,5-diBr-salo)$_2$(CH$_3$OH)$_2$] (Complex **1**)

A methanolic solution containing 3,5-diBr-saloH (0.5 mmol, 140 mg) and CH_3ONa (0.5 mmol, 27 mg) was stirred for 1 h and afterwards was added into a methanolic solution of $MnCl_2{\cdot}4H_2O$ (0.25 mmol, 49 mg) at RT. The reaction mixture was stirred for an additional 60 min and left to slowly evaporate. After two weeks, a beige microcrystalline product (yield: 85 mg, 50%) of complex **1** was collected with filtration. *Anal.* Calc. for [Mn(3,5-diBr-salo)$_2$(MeOH)$_2$] ($C_{16}H_{14}Br_4MnO_6$) (MW = 676.86). C: 28.39, H: 2.08%; found: C: 28.55, H: 2.17%. IR (KBr disk), ν_{max} (in cm^{-1}): $(O\text{–}H)_{MeOH}$, 3400(m); υ(C=O), 1641(s); υ(C–O → Mn), 1312(m). UV–vis: as nujol mull, λ (in nm): 419 (sh (shoulder)); in DMSO, λ (in nm) (ε, in M^{-1} cm^{-1}): 425 (10200), 268 (9140). μ_{eff} at RT = 5.90 BM. The complex is soluble in DMF and DMSO (Λ_M = 10 $S{\cdot}cm^2{\cdot}mol^{-1}$, in 1 mM DMSO solution) and partially soluble in methanol.

3.2.2. Synthesis of Complexes [Mn(3,5-diBr-salo)$_2$(*N,N′*-donor) (Complexes **2**–**5**)

Complexes **2**–**5** were prepared at RT according to the following procedure: a methanolic solution (10 mL) of CH_3ONa (0.5 mmol, 27 mg) and 3,5–diBr–saloH (0.5 mmol, 70 mg) was stirred for 30 min in order to deprotonate 3,5–dibromo–salicylaldehyde. Afterwards, the resultant solution was added dropwise and simultaneously with a solution of the corresponding *N,N′*-donor (0.25 mmol) into a methanolic solution of $MnCl_2{\cdot}4H_2O$ (0.25 mmol, 49 mg). The final solution was stirred for an additional 45 min and was left to evaporate slowly. The formation of the desired product was observed after a few days.

[Mn(3,5-diBr-salo)$_2$(bipyam)] (complex **2**): For the synthesis of complex **2**, bipyam (0.25 mmol, 43 mg) was used as the *N,N′*-donor. Light-brown crystals of **2** (90 mg, 45%) suitable for X-ray crystallography were collected after ten days. *Anal.* Calc. for [Mn(3,5-diBr-salo)$_2$(bipyam)] ($C_{24}H_{15}Br_4MnN_3O_4$) (MW = 783.94). C: 36.77, H: 1.93, N: 5.36%; found: C: 36.60, H: 1.83, N: 5.13%. IR (KBr disk), ν_{max} (in cm^{-1}): $\nu(C{=}O)_{aldehydo}$, 1641 (s); $\nu(C\text{–}O)_{phenolato}$, 1306 (sm); $\rho(C\text{–}H)_{bipyam}$, 767 (m). UV–vis: as nujol mull, λ (in nm): 422 (sh); in DMSO, λ (in nm) (ε, in M^{-1} cm^{-1}): 425 (6250), 315 (7500), 287 (9500). μ_{eff} at RT = 5.96 BM. The complex is soluble in DMF and DMSO (Λ_M = 8 $S{\cdot}cm^2{\cdot}mol^{-1}$, in 1 mM DMSO solution) and partially soluble in methanol and acetonitrile.

CCDC deposition number 2,468,794 contains the supplementary crystallographic data for complex **2**. These data can be obtained free of charge via www.ccdc.cam.ac.uk (or from the Cambridge Crystallographic Data Centre, 12 Union Road, Cambridge CB21EZ, UK; fax: (+44) 1223–336–033; or deposit@ccdc.cam.ac.uk).

[Mn(3,5-diBr-salo)$_2$(bipy)] (complex **3**): For the synthesis of complex **3**, bipy (0.25 mmol, 39 mg) was used as the *N,N′*-donor. The orange product of complex **3** (yield: 95 mg, 50%)

was collected with filtration after two weeks. *Anal.* Calc. for [Mn(3,5-diBr-salo)$_2$(bipy)] ($C_{24}H_{14}Br_4MnN_2O_4$) (MW = 768.96). C: 37.49, H: 1.83, N: 3.64%; found: C: 37.65, H: 1.69, N: 3.46%. IR (KBr disk), ν_{max} (in cm^{-1}): $\nu(C{=}O)_{aldehydo}$, 1625 (s); $\nu(C{-}O)_{phenolato}$, 1324 (sm); $\rho(C{-}H)_{bipy}$, 760 (m). UV–vis: as nujol mull, λ (in nm): 415 (sh); in DMSO, λ (in nm) (ε, in $M^{-1}cm^{-1}$): 425 (7450), 283 (8040). μ_{eff} at RT = 5.88 BM. The complex is soluble in DMF and DMSO (Λ_M = 15 S·cm^2·mol^{-1}, in 1 mM DMSO solution) and partially soluble in acetonitrile.

[Mn(3,5-diBr-salo)$_2$(phen)] (complex **4**): For the synthesis of complex **4**, phen (0.25 mmol, 45 mg) was used as the *N,N′*-donor. The orange product of complex **4** (yield: 105 mg, 52%) was collected with filtration after two weeks. *Anal.* Calc. for [Mn(3,5-diBr-salo)$_2$(phen)] ($C_{26}H_{14}Br_4MnN_2O_4$) (MW = 792.98). C: 39.38, H: 1.78, N: 3.53%; found: C: 39.56, H: 1.67, N: 3.33%. IR (KBr disk), ν_{max} (in cm^{-1}): $\nu(C{=}O)_{aldehydo}$, 1632 (s); $\nu(C{-}O)_{phenolato}$, 1325 (sm); $\rho(C{-}H)_{phen}$, 729 (m). UV–vis: as nujol mull, λ (in nm): 421 (sh); in DMSO, λ (in nm) (ε, in M^{-1} cm^{-1}): 425 (6200), 300 (11500). μ_{eff} at RT = 5.87 BM. The complex is soluble in DMF and DMSO (Λ_M = 11 S·cm^2·mol^{-1}, in 1 mM DMSO solution) and partially soluble in methanol and acetonitrile.

[Mn(3,5-diBr-salo)$_2$(neoc)] (complex **5**): For the synthesis of complex **5**, neoc (0.25 mmol, 52 mg) was used as the *N,N′*-donor. The orange product of complex **5** (yield: 95 mg, 45%) was collected with filtration after twenty days. *Anal.* Calc. for [Mn(3,5-diBr-salo)$_2$(neoc)] ($C_{28}H_{18}Br_4MnN_2O_4$) (MW = 821.04). C: 40.96, H: 2.21, N: 3.41%; found: C: 40.75, H: 2.37, N: 3.26%. IR (KBr disk), ν_{max} (in cm^{-1}): $\nu(C{=}O)_{aldehydo}$, 1632 (s); $\nu(C{-}O)_{phenolato}$, 1328 (sm); $\rho(C{-}H)_{neoc}$, 735 (m). UV–vis: as nujol mull, λ (in nm): 419 (sh); in DMSO, λ (in nm) (ε, in M^{-1} cm^{-1}): 425 (6430), 311 (5500), 279 (6120). μ_{eff} at RT = 5.85 BM. The complex is soluble in DMF and DMSO (Λ_M = 10 S·cm^2·mol^{-1}, in 1 mM DMSO solution) and partially soluble in methanol.

4. Conclusions

Five neutral mononuclear manganese(II) complexes with 3,5–dibromo–salicylaldehyde (3,5–diBr–saloH) were synthesized in the absence or the presence of the *N,N′*-donors 2,2′–bipyridylamine (bipyam), 2,2′–bipyridine (bipy), 1,10–phenanthroline (phen), and 2,9–dimethyl–1,10–phenanthroline (neoc) and were characterized with various techniques. The 3,5–diBr–salo$^-$ ligands are bound to the manganese(II) ion in a bidentate fashion via the deprotonated phenolato and the carbonyl oxygen atoms. The manganese(II) ions are six-coordinated with a distorted octahedral geometry.

The complexes interact with linear CT DNA via intercalation as derived from the techniques employed, and complex [Mn(3,5-diBr-salo)$_2$(bipy)] exhibits the highest DNA-binding constant (K_b = 6.58 × 10^5 M^{-1}) among the compounds studied herein. Furthermore, the complexes tightly and reversibly bind to human and serum albumins, as concluded from the values of the corresponding binding constants calculated. The DNA- and albumin-binding constants of complexes **1–5** are among the highest constants of the reported manganese(II) complexes with di- or mono-substituted salicylaldehydes. In conclusion, complexes **1–5** showed promising features regarding the interaction with biomolecules, deserving further bioactivity studies in future projects.

Supplementary Materials: The following supporting information can be downloaded at: https://www.mdpi.com/article/10.3390/inorganics13080263/s1; Cif file for complex **2**; Checkcif file for complex **2**; Information for materials, instrumentation, and physical measurements (Section S1); Procedure for single-crystal X-ray crystallography (Section S2); Protocols and equations regarding interaction studies with CT DNA (Section S3); Albumin-binding studies (Section S4). Table S1 and Figures S1–S13 are included in the ESI file.

Author Contributions: Conceptualization, V.T., A.Z., and G.P.; methodology, V.T., A.Z., and G.P.; formal analysis, V.T., A.Z., A.G.H., and G.P.; investigation, V.T., A.Z., and A.G.H.; resources, A.G.H. and G.P.; data curation, V.T., A.Z., A.G.H., and G.P.; writing—original draft preparation, V.T., A.Z., A.G.H., and G.P.; writing—review and editing, G.P.; supervision, G.P.; project administration, G.P.; funding acquisition, G.P. All authors have read and agreed to the published version of the manuscript.

Funding: This research received no external funding.

Institutional Review Board Statement: Not applicable.

Informed Consent Statement: Not applicable.

Data Availability Statement: Data are contained within the article and Supplementary Materials.

Conflicts of Interest: The authors declare no conflicts of interest.

Abbreviations

The following abbreviations are used in this manuscript:

3,5-diBr–saloH	3,5–dibromo–salicylaldehyde
4-OMe–saloH	4–methoxy–salicylaldehyde
5-Br–saloH	5–bromo–salicylaldehyde
5-Cl–saloH	5–chloro–salicylaldehyde
5-NO_2–saloH	5–nitro–salicylaldehyde
bipy	2,2′–bipyridine
bipyam	2,2′–bipyridylamine
BSA	bovine serum albumin
CT	calf-thymus
EB	ethidium bromide
HSA	human serum albumin
K	SA-binding constant
K_b	DNA-binding constant
K_q	quenching constant
K_{SV}	Stern–Volmer constant
neoc	2,9–dimethyl–1,10–phenanthroline
phen	1,10–phenanthroline
RT	room-temperature
SA	serum albumin
saloH	salicylaldehyde
X–saloH	substituted salicylaldehyde

References

1. Mullins, C.S.; Pecoraro, V.L. Reflections on Small Molecule Manganese Models That Seek to Mimic Photosynthetic Water Oxidation Chemistry. *Coord. Chem. Rev.* **2008**, *252*, 416–443. [CrossRef]
2. Wu, A.J.; Penner-Hahn, J.E.; Pecoraro, V.L. Structural, Spectroscopic, and Reactivity Models for the Manganese Catalases. *Chem. Rev.* **2004**, *104*, 903–938. [CrossRef] [PubMed]
3. Horning, K.J.; Caito, S.W.; Tipps, K.G.; Bowman, A.B.; Aschner, M. Manganese Is Essential for Neuronal Health. *Annu. Rev. Nutr.* **2015**, *35*, 71–108. [CrossRef] [PubMed]
4. Li, L.; Yang, X. The Essential Element Manganese, Oxidative Stress, and Metabolic Diseases: Links and Interactions. *Oxid. Med. Cell Longev.* **2018**, *2018*, 7580707. [CrossRef] [PubMed]
5. Krstic, N.S.; Nikolic, R.S.; Stankovic, M.N.; Nikolic, N.G.; Djordjevic, D.M. Coordination Compounds of M(II) Biometal Ions with Acid-Type Anti-Inflammatory Drugs as Ligands—A Review. *Trop. J. Pharm. Res.* **2015**, *14*, 337–349. [CrossRef]
6. Peres, T.V.; Schettinger, M.R.C.; Chen, P.; Carvalho, F.; Avila, D.S.; Bowman, A.B.; Aschner, M. Manganese-Induced Neurotoxicity: A Review of Its Behavioral Consequences and Neuroprotective Strategies. *BMC Pharmacol. Toxicol.* **2016**, *17*, 57. [CrossRef]
7. Eremina, J.A.; Ermakova, E.A.; Smirnova, K.S.; Klyushova, L.S.; Berezin, A.S.; Sukhikh, T.S.; Zubenko, A.A.; Fetisov, L.N.; Kononenko, K.N.; Lider, E.V. Cu(II), Co(II), Mn(II) Complexes with 5-Phenyltetrazole and Polypyridyl Ligands: Synthesis, Characterization and Evaluation of the Cytotoxicity and Antimicrobial Activity. *Polyhedron* **2021**, *206*, 115352. [CrossRef]

8. Abdallah, A.M.; Zaki, N.G.; Mahmoud, W.H.; El Kerdawy, A.M.; Mohamed, G.G. Synthesis, Structural Characterization, Density Functional Theory Calculations, and Antimicrobial, Anticancer, and Antimetastatic Properties of Nanosized Heteroleptic Complexes of Cocaine/TMEDA with d-Block Metal Ions. *Appl. Organomet. Chem.* **2021**, *35*, e6441. [CrossRef]
9. Archana, B.; Sreedaran, S. Synthesis, Characterization, DNA Binding and Cleavage Studies, in-Vitro Antimicrobial, Cytotoxicity Assay of New Manganese(III) Complexes of N-Functionalized Macrocyclic Cyclam Based Schiff Base Ligands. *Polyhedron* **2023**, *231*, 116269. [CrossRef]
10. Bourouai, M.A.; Bouchoucha, A.; Si Larbi, K.; Cosnier, S.; Djebbar, S. Novel Mn(II) and Cu(II) Metal Complexes with Sulfa Drug-Derived Ligands as Potent Antimicrobial and Anticancer Agents: In Vitro Studies, ADMET Profile and Molecular Docking. *Polyhedron* **2024**, *253*, 116914. [CrossRef]
11. Friaes, S.; Trigueiros, C.; Gomes, C.S.B.; Fernandes, A.R.; Lenis-Rojas, O.A.; Martins, M.; Royo, B. Antimicrobial Activity of Manganese(I) Tricarbonyl Complexes Bearing 1,2,3-Triazole Ligands. *Molecules* **2023**, *28*, 7453. [CrossRef]
12. Saleem, S.; Parveen, B.; Abbas, K.; Iqbal, S.; Altaf, A.A.; Kausar, S. Synthesis, Structural Elucidation, Molecular Modeling and Antimicrobial Studies of Mn(II), Co(II), Ni(II), and Cu(II) Complexes Containing NO Donor Bidentate Schiff Base. *Appl. Organomet. Chem.* **2023**, *37*, e7234. [CrossRef]
13. Swiderski, G.; Wojtulewski, S.; Kalinowska, M.; Swisłocka, R.; Wilczewska, A.Z.; Pietryczuk, A.; Cudowski, A.; Lewandowski, W. The Influence of Selected Transition Metal Ions on the Structure, Thermal and Microbiological Properties of Pyrazine-2-Carboxylic Acid. *Polyhedron* **2020**, *175*, 114173. [CrossRef]
14. Jablonska-Wawrzycka, A.; Rogala, P.; Czerwonka, G.; Michalkiewicz, S.; Hodorowicz, M.; Galczynska, K.; Cieslak, B.; Kowalczyk, P. Tuning Anti-Biofilm Activity of Manganese(Ii) Complexes: Linking Biological Effectiveness of Heteroaromatic Complexes of Alcohol, Aldehyde, Ketone, and Carboxylic Acid with Structural Effects and Redox Activity. *Int. J. Mol. Sci.* **2021**, *22*, 4847. [CrossRef] [PubMed]
15. Dimiza, F.; Hatzidimitriou, A.G.; Psomas, G. Manganese(II) Complexes with Non-Steroidal Anti-Inflammatory Drugs: Structure and Biological Activity. *Int. J. Mol. Sci.* **2024**, *25*, 13457. [CrossRef] [PubMed]
16. Sakthikumar, K.; Kabuyaya Isamura, B.; Krause, R.W.M. Exploring the Antioxidant, Antimicrobial, Cytotoxic and Biothermodynamic Properties of Novel Morpholine Derivative Bioactive Mn(II), Co(II) and Ni(II) Complexes—Combined Experimental and Theoretical Measurements towards DNA/BSA/SARS-CoV-2 3CLPro. *RSC Med. Chem.* **2023**, *14*, 1667–1697. [CrossRef] [PubMed]
17. Guo, Z.; Sadler, P.J. Metals in Medicine. *Angew. Chem. Int. Ed.* **1999**, *38*, 1512–1531. [CrossRef]
18. Elo, H.; Kuure, M.; Pelttari, E. Correlation of the Antimicrobial Activity of Salicylaldehydes with Broadening of the NMR Signal of the Hydroxyl Proton. Possible Involvement of Proton Exchange Processes in the Antimicrobial Activity. *Eur. J. Med. Chem.* **2015**, *92*, 750–753. [CrossRef]
19. Pelttari, E.; Karhumaki, E.; Langshaw, J.; Perakyla, H.; Elo, H. Antimicrobial Properties of Substituted Salicylaldehydes and Related Compounds. *Z. Naturforschung C* **2007**, *62*, 487–497. [CrossRef]
20. Bountagkidou, O.G.; Ordoudi, S.A.; Tsimidou, M.Z. Structure–Antioxidant Activity Relationship Study of Natural Hydroxybenzaldehydes Using in Vitro Assays. *Food Res. Int.* **2010**, *43*, 2014–2019. [CrossRef]
21. Kordestani, N.; Rudbari, H.A.; Fernandes, A.R.; Raposo, L.R.; Baptista, P.V.; Ferreira, D.; Bruno, G.; Bella, G.; Scopelliti, R.; Braun, J.D.; et al. Antiproliferative Activities of Diimine-Based Mixed Ligand Copper(II) Complexes. *ACS Comb. Sci.* **2020**, *22*, 89–99. [CrossRef]
22. Xu, Z.; Yang, Z.; Liu, Y.; Lu, Y.; Chen, K.; Zhu, W. Halogen Bond: Its Role beyond Drug–Target Binding Affinity for Drug Discovery and Development. *J. Chem. Inf. Model.* **2014**, *54*, 69–78. [CrossRef]
23. Varadwaj, P.R.; Varadwaj, A.; Marques, H.M. Halogen Bonding: A Halogen-Centered Noncovalent Interaction Yet to Be Understood. *Inorganics* **2019**, *7*, 40. [CrossRef]
24. Lu, Y.; Wang, Y.; Zhu, W. Nonbonding Interactions of Organic Halogens in Biological Systems: Implications for Drug Discovery and Biomolecular Design. *Phys. Chem. Chem. Phys.* **2010**, *12*, 4543–4551. [CrossRef] [PubMed]
25. Ntanatsidis, S.; Perontsis, S.; Konstantopoulou, S.; Kalogiannis, S.; Hatzidimitriou, A.G.; Papadopoulos, A.N.; Psomas, G. Manganese(II) Complexes of Substituted Salicylaldehydes and α-Diimines: Synthesis, Characterization and Biological Activity. *J. Inorg. Biochem.* **2022**, *227*, 111693. [CrossRef] [PubMed]
26. Stamou, P.; Hatzidimitriou, A.G.; Psomas, G. Manganese(II) Complexes with 5–Nitro–2–Hydroxy–Benzaldehyde or Substituted 2–Hydroxy–Phenones: Structure and Interaction with Bovine Serum Albumin and Calf–Thymus DNA. *J. Inorg. Biochem.* **2022**, *235*, 111923. [CrossRef]
27. Feng, D.; Wang, B.; Wang, L.; Abraham, N.; Tao, K.; Huang, L.; Shi, W.; Dong, Y.; Qu, Y. Pre-Ischemia Melatonin Treatment Alleviated Acute Neuronal Injury after Ischemic Stroke by Inhibiting Endoplasmic Reticulum Stress-Dependent Autophagy via PERK and IRE1 Signalings. *J. Pineal Res.* **2017**, *62*, e12395. [CrossRef]
28. Thapa, K.; Khan, H.; Singh, T.G.; Kaur, A. Traumatic Brain Injury: Mechanistic Insight on Pathophysiology and Potential Therapeutic Targets. *J. Mol. Neurosci.* **2021**, *71*, 1725–1742. [CrossRef]

29. Zianna, A.; Geromichalou, E.; Geromichalos, G.; Fiotaki, A.M.; Hatzidimitriou, A.G.; Kalogiannis, S.; Psomas, G. Zinc(II) Complexes of 3,5–Dibromo–Salicylaldehyde and α–Diimines: Synthesis, Characterization and in Vitro and in Silico Biological Profile. *J. Inorg. Biochem.* **2022**, *226*, 111659. [CrossRef]
30. Christidou, A.; Zavalani, K.; Hatzidimitriou, A.G.; Psomas, G. Copper(II) Complexes with 3,5–Dihalogeno–Salicylaldehydes: Synthesis, Structure and Interaction with DNA and Albumins. *J. Inorg. Biochem.* **2023**, *238*, 112049. [CrossRef]
31. Psarras, G.I.; Zianna, A.; Hatzidimitriou, A.G.; Psomas, G. Coordination Compounds of Nickel(II) with 3,5–Dibromo–Salicylaldehyde: Structure and Interaction with Biomolecules. *Inorganics* **2024**, *12*, 138. [CrossRef]
32. Papadopoulos, Z.; Hatzidimitriou, A.G.; Psomas, G. Iron(III) Complexes with Substituted Salicylaldehydes: Synthesis, Interaction with DNA and Serum Albumins, and Antioxidant Activity. *Molecules* **2025**, *30*, 2383. [CrossRef]
33. Zianna, A.; Geromichalos, G.; Fiotaki, A.M.; Hatzidimitriou, A.G.; Kalogiannis, S.; Psomas, G. Palladium(II) Complexes of Substituted Salicylaldehydes: Synthesis, Characterization and Investigation of Their Biological Profile. *Pharmaceuticals* **2022**, *15*, 886. [CrossRef] [PubMed]
34. Aryaeifar, M.; Rudbari, H.A.; Moreno-Pineda, E.; Cuevas-Vicario, J.V.; Paul, S.; Schulze, M.; Wernsdorfer, W.; Lloret, F.; Moini, N.; Blacque, O. Synthesis, Characterization and Magnetic Properties of Halogenated Tetranuclear Cubane-like Nickel(II) Complexes. *New J. Chem.* **2024**, *48*, 3603–3613. [CrossRef]
35. Geary, W.J. The Use of Conductivity Measurements in Organic Solvents for the Characterisation of Coordination Compounds. *Coord. Chem. Rev.* **1971**, *7*, 81–122. [CrossRef]
36. Chiswell, B.; McKenzie, E.D.; Lindoy, L.F. 41 Manganese. *Compr. Coord. Chem.* **1987**, *4*, 1–122.
37. Nakamoto, K. Infrared and Raman Spectra of Inorganic and Coordination Compounds: Part B: Applications in Coordination, Organometallic, and Bioinorganic Chemistry. In *Infrared and Raman Spectra of Inorganic and Coordination Compounds: Part B: Applications in Coordination, Organometallic, and Bioinorganic Chemistry*; John Wiley & Sons: Hoboken, NJ, USA, 2008; pp. 1–408. [CrossRef]
38. Papadopoulos, C.D.; Lalia-Kantouri, M.; Jaud, J.; Hatzidimitriou, A.G. Substitution Effect on New Co(II) Addition Compounds with Salicylaldehydes and the Nitrogenous Bases Phen or Neoc: Crystal and Molecular Structures of [Co^{II}(5-NO_2-Salicylaldehyde)$_2$(Phen)], [Co^{II}(5-CH_3-Salicylaldehyde)$_2$(Neoc)] and [Co^{II}(5-Cl-Salicylaldehyde)$_2$(Neoc)]. *Inorganica Chim. Acta* **2007**, *360*, 3581–3589. [CrossRef]
39. Zianna, A.; Vradi, E.; Hatzidimitriou, A.G.; Kalogiannis, S.; Psomas, G. Zinc(II) Complexes of 3-Bromo-5-Chloro-Salicylaldehyde: Characterization and Biological Activity. *Dalton Trans.* **2022**, *51*, 17629–17641. [CrossRef]
40. Zianna, A.; Sumar Ristovic, M.; Psomas, G.; Hatzidimitriou, A.; Coutouli-Argyropoulou, E.; Lalia-Kantouri, M. Cadmium(II) Complexes of 5-Nitro-Salicylaldehyde and α -Diimines: Synthesis, Structure and Interaction with Calf-Thymus DNA. *J. Coord. Chem.* **2015**, *68*, 4444–4463. [CrossRef]
41. Hadjiliadis, N.D.; Sletten, E. *Metal Complex-DNA Interactions*; Hadjiliadis, N., Sletten, E., Eds.; Wiley: New York, NY, USA, 2009; ISBN 978-1-405-17629-3.
42. Rehman, S.U.; Sarwar, T.; Husain, M.A.; Ishqi, H.M.; Tabish, M. Studying Non-Covalent Drug–DNA Interactions. *Arch. Biochem. Biophys.* **2015**, *576*, 49–60. [CrossRef]
43. Zeglis, B.M.; Pierre, V.C.; Barton, J.K. Metallo-Intercalators and Metallo-Insertors. *Chem. Commun.* **2007**, *44*, 4565–4579. [CrossRef]
44. Pyle, A.M.; Rehmann, J.P.; Meshoyrer, R.; Kumar, C.V.; Turro, N.J.; Barton, J.K. Mixed-Ligand Complexes of Ruthenium(II): Factors Governing Binding to DNA. *J. Am. Chem. Soc.* **2002**, *111*, 3051–3058. [CrossRef]
45. Wolfe, A.; Shimer, G.H.; Meehan, T. Polycyclic Aromatic Hydrocarbons Physically Intercalate into Duplex Regions of Denatured DNA. *Biochemistry* **1987**, *26*, 6392–6396. [CrossRef] [PubMed]
46. Dimitrakopoulou, A.; Dendrinou-Samara, C.; Pantazaki, A.A.; Alexiou, M.; Nordlander, E.; Kessissoglou, D.P. Synthesis, Structure and Interactions with DNA of Novel Tetranuclear, [Mn_4(II/II/II/IV)] Mixed Valence Complexes. *J. Inorg. Biochem.* **2008**, *102*, 618–628. [CrossRef] [PubMed]
47. Bravo-Anaya, L.; Rinaudo, M.; Martinez, F. Conformation and Rheological Properties of Calf-Thymus DNA in Solution. *Polymers* **2016**, *8*, 51. [CrossRef] [PubMed]
48. Lakowicz, J.R. *Principles of Fluorescence Spectroscopy*; Springer: Berlin/Heidelberg, Germany, 2006; ISBN 0387312781.
49. Heller, D.P.; Greenstock, C.L. Fluorescence Lifetime Analysis of DNA Intercalated Ethidium Bromide and Quenching by Free Dye. *Biophys. Chem.* **1994**, *50*, 305–312. [CrossRef]
50. He, X.M.; Carter, D.C. Atomic Structure and Chemistry of Human Serum Albumin. *Nature* **1992**, *358*, 209–215. [CrossRef]
51. Olson, R.E.; Christ, D.D. Chapter 33. Plasma Protein Binding of Drugs. *Annu. Rep. Med. Chem.* **1996**, *31*, 327–336. [CrossRef]
52. Stella, L.; Capodilupo, A.L.; Bietti, M. A Reassessment of the Association between Azulene and [60]Fullerene. Possible Pitfalls in the Determination of Binding Constants through Fluorescence Spectroscopy. *Chem. Commun.* **2008**, *39*, 4744–4746. [CrossRef]
53. Laitinen, O.H.; Hytonen, V.P.; Nordlund, H.R.; Kulomaa, M.S. Genetically Engineered Avidins and Streptavidins. *Cell. Mol. Life Sci.* **2006**, *63*, 2992–3017. [CrossRef]

54. Reichmann, M.E.; Rice, S.A.; Thomas, C.A.; Doty, P. A Further Examination of the Molecular Weight and Size of Desoxypentose Nucleic Acid. *J. Am. Chem. Soc.* **1954**, *76*, 3047–3053. [CrossRef]
55. *Apex2, Version 2 User Manual, M86–E01078*; Bruker Analytical X–Ray Systems, Inc.: Madison, WI, USA, 2006.
56. *SADABS: Area–Detector Absorption Correction*; Siemens Industrial Automation, Inc.: München, Germany, 1996.
57. Palatinus, L.; Chapuis, G. SUPERFLIP—A Computer Program for the Solution of Crystal Structures by Charge Flipping in Arbitrary Dimensions. *J. Appl. Crystallogr.* **2007**, *40*, 786–790. [CrossRef]
58. Betteridge, P.W.; Carruthers, J.R.; Cooper, R.I.; Prout, K.; Watkin, D.J. CRYSTALS Version 12: Software for Guided Crystal Structure Analysis. *J. Appl. Crystallogr.* **2003**, *36*, 1487. [CrossRef]
59. de Meulenaer, J.; Tompa, H. The Absorption Correction in Crystal Structure Analysis. *Acta Crystallogr.* **1965**, *19*, 1014–1018. [CrossRef]
60. Watkin, D.J.; Cooper, R.I. Why Direct and Post-Refinement Determinations of Absolute Structure May Give Different Results. *Acta Crystallogr. B Struct. Sci. Cryst. Eng. Mater.* **2016**, *72*, 661–683. [CrossRef]
61. Wang, Y.-Q.; Zhang, H.-M.; Zhang, G.-C.; Tao, W.-H.; Tang, S.-H. Interaction of the Flavonoid Hesperidin with Bovine Serum Albumin: A Fluorescence Quenching Study. *J. Lumin.* **2007**, *126*, 211–218. [CrossRef]

inorganics

MDPI

Article

Optimized $NaYF_4$: Er^{3+}/Yb^{3+} Upconversion Nanocomplexes via Oleic Acid for Biomedical Applications

Ha Thi Phuong [1], Le Thi Vinh [2], Tong Quang Cong [3], Tran Quoc Tien [3], Nguyen Duc Van [3], Vu Thi Hong Ha [4,*], Vu Ngoc Phan [4], Le Thi Hoi [5], Pham Duc Thang [6], Do Thi Thao [7] and Tran Thu Huong [3,*]

1 Department of Chemistry, Hanoi Medical University, 1 Ton That Tung, Hanoi 100000, Vietnam; hathiphuong@hmu.edu.vn
2 Faculty of Basic Science, Hanoi University of Mining and Geology, 18 Pho Vien, Hanoi 100000, Vietnam; levinhmdc@gmail com
3 Institute of Materials Science, Vietnam Academy of Science and Technology, 18 Hoang Quoc Viet, Hanoi 100000, Vietnam; congtq2004@gmail.com (T.Q.C.); tientq@ims.vast.ac.vn (T.Q.T.); vannguyenduc1972@gmail.com (N.D.V.)
4 Faculty of Biotechnology, Chemistry and Environmental Engineering, Phenikaa University, Yen Nghia, Ha Dong, Hanoi 100000, Vietnam; phan.vungoc@phenikaa-uni.edu.vn
5 Faculty of Medical Technology, Hanoi Medical University, 1 Ton That Tung, Hanoi 100000, Vietnam; lethihoi@hmu.edu.vn
6 Faculty of Physics, VNU University of Science, Vietnam National University, Hanoi, 334 Nguyen Trai, Hanoi 100000, Vietnam; pdthang@vnu.edu.vn
7 Institute of Biology, Vietnam Academy of Science and Technology, 18 Hoang Quoc Viet, Hanoi 100000, Vietnam; thaodo@ibt.ac.vn
* Correspondence: ha.vuthihong@phenikaa-uni.edu.vn (V.T.H.H.); tthuongims@gmail.com (T.T.H.)

Abstract: This study presents the synthesis of $NaYF_4$: Er^{3+}/Yb^{3+} upconversion luminescent nanomaterials using a wet chemistry method. The role of oleic acid in influencing the size, shape, and luminescent properties of the materials was also investigated. The results showed that, at a suitable oleic acid concentration of 10^{-3} M, the obtained nanoparticles exhibited a nearly spherical morphology with diameters ranging from 150 to 250 nm and predominantly display a hexagonal (β-$NaYF_4$) crystalline phase. Photoluminescence measurements under 980 nm laser excitation reveal that these nanoparticles emit strong, stable luminescence with narrow emission bands characteristic of Er^{3+} transitions. Subsequently, the nanoparticles were coated with a silica shell, functionalized with amine groups, and conjugated with IgG antibodies via glutaraldehyde (GA) to form the bio-nano complex β-$NaYF_4$: Er^{3+}/Yb^{3+}@SNGA-IgG. In vitro experiments using fluorescence microscopy demonstrated that the complex effectively labels HeLa cervical cancer cells. With its robust upconversion luminescence and excellent biocompatibility, the developed nanocomplex shows promising potential for rapid pathogen detection and other biomedical applications.

Keywords: $NaYF_4$: Er^{3+}/Yb^{3+}; upconversion; oleic acid; nanoparticles; wet chemistry

1. Introduction

Rare-earth ion-doped luminescent nanomaterials have attracted significant research interest due to their ability to exhibit both downconversion (DC) and upconversion (UC) luminescence. These properties are suitable for applications such as photocatalysis, security printing, solar cells, drug delivery, and cell labeling for biomedical applications [1–8].

Among these effects, upconversion luminescence has several advantages, including high sensitivity, generating a strong optical signal to enhance biological detection when using an infrared excitation source (e.g., 980 nm laser). Using infrared excitation sources

https://doi.org/10.3390/inorganics13050140

that are non-damaging to cells, friendly to humans, and capable of penetrating several millimeters into human tissue allows deeper interaction with damaged areas. Consequently, these materials have attracted interest in medical applications, especially for pathogen detection and cancer cell labeling [9–15].

$NaYF_4$ luminescent materials doped with Er^{3+} and Yb^{3+} ions are especially promising due to their upconversion effects. However, in order to achieve the target of using these materials for biomedical applications, numerous challenges still remain to be solved such as relatively low luminescence intensity compared to other materials and the need to optimize surface functionalization for effective biomolecule binding [16–19]. Researchers are, therefore, focusing on enhancing luminescence intensity, developing more stable and safer synthesis methods, and expanding applications for rapid and accurate pathogen detection. Controlled synthesis of nanoparticle size and shape is crucial, as these parameters directly influence optical properties and overall luminescence efficiency.

Oleic acid, a surfactant with the chemical formula $C_{17}H_{33}COOH$, plays a pivotal role in nanoparticle synthesis. Its *cis* C=C double bond facilitates the formation of a protective layer around nanoparticles, reducing energy loss and preventing aggregation. This protective effect allows for precise control over nanoparticle size and shape. Given that nanoparticle size directly affects optical performance, achieving optimal size control is essential for maximizing luminescence efficiency [20].

In this study, we investigate the effect of surfactant concentration—specifically, oleic acid (OA)—on the size, shape, and luminescent properties of $NaYF_4$: Er^{3+}/Yb^{3+} nanomaterials. Our goal is to determine the suitable synthesis conditions for producing highly luminescent materials for diagnostic applications, particularly in rapid pathogen detection. Additionally, we evaluated the ability of the nanocomplex containing $NaYF_4$:Yb^{3+}/Er^{3+} to detect HeLa cervical cancer cells, thereby confirming its potential for biomedical diagnostics.

2. Results and Discussion

2.1. Morphological Characterization

Figure 1 presents the FESEM images of the nanomaterial samples: (a) $NaYF_4$: Er^{3+}/Yb^{3+}, (b) $NaYF_4$: Er^{3+}/Yb^{3+}–OA (10^{-1} M), (c) $NaYF_4$: Er^{3+}/Yb^{3+}–OA (10^{-2} M), and (d) $NaYF_4$: Er^{3+}/Yb^{3+}–OA (10^{-3} M), respectively. The results indicate that the $NaYF_4$: Er^{3+}/Yb^{3+} sample that was synthesized without using OA (Figure 1a) exhibits an oval block shape with a diameter ranging from 800 to 900 nm. When oleic acid (OA) is introduced at a concentration of OA (10^{-1} M) (Figure 1b), the material tends to transition into a rod-like form with a length of 800–900 nm and a diameter of 400–500 nm. As the OA concentration decreases to OA (10^{-2} M), the particles become completely separated and tend to be hexagonal in shape with sizes of 400–500 nm. Further dilution to OA (10^{-3} M) results in progressively smaller spherical-like particles, ranging from 150 to 250 nm.

The gradual reduction in particle size can be probably attributed to the dilution process, which decreases the extent of reaction participation and promotes increased nucleation relative to individual particle growth. Additionally, a more diluted environment reduces particle collisions, thereby limiting aggregation and ensuring both small size and uniform dispersion.

These findings demonstrate that OA plays a crucial role in controlling the size and morphology of the material. At high OA concentrations, the material adopts a rod-like morphology, whereas at lower concentrations, it shifts towards a more spherical-like form and becomes progressively smaller. The controlled addition and dilution of OA provide an effective means of tailoring particle size and morphology, making the material suitable for various applications, particularly in biomedicine. This demonstrates the role of oleic acid as a morphological controller of the material, where dilution of its concentration leads to a gradual decrease in particle size.

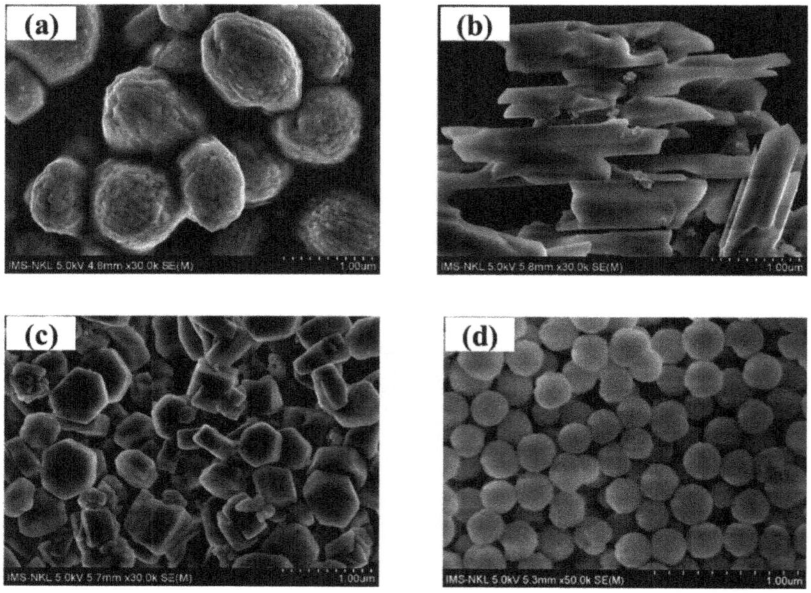

Figure 1. FESEM images of (**a**) $NaYF_4$: Er^{3+}/Yb^{3+}, (**b**) $NaYF_4$: Er^{3+}/Yb^{3+}–OA (10^{-1} M), (**c**) $NaYF_4$: Er^{3+}/Yb^{3+}–OA (10^{-2} M), and (**d**) $NaYF_4$: Er^{3+}/Yb^{3+}–OA (10^{-3} M) samples.

2.2. Structure Characterization

The XRD patterns of the nanoparticle samples, namely, $NaYF_4$: Er^{3+}/Yb^{3+}, $NaYF_4$: Er^{3+}/Yb^{3+}–OA (10^{-1} M), $NaYF_4$: Er^{3+}/Yb^{3+}–OA (10^{-2} M), and $NaYF_4$: Er^{3+}/Yb^{3+}–OA (10^{-3} M) are shown in Figure 2. The results reveal that characteristic diffraction peaks appear in the $NaYF_4$: Er^{3+}/Yb^{3+} samples synthesized by using OA at 2θ angles of 17.2, 29.8, 30.8, 34.7, 43.5, 46.5, 53.6, 55.2, and 62.3°. These peaks correspond to those in the JCPDS reference card No. 28-1192 for the β-$NaYF_4$ (hexagonal) crystal phase, which is the preferred phase for achieving high luminescence efficiency. For the $NaYF_4$: Er^{3+}/Yb^{3+} sample with the added OA concentration of 10^{-1} M, additional diffraction peaks at 28.2, 32.6, 46.9, and 55.6° were observed alongside those of the β-$NaYF_4$ (hexagonal) phase. These peaks correspond to the α-$NaYF_4$ (cubic) phase, as indicated by JCPDS reference card no. 01-077-2042. Importantly, for the case of using the suitable OA concentrations in the range of 10^{-2}–10^{-3} M, no impurity peaks are observed, confirming that the β-$NaYF_4$: Er^{3+}/Yb^{3+} material was synthesized readily and purely with a high crystallinity.

By combining the results obtained from FESEM and XRD measurements, it can be concluded that, for the $NaYF_4$: Er^{3+}/Yb^{3+}–OA (10^{-1} M) sample (Figure 2, curve 2), the material exhibits a rod-like morphology with a mixed-phase structure (hexagonal and cubic). However, as the OA concentration is diluted to 10^{-3} M, the material adopts a spherical-like morphology with only the hexagonal phase presents.

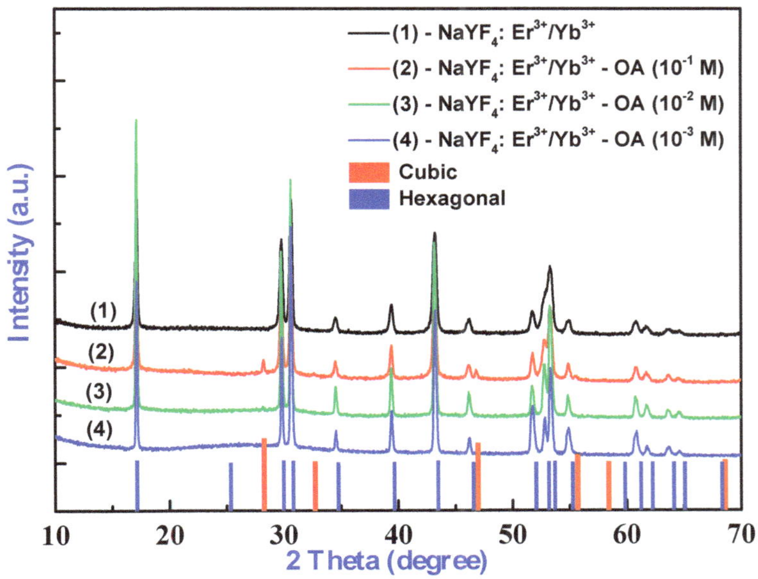

Figure 2. X-ray diffraction pattern of (1) $NaYF_4$: Er^{3+}/Yb^{3+}, (2) $NaYF_4$: Er^{3+}/Yb^{3+}–OA (10^{-1} M), (3) $NaYF_4$: Er^{3+}/Yb^{3+}–OA (10^{-2} M), and (4) $NaYF_4$: Er^{3+}/Yb^{3+}–OA (10^{-3} M) samples.

This shows that the concentration of oleic acid (OA) plays a crucial role not only in controlling particle size and morphology but also in determining the crystal phase of the synthesized $NaYF_4$:Yb^{3+}/Er^{3+} nanomaterials. At low OA concentrations, the surface of crystal nuclei is only partially covered, allowing for anisotropic growth that favors the formation of the thermodynamically stable hexagonal β-phase, which is typically associated with enhanced upconversion luminescence. However, when the OA concentration increases, the nanocrystal surfaces become densely coated with OA molecules, significantly altering the surface energy and growth dynamics. This hinders anisotropic crystal growth, thereby kinetically stabilizing the formation of the cubic α-phase, which is known to form during early nucleation or under growth-restricted conditions.

Additionally, excess OA can suppress the transition from cubic to hexagonal phase by reducing the crystallization rate, ultimately leading to the coexistence of both α- and β-phases in the final product. However, the hexagonal β-$NaYF_4$ phase features a larger separation between Er^{3+} ions, which helps to reduce fluorescence quenching. This structural advantage is highly beneficial for achieving strong luminescence, making the material well-suited for various applications [21–24].

Figure 3 presents the Fourier transform infrared (FTIR) spectrum of the β-$NaYF_4$: Er^{3+}/Yb^{3+}–OA (10^{-3} M) sample. The infrared spectrum displays a broad and intense band in the 3600–3300 cm^{-1} region, corresponding to O–H bond vibrations caused by the absorption of H_2O. Despite multiple washing steps with water and ethanol, organic citrate molecules remain on the particle surface. The peaks at approximately 2950 and 2854 cm^{-1} are attributed to the asymmetric and symmetric stretching vibrations of the methylene group (–CH_2), respectively [25,26]. A distinct peak at ~1614 cm^{-1} corresponds to the carboxylate (COO^-) group, confirming the attachment of oleic acid (OA) to the particle surface and its interaction with metal ions in the $NaYF_4$ lattice. Additionally, the band at ~1011 cm^{-1} is assigned to the C–O stretching vibration, indicating coordination with metal cations. The presence of rare earth elements bonded to the $NaYF_4$ matrix is reflected in the lower-frequency vibrations observed at 792, 576, and 454 cm^{-1}. These results confirm the successful formation of $NaYF_4$: Er^{3+}/Yb^{3+} nanoparticles.

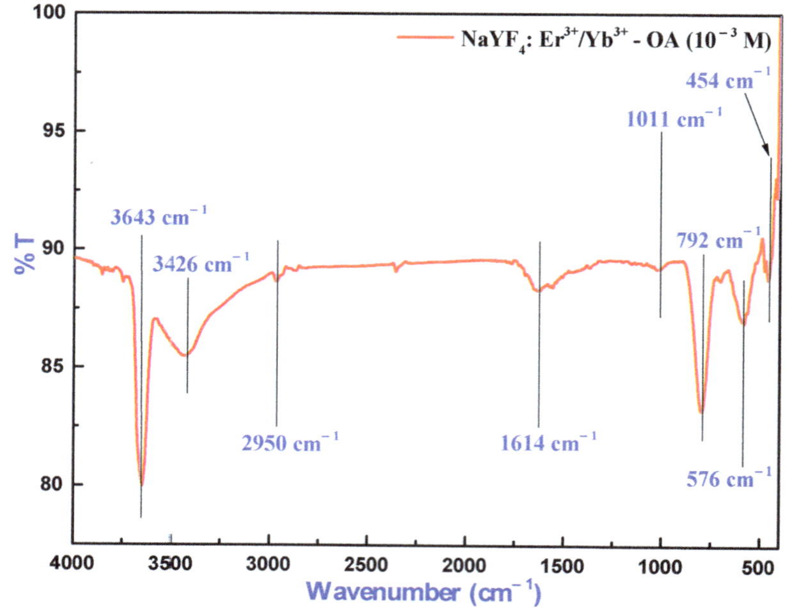

Figure 3. FTIR spectrum of the $NaYF_4$: Er^{3+}/Yb^{3+}–OA (10^{-3} M) samples.

By integrating the findings from XRD and FTIR measurements, it is evident that the incorporation of oleic acid in the synthesis process plays a crucial role in controlling particle size and morphology, ultimately enhancing the optical properties of the material.

2.3. Energy Dispersive X-Ray Analysis

Figure 4 presents the energy dispersive X-ray (EDX) spectrum of the β-$NaYF_4$: Er^{3+}/Yb^{3+}–OA (10^{-3} M) nanoparticles. The spectrum reveals strong peaks corresponding to the primary elements Na, Y, and F, along with weaker peaks for Yb and Er, which are consistent with the expected composition of $NaYF_4$: Er^{3+}/Yb^{3+}. Notably, no foreign elements are detected, confirming the successful synthesis of the β-$NaYF_4$: Er^{3+}/Yb^{3+}–OA (10^{-3} M) material.

To further analyze the elemental distribution and verify the uniform dispersion of dopant elements within the host matrix, energy dispersive X-ray spectroscopy (EDS) mapping was conducted (Figure 5). The EDS mapping images of the β-$NaYF_4$: Er^{3+}/Yb^{3+}–OA (10^{-3} M) nanoparticles illustrate the distribution of: (I) sodium (Na); (II) yttrium (Y); (III) fluorine (F); (IV) ytterbium (Yb); (V) erbium (Er) elements.

The mapping results confirm that Na, F, and Y are evenly distributed throughout the sample, indicating the successful incorporation of these elements into the host lattice. Additionally, the distributions of Yb and Er are also homogeneous, which plays a crucial role in reducing concentration quenching and enhancing luminescence intensity. The composite (merged color) image (VI) demonstrates the overlap of elemental signals, verifying that all components are present and uniformly distributed within the system, with no signs of phase separation or aggregation. This finding underscores the high potential of this material for optical applications, particularly in the biomedical field.

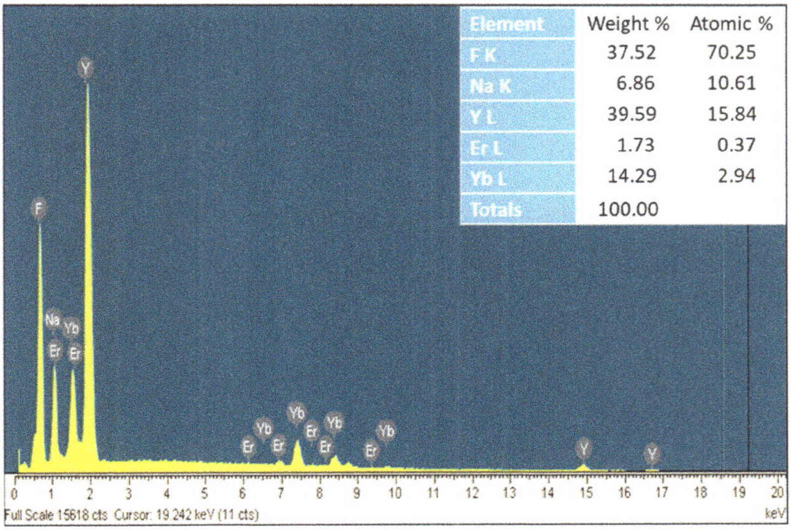

Figure 4. EDX spectrum of the β-$NaYF_4$: Er^{3+}/Yb^{3+}—oleic acid 10^{-3} M sample.

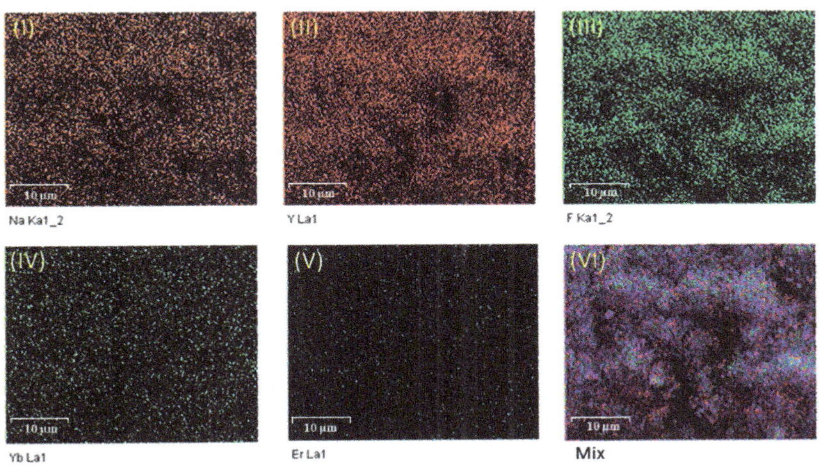

Figure 5. Mapping EDS images of β-$NaYF_4$: Er^{3+}/Yb^{3+}–OA (10^{-3} M) nanoparticles, the distribution of: **(I)** sodium (Na) element; **(II)** yttrium (Y) element; **(III)** fluorine (F) element; **(IV)** ytterbium (Yb) element; **(V)** erbium (Er) element, and **(VI)** merged color.

2.4. Luminescence Properties

Figure 6 presents the upconversion photoluminescence spectra of the nanoparticle samples: (a) $NaYF_4$: Er^{3+}/Yb^{3+}, (b) $NaYF_4$: Er^{3+}/Yb^{3+}–OA (10^{-1} M), (c) $NaYF_4$: Er^{3+}/Yb^{3+}–OA (10^{-2} M), and (d) $NaYF_4$: Er^{3+}/Yb^{3+}–OA (10^{-3} M), under 980 nm excitation, at 450 mW.

The spectra exhibit three broad emission bands ranging from green to red, centered at 520, 540, and 650 nm. These emissions correspond to the ${}^2H_{11/2} \rightarrow {}^4I_{15/2}$, ${}^4S_{3/2} \rightarrow {}^4I_{15/2}$, and ${}^4F_{9/2} \rightarrow {}^4I_{15/2}$ energy transitions in Er^{3+}, respectively [7,27,28]. The upconversion emission of $NaYF_4$: Er^{3+}/Yb^{3+} nanoparticles clearly demonstrates these typical energy transitions. A significant enhancement in luminescence intensity was observed in the $NaYF_4$: Er^{3+}/Yb^{3+}–OA (10^{-3} M) sample compared to the $NaYF_4$: Er^{3+}/Yb^{3+}–OA (10^{-1} M) sample. The emission intensity of the β-$NaYF_4$: Er^{3+}/Yb^{3+}–OA (10^{-3} M) sample is 8.2 times higher than that of the mixed phase $NaYF_4$: Er^{3+}/Yb^{3+}–OA (10^{-1} M) sample. This enhancement can be attributed to the small particle size and uniform distribution of dopant ions in the material, which further contributes to reducing the concentration quenching effect, thereby enhancing the luminescence intensity.

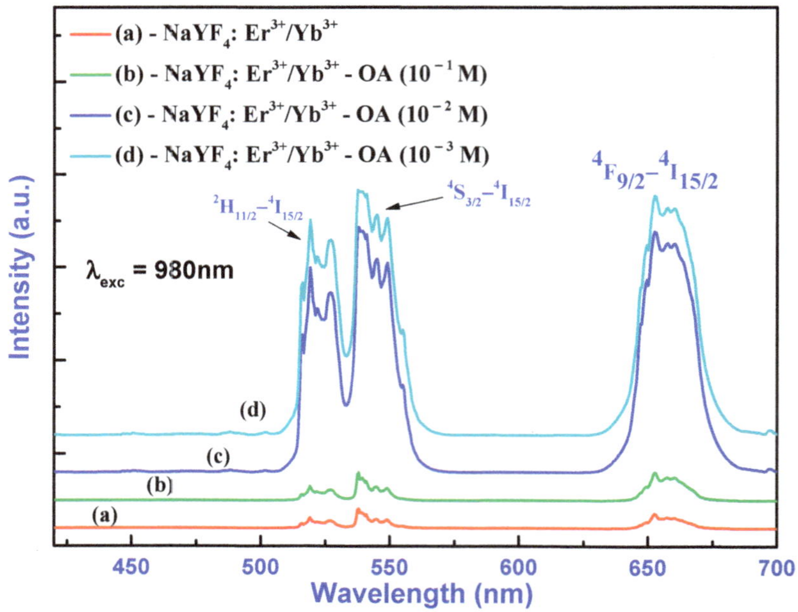

Figure 6. The upconversion photoluminescence spectra of (a) $NaYF_4$: Er^{3+}/Yb^{3+}, (b) $NaYF_4$: Er^{3+}/Yb^{3+}–OA (10^{-1} M), (c) $NaYF_4$: Er^{3+}/Yb^{3+}–OA (10^{-2} M), and (d) $NaYF_4$: Er^{3+}/Yb^{3+}–OA (10^{-3} M) samples under 980 nm excitation.

Some reports have demonstrated that $NaYF_4$-based upconversion nanocrystals can be synthesized with significantly smaller sizes (e.g., 20–50 nm) by employing different synthetic strategies [29,30]. To further reduce particle size, several optimization approaches can be considered, such as: decreasing the OA concentration; adjusting the reaction temperature and time; employing a co-surfactant or alternative ligands using microwave-assisted synthesis. In our work, the primary focus was to ensure strong upconversion luminescence and phase purity (especially stabilizing the β-phase). We have investigated the effects of three representative oleic acid (OA) concentrations (10^{-1}, 10^{-2}, and 10^{-3} M) on the morphology and upconversion luminescence of $NaYF_4$:Er^{3+}/Yb^{3+} nanomaterials. Among these, the 10^{-3} M concentration exhibited the very good performance in terms of emission intensity, particle uniformity, and size. Therefore, this sample was selected for cancer cell labeling experiments. Further investigation of these material characteristics will be carried out in the next study.

2.5. In Vitro Cellular Imaging

We used fluorescence microscopy to evaluate the binding ability between $NaYF_4$: Er^{3+}/Yb^{3+}@SNGA-IgG conjugates and HeLa cervical cancer cells after the incubation process. Figure 7a,b show the fluorescent images of HeLa cervical cancer cells (negative control) and HeLa cervical cancer cells incubated with $NaYF_4$: Er^{3+}/Yb^{3+}@SNGA-IgG, respectively.

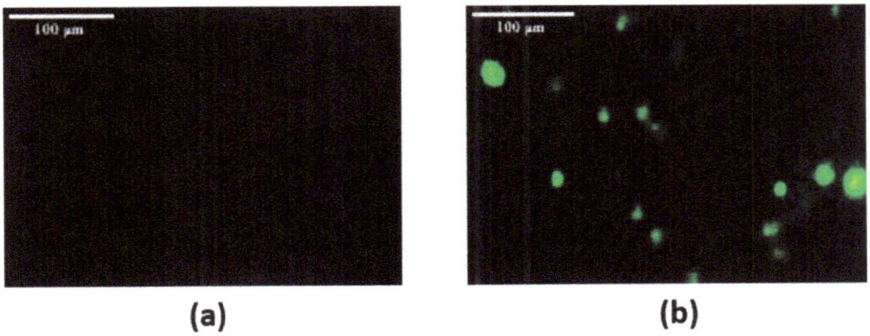

Figure 7. Fluorescence microscopy images of HeLa cervical cancer cells after 3 h of incubation with (**a**) HeLa cervical cancer cells (negative control) and (**b**) β-$NaYF_4$: Er^{3+}/Yb^{3+}@SNGA-IgG conjugates at a concentration of 20 μg/mL.

In the first case, no photoluminescence (PL) emission was observed in the reference sample (Figure 7a). In the second case, the $NaYF_4$: Er^{3+}/Yb^{3+}@SNGA-IgG sample exhibited strong green luminescence (Figure 7b). This might be attributed to the formation of bonds between $NaYF_4$: Er^{3+}/Yb^{3+}@SNGA-IgG and HeLa cervical cancer cells, demonstrating a strong interaction between the conjugates and HeLa cancer cells due to biological conjugation. Furthermore, it can be seen that the $NaYF_4$: Er^{3+}/Yb^{3+}@SNGA-IgG particles are localized within the cell cytoplasm. The Glutaraldehyde facilitates ligand binding between the cell and the luminescent labeling particles, and, subsequently, the conjugates are internalized into the cell via the invagination process. Therefore, β-$NaYF_4$: Er^{3+}/Yb^{3+}@SNGA-IgG could serve as a potential bio-label for cancer cells.

3. Experimental

3.1. Materials and Methods

Yttrium (III) nitrate hexahydrate, $Y(NO_3)_3{\cdot}6H_2O$ (99.9%), ytterbium (III) nitrate pentahydrate, $Yb(NO_3)_3{\cdot}5H_2O$ (99.9%), erbium (III) nitrate pentahydrate, $Er(NO_3)_3{\cdot}5H_2O$ (99.9%), NaF (99%), and tetraethyl orthosilicate (TEOS) (99.9%) were purchased from Sigma-Aldrich, St. Louis, MO, USA. NaOH (99%), trisodium citrate dehydrate HOC(COONa)$(CH_2COONa)_2{\cdot}2H_2O$ (99%) were provided from Merck, Darmstadt, Germany, while oleic acid ($C_{18}H_{34}O_2$) (92%) was provided from Fisher, London, England. All chemicals were of analytical grade without any purification.

The crystalline phase identification of $NaYF_4$:Yb^{3+}/Er^{3+} nanomaterials were carried out by using X-ray diffraction (XRD) (D8 Advance, Bruker, Karlsruhe, Germany). The morphology and energy dispersive X-ray spectra (EDS) of the $NaYF_4$:Yb^{3+}/Er^{3+} nanoparticles were analyzed using a field emission scanning electron microscope (FE-SEM, S-4800, Hitachi, Tokyo, Japan) and high-resolution transmission electron microscopy (HRTEM, JEM 2100, Jeol, Tokyo, Japan). To investigate the chemical bonding of the material, infrared absorption spectra (IR) were recorded in the wavenumber range of 4000 to 400 cm^{-1} using a Fourier transform infrared (FT-IR) spectrometer (Spectrum Two, PerkinElmer, Waltham, MA, USA). Photoluminescence (PL) measurements were conducted to evaluate the optical properties of $NaYF_4$:Yb^{3+}/Er^{3+} nanomaterials.

The upconversion photoluminescence properties were analyzed using an iHR320 photoluminescence measurement system (Horiba, Kyoto, Japan), under 980 nm excitation (the dried nanopowder samples were placed directly onto flat copper slides). The cells were observed under an Olympus ScanR fluorescence microscope (Olympus Europa SE and Co.KG, Hamburg, Germany).

3.2. Preparation of $NaYF_4$: Er^{3+}/Yb^{3+} Nanoparticles

The synthesis process of $NaYF_4$: Er^{3+}/Yb^{3+} nanomaterials is as follows:

First, a mixed solution of rare-earth metal salts, including $Y(NO_3)_3{\cdot}6H_2O$, $Yb(NO_3)_3{\cdot}5H_2O$, and $Er(NO_3)_3{\cdot}5H_2O$, is prepared with a molar ratio of $NaYF_4$:Yb^{3+}:Er^{3+} = 79:19:02. The solution is stirred for 15 min to obtain Solution 1.

Next, a solution of NaF and sodium hydroxide is slowly added to Solution 1, followed by continuous stirring for 120 min to obtain Solution 2. Then, trisodium citrate dihydrate is gradually added to Solution 2 and stirred for 15 min to obtain Solution 3.

After that, oleic acid solutions (10^{-1}, 10^{-2}, 10^{-3} M, etc.) are added to Solution 3 and stirred for 120 min. Finally, the resulting solution is transferred into an autoclave and heated at 200 °C for 24 h.

After completion of the hydrothermal reaction, the resulting products were collected by centrifugation and washed several times with deionized water to remove unreacted precursors and soluble byproducts. The washed samples were then dried at 70 °C for 48 h to obtain dry nanopowders. Notably, although the nanocrystals were surface-modified with oleic acid (OA), the washing step did not involve redispersion in an aqueous medium. As a result, the dried OA-coated nanocrystals retained their hydrophobic nature and were not dispersed in water during any of the characterization steps. The obtained product is centrifuged, thoroughly washed with deionized water to remove residual solvents and dried at 70 °C for 48 h. The final product is a $NaYF_4$:Er^{3+}/Yb^{3+} material in powder form.

3.3. Preparation of $NaYF_4$: Er^{3+}/Yb^{3+}@SNGA-IgG Bio-NanoComplexes

After synthesis, the $NaYF_4$: Er^{3+}/Yb^{3+} material was coated with a layer of silica via the hydrolysis of tetraethyl orthosilicate (TEOS). To functionalize the surface with amine groups, we used a solution of 3-aminopropyltrimethoxysilane [31]. The surface functionalization process was carried out for 24 h. Next, a solution of 50% glutaraldehyde (GA) was mixed with the functionalized $NaYF_4$: Er^{3+}/Yb^{3+}material; this mixture was then dispersed in a PBS solution (0.1 M, pH = 5) at a concentration of 5 g L^{-1} and agitated for 30 min. Subsequently, the compound was combined with various concentrations of IgG. The reaction mixtures were incubated in the presence of glycerol at room temperature for 4 h. Finally, the $NaYF_4$: Er^{3+}/Yb^{3+}@SNGA-IgG products were collected by centrifugation at 5900 rpm, washed three times with water, and stored at 4 °C in a sealed container.

3.4. Experiment on Incubating the Biomedical $NaYF_4$: Er^{3+}/Yb^{3+}@SNGA-IgG Nanocomplex with HeLa Cervical Cancer Cells

3.4.1. In Vitro Cell Culture

HeLa cervical cancer cells were cultured at 37 °C in DMEM medium supplemented with 10% fetal bovine serum (FBS), 1% antibiotic–antimycotic, and 50 μg/mL gentamicin. The cells were subcultured after 3–5 days at a ratio of 1:3 and incubated at 37 °C in an atmosphere containing 5% CO_2.

3.4.2. Experiment for Labeling HeLa Cervical Cancer Cells Using the Biomedical Nanocomplex $NaYF_4$: Er^{3+}/Yb^{3+}@SNGA-IgG

To evaluate cell detection capabilities, we performed cell staining. The procedure was as follows: HeLa cervical cancer cells were seeded into a 24-well plate at a density of 5×10^4 cells/mL and allowed to stabilize in an incubator at 37 °C with 5% CO_2 for 24 h. Next, PEG 1500 and the nanocomplex of β-$NaYF_4$: Er^{3+}/Yb^{3+}@SNGA-IgG (at a concentration of 20 μg/mL) were added to the cell culture wells, and the cells were incubated for further 3 h.

After incubation, the culture medium (including nanoparticles) was discarded. The cells were then washed three times with sterile phosphate-buffered saline (pH 7.4) before being observed under an Olympus ScanR fluorescence microscope.

4. Conclusions

We have successfully synthesized the $NaYF_4$: Er^{3+}/Yb^{3+} material using a wet chemistry method. The influence of oleic acid on the morphology and luminescent properties of the material was investigated. The resulting $NaYF_4$: Er^{3+}/Yb^{3+} nanomaterials exhibit upconversion luminescence in the green region, and our results indicate that a suitable oleic acid concentration of 10^{-3} M is required. The synthesized nanoparticles are spherical in shape, with diameters ranging from 150 to 250 nm, and possess a single-phase β-$NaYF_4$ (hexagonal phase) crystalline structure.

The β-$NaYF_4$: Er^{3+}/Yb^{3+} materials synthesized with an oleic acid concentration of 10^{-3} M demonstrate good and stable upconversion luminescence. They were subsequently coated with silica, functionalized with amino-silane, and conjugated with IgG antibodies. The pairing ability of the biomedical β-$NaYF_4$: Er^{3+}/Yb^{3+}@SNGA-IgG nanocomplex for labeling HeLa cervical cancer cells was successfully evaluated using an inverted fluorescence microscope. The results show that the nanocomplex effectively binds to HeLa cervical cancer cells under in vitro conditions.

Initial test results suggest that the biomedical nanocomplex containing the β-$NaYF_4$: Er^{3+}/Yb^{3+} material shows promising potential for rapid pathogen detection.

Author Contributions: Conceptualization, T.T.H.; data curation, L.T.V., T.Q.T., V.T.H.H., L.T.H., P.D.T. and T.T.H.; formal analysis, H.T.P., V.N.P., L.T.H., P.D.T., D.T.T. and T.T.H.; investigation, H.T.P., L.T.V., T.Q.C., T.Q.T., V.T.H.H. and D.T.T.; methodology, H.T.P., L.T.V., T.Q.C., T.Q.T., N.D.V., V.T.H.H. and V.N.P.; writing—original draft, T.Q.C., T.Q.T., V.T.H.H., V.N.P., P.D.T., D.T.T. and T.T.H.; writing—review and editing, N.D.V. and T.T.H. All authors have read and agreed to the published version of the manuscript.

Funding: This research was funded by the ĐTĐL.CN-26/23 project of the Ministry of Science and Technology.

Institutional Review Board Statement: Not applicable.

Informed Consent Statement: Not applicable.

Data Availability Statement: All data are available in this publication.

Conflicts of Interest: The authors declare no conflicts of interest.

References

1. Yaoguang, Y.; Gang, C.; Yansong, Z.; Zhonghui, H. Recent advances in rare-earth elements modification of inorganic semiconductor-based photocatalysts for efficient solar energy conversion: A review. *J. Rare Earths* **2015**, *33*, 453–462. [CrossRef]
2. Guo, X.; Chen, C.; Zhang, D.; Tripp, C.P.; Yin, S.; Qin, W. Photocatalysis of $NaYF_4$:Yb,Er/CdSe composites under 1560 nm laser excitation. *RSC Adv.* **2016**, *6*, 8127–8133. [CrossRef]
3. You, M.; Zhong, J.; Hong, Y.; Duan, Z.; Lin, M.; Xu, F. Inkjet printing of upconversion nanoparticles for anti-counterfeit applications. *Nanoscale* **2015**, *7*, 4423–4431. [CrossRef]
4. Shao, J.; Yan, J.; Xiaoguang, L.; Li, S.; Hu, T. Novel fluorescent label based on YVO_4: Bi^{3+}, Eu^{3+} for latent fingerprint detection. *Dye. Pigment.* **2019**, *160*, 555–562. [CrossRef]
5. Huang, X.; Han, S.; Huang, W.; Liu, X. Enhancing solar cell efficiency: The search for luminescent materials as spectral converters. *Chem. Soc. Rev.* **2013**, *42*, 173–201. [CrossRef]
6. Akbarzadeh, I.; Yaraki, M.T.; Ahmadi, S.; Chiani, M.; Nourouzian, D. Folic acid-functionalized niosomal nanoparticles for selective dual-drug delivery into breast cancer cells: An in-vitro investigation. *Adv. Powder Technol.* **2020**, *31*, 4064–4071. [CrossRef]

7. Zhou, J.; Liu, Q.; Feng, W.; Sun, Y.; Li, F. Upconversion luminescent materials: Advances and applications. *Chem. Rev.* **2015**, *115*, 395–465. [CrossRef]
8. Chen, G.; Qiu, H.; Prasad, P.N.; Chen, X. Upconversion Nanoparticles: Design, Nanochemistry, and Applications in Theranostics. *Chem. Rev.* **2014**, *114*, 5161–5214. [CrossRef]
9. Chitta, R.P.; Resham, B.; Sujata, P.; Sujit, B.; Priyabrata, M.; Debabrata, M. Inorganic Phosphate nanorods are a novel fluorescent label in cell biology. *J. Nanobiotechnol.* **2006**, *4*, 11.
10. Xu, R.; Cao, H.; Lin, D.; Yu, B.; Junle, Q. Lanthanide-doped upconversion nanoparticles for biological super-resolution fluorescence imaging. *Cell Rep. Phys. Sci.* **2022**, *3*, 100922. [CrossRef]
11. Chen, G.; Ågren, H.; Ohulchanskyy, T.Y.; Prasad, P.N. Light upconverting core–shell nanostructures: Nanophotonic control for emerging applications. *Chem. Soc. Rev.* **2015**, *44*, 1680–1713. [CrossRef]
12. Liang, X.; Fan, J.; Zhao, Y.; Jin, R. Core–Shell Structured $NaYF_4$:Yb,Er Nanoparticles with Excellent Upconversion Luminescent for Targeted Drug Delivery. *J. Clust. Sci.* **2021**, *32*, 1683–1691. [CrossRef]
13. Lei, Y.; Huang, T.; Su, M.; Luo, J.; Korteweg, C.; Li, J.; Chen, Z.; Qiu, Y.; Liu, X.; Yan, M.; et al. Expression and distribution of immunoglobulin G in the normal liver, hepatocarcinoma and postpartial hepatectomy liver. *Lab. Investig.* **2014**, *94*, 1283–1295. [CrossRef]
14. Chinen, A.B.; Guan, C.M.; Ferrer, J.R.; Barnaby, S.N.; Merkel, T.J.; Mirkin, C.A. Nanoparticle Probes for the Detection of Cancer Biomarkers, Cells, and Tissues by Fluorescence. *Chem. Rev.* **2015**, *115*, 10530–10574. [CrossRef]
15. Feng, W.; Zhu, X.; Li, F. Recent advances in the optimization and functionalization of upconversion nanomaterials for in vivo bioapplications. *NPG Asia Mater.* **2013**, *5*, e75. [CrossRef]
16. Oleksa, V.; Macková, H.; Engstová, H.; Patsula, V.; Shapoval, O.; Velychkivska, N.; Ježek, P.; Horák, D. Poly (N,N-di met hyl acr yla mid e)-coated upconverting NaYF4:Yb,Er@ $NaYF_4$:Nd core–shell nanoparticles for fluorescent labeling of carcinoma cells. *Sci. Rep.* **2021**, *11*, 21373. [CrossRef]
17. Doronkina, A.A.; Kochubey, V.I.; Maksutova, A.V.; Pravdin, A.B.; Mylnikov, A.M.; Navolokin, N.A.; Yanina, I.Y. $NaYF_4$: Yb,Er Upconversion Nanoparticles for Imaging: Effect on Red Blood Cells. *Photonics* **2023**, *10*, 1386. [CrossRef]
18. Skaptsov, A.A.; Ustalkov, S.O.; Mohammed, A.H.; Zakharevich, A.M.; Kozyrev, A.A.; Sagaidachnaya, E.A.; Kochubey, V.I. Application of Upconversion Luminescence of NaYF4:Yb,Er Nanoparticles to Study Protein Coagulation Dynamics. *Opt. Spectrosc.* **2020**, *128*, 952–958. [CrossRef]
19. Hua, L.; Xiaobo, S.; Xia, L.; Lingbo, Z. Size-tunable β-$NaYF_4$:Yb/Er up-converting nanoparticles with a strong green emission synthesized by thermal decomposition. *Opt. Mater.* **2020**, *108*, 110144. [CrossRef]
20. Takahashi, N.; Gubarevich, A.; Sakurai, J.; Hata, S.; Tsuge, T.; Kitamoto, Y.; Yamazaki, Y.; Odawara, O.; Wada, H. Preparation and Optical Properties of Rare Earth Doped Y_2O_3 Nanoparticles Synthesized by Thermal Decomposition with Oleic Acid. *Adv. Mater. Res.* **2011**, *332–334*, 1974–1978. [CrossRef]
21. Gunaseelan, M.; Yamini, S.; Kumar, G.A.; Senthilselvan, J. Highly efficient upconversion luminescence in hexagonal $NaYF_4$:Yb^{3+}, Er^{3+} nanocrystals synthesized by a novel reverse microemulsion method. *Opt. Mater.* **2018**, *75*, 174–186. [CrossRef]
22. Wang, J.; Song, H.; Xu, W.; Dong, B.; Xu, S.; Chen, B.; Yu, W.; Zhang, S. Phase transition, size control and color tuning of $NaREF_4$:Yb^{3+}, Er^{3+} (RE = Y, Lu) nanocrystals. *Nanoscale* **2013**, *5*, 3412–3420. [CrossRef]
23. Ding, M.; Yin, S.; Ni, Y.; Lu, C.; Chen, D.; Zhong, J.; Ji, Z.; Xu, Z. Controlled synthesis of β-$NaYF_4$:Yb^{3+}/Er^{3+} microstructures with morphology-and size dependent up conversion luminescence. *Ceram. Int.* **2015**, *41*, 7411–7420. [CrossRef]
24. Xie, J.; Hu, W.; Tian, D.; Wei, Y.; Zheng, G.; Huang, L.; Liang, E. Selective growth and upconversion photoluminescence of Y-based fluorides: From $NaYF_4$: Yb/Er to YF_3: Yb/Er crystals. *Nanotechnology* **2020**, *31*, 505605. [CrossRef]
25. Stauffer, M.T. Fourier Transform Infrared and Raman Characterization of Silica-Based Materials. In *Applications of Molecular Spectroscopy to Current Research in the Chemical and Biological Sciences*; Capeletti, L.B., Zimnoch, J.H., Eds.; IntechOpen: London, UK, 2016. [CrossRef]
26. Haase, M.; Schäfer, H. Upconverting nanoparticles. *Angew. Chem. Int. Ed.* **2011**, *50*, 5808–5829. [CrossRef]
27. Feng, Y.; Li, Z.; Li, Q.; Yuan, J.; Tu, L.; Ning, L.; Zhang, H. Internal OH^- induced cascade quenching of upconversion luminescence in $NaYF_4$:Yb/Er nanocrystals. *Light Sci. Appl.* **2021**, *10*, 105. [CrossRef]
28. Huong, T.T.; Phuong, H.T.; Vinh, L.T.; Khuyen, H.T.; Thao, D.T.; Tuyen, L.D.; Anh, T.K.; Minh, L.Q. Upconversion $NaYF_4$:Yb^{3+}/Er^{3+}@silica-TPGS bio-nano complexes: Synthesis, characterization, and in vitro test for labelling cancer cells. *J. Phys. Chem. B* **2021**, *125*, 9768–9775. [CrossRef]
29. Liu, S.; De, G.; Xu, Y.; Wang, X.; Liu, Y.; Cheng, C.; Wang, J. Size, phase-controlled synthesis, the nucleation and growth mechanisms of $NaYF_4$: Yb/Er nanocrystals. *J. Rare Earths* **2018**, *36*, 1060–1066. [CrossRef]

30. Wang, F.; Han, Y.; Lim, C.S.; Lu, Y.; Wang, J.; Xu, J.; Chen, H.; Zhang, C.; Hong, M.; Liu, X. Simultaneous phase and size control of upconversion nanocrystals through lanthanide doping. *Nature* **2010**, *463*, 1061–1065. [CrossRef]
31. Huong, T.T.; Vinh, L.T.; Khuyen, H.T.; Tuyen, L.D.; Van, N.D.; Thao, D.T.; Phuong, H.T. Synthesis and in vitro testing of $YVO_4:Eu^{3+}$@silica- NH-GDA-IgG bio-nano complexes for labelling MCF-7 breast cancer cells. *Molecules* **2023**, *28*, 280. [CrossRef]

inorganics

MDPI

Article

Novel Tricarbonylrhenium-Anthrapyrazole Complexes with DNA-Binding and Antitumor Properties: In Vitro and In Vivo Pharmacokinetic Studies with ^{99m}Tc-Analogue

Georgios Papaparidis [1], Melpomeni Akrivou [2], George Psomas [3], Ioannis S. Vizirianakis [2,4], Antonios Hatzidimitriou [3], Catherine Gabriel [5,6], Dimosthenis Sarigiannis [5,6] and Dionysia Papagiannopoulou [1,*]

1 Laboratory of Pharmaceutical Chemistry, School of Pharmacy, Faculty of Health Sciences, Aristotle University of Thessaloniki, 54124 Thessaloniki, Greece
2 Laboratory of Pharmacology, School of Pharmacy, Faculty of Health Sciences, Aristotle University of Thessaloniki, 54124 Thessaloniki, Greece
3 Department of General and Inorganic Chemistry, Faculty of Chemistry, Aristotle University of Thessaloniki, 54124 Thessaloniki, Greece
4 Department of Health Sciences, School of Life & Health Sciences, University of Nicosia, Nicosia 2417, Cyprus
5 Environmental Engineering Laboratory, Department of Chemical Engineering, Aristotle University of Thessaloniki, 54124 Thessaloniki, Greece
6 HERACLES Research Center on the Exposome and Health, Center for Interdisciplinary Research and Innovation, Balkan Center, Bldg. B, 10th km Thessaloniki-Thermi Road, 57001 Thessaloniki, Greece
* Correspondence: papagd@pharm.auth.gr

Abstract: Organometallic complexes of *fac*-tricarbonylrhenium have been shown to exhibit anticancer properties. Anthrapyrazole anticancer agents act as DNA intercalators and topoisomerase IIα inhibitors, leading to double-strand breaks (DBS) and cell cycle arrest. This work involves the synthesis and biological evaluation of novel *fac*-tricarbonyl-rhenium complexes with anthrapyrazole derivatives. The anthrapyrazole moiety was synthesized from 1,8-dihydroxyanthraquinone, and three ligands **L1**, **L2** and **L3** were prepared. Ligand **L1** coordinates via the phenolic O and pyrazole N as bidentate chelator forming the *fac*-[Re(CO)$_3$(κ^2-N,O)(MeOH)]-type complex, **ReL1**. Ligand **L2** contains a pendant picolylamine N,N′-chelating system, forming the bidentate *fac*-[Re(CO)$_3$(κ^2-N,N′)Br]-type complex, **ReL2**. Ligand **L3** contains a pendant picolylaminomonoacetic acid chelating system, forming a tridentate *fac*-[Re(CO)$_3$(κ^3-N,N′,O)]-type complex, **ReL3**. Complex **ReL4** contains a picolylamine chelator, forming a complex with structure *fac*-[Re(CO)$_3$(κ^2-N,N′)Br], which was synthesized as a model for **ReL2**, and its coordination mode was resolved by X-ray crystallography. The complexes were characterized spectroscopically, and their biological properties were evaluated in vitro, in terms of DNA binding as well as for the cytotoxicity against CT-26 tumor cell line. Tumor cell cytotoxicity was high for ligand **L2** and complex **ReL2**, exhibiting IC$_{50}$ values of 0.36 and 0.64 μM, respectively. The most promising complex **ReL2** was evaluated further by the preparation of its congener γ-emitting technetium-99m radio-complex, ^{99m}**TcL2**. The in vitro uptake in CT26 tumor cells and the in vivo uptake in CT26 tumor-bearing mice of ^{99m}**TcL2** was determined, and its pharmacokinetic profile was established. These data indicate that the ^{99m}Tc complex has suitable properties to enter tumor cells in vitro and in vivo, and therefore **ReL2** is promising for further evaluation.

Keywords: rhenium; tricarbonylrhenium; anthrapyrazole; anthraquinone; DNA-binding studies; cytotoxicity; technetium-99m

1. Introduction

The development of metal complexes in therapeutics, especially in the treatment of cancer, has made significant advances to date [1–7]. Platinum-based drugs have been used as first-line anticancer agents for half a century now [8]. In this area of research, organometallic complexes of *fac*-tricarbonylrhenium have attracted attention in terms of their potential pharmaceutical applications for the development of novel antitumor and antimicrobial agents, as well as of luminescent probes for optical imaging [9–13]. The utilization of the *fac*-tricarbonylrhenium core is attractive for pharmaceutical applications due to its thermodynamic and kinetic stability, as well as for its versatility in drug design, with the existence of a variety of suitable chelating strategies. In addition, a number of research results show strong biological properties, which warrants their future exploration [13].

DNA intercalators are small molecules that can reversibly bind in between adjacent base pairs of double-stranded DNA (dsDNA). By binding to DNA, intercalators may cause fatal perturbations in essential DNA-associated processes such as replication, transcription and repair. Such polyaromatic frameworks include anthracenes, acridines, anthraquinones, phenazines, quinolones, phenanthridines, etc. [14]. Metal complexes may also intercalate with nucleic acid sequences either by conjugation to an organic intercalator or by coordination with a suitable chelating system, also known in the literature as metallointercalators, containing phenanthroline, phenanthrenequinone diimine and other ligands with extended aromatic systems [15]. Tricarbonylrhenium complexes with polyaromatic ligands such as phenanthrolines [16], quinolones [17] and bipyridine [18] have also been reported as antitumor agents.

Drug molecules containing the anthraquinone group are known to have clinical applications in therapeutic regiments against cancer. The anthraquinone scaffold provides flatness, due to the aromatic polycyclic system, causing intercalation between the double helix of DNA and the inhibition of topoisomerase II enzyme. Anthraquinone-based antitumor cytotoxic agents such as daunorubicin, doxorubicin, mitoxantrone and pixantrone, used in tumor therapy, are both Topo-II inhibitors and intercalators [19,20]. In this light, Imstepf et al. developed tricarbonylrhenium-doxorubicin complexes that were evaluated for their ability to inhibit Topo-IIα [21].

The anthraquinone scaffold participates in redox processes, leading to the production of free radicals, which enhances the cytotoxic profile of the compound in both cancer and normal cells. The cardiotoxicity of anthraquinone derivatives led to the design and development of an alternative scaffold [22]. Anthrapyrazoles, such as losoxantrone, contain a pyrazole ring in the chromophore site instead of the quinone group, and they exhibit significantly reduced cardiotoxicity, as shown in preclinical and clinical studies [23].

In this work, we developed new anthrapyrazole-based ligands suitable for complexation with tricarbonylrhenium as potential antitumor agents, taking into consideration the reduced cardiotoxicity as well as the presence of coordinating atoms in this moiety. Therefore, we designed ligands to coordinate with *fac*-tricarbonylrhenium core either directly with the (N,O)-donor atoms of the anthrapyrazole pharmacophore, ligand **L1** or indirectly via the pendant bidentate (N,N)- and tridentate (N,N,O)-donor atom systems, ligands **L2** and **L3**, respectively (Figure 1). The ligands and the rhenium complexes were evaluated in vitro for their ability to interact with calf-thymus (CT) DNA as well as for their cytotoxic properties in a tumor cell line. Complex **ReL2,** which exhibited the best biological properties, was translated to the analogous γ-emitting **^{99m}TcL2** complex, which was evaluated in vitro and in vivo for its tumor uptake and distribution properties. Also, in order to simulate the structure of **ReL2**, the model complex **ReL4**, which also contains a picolylamine chelator and shares the same coordination mode of *fac*-[Re(CO)$_3$(κ^2-N,N′)Br], was synthesized and characterized by X-ray crystallography.

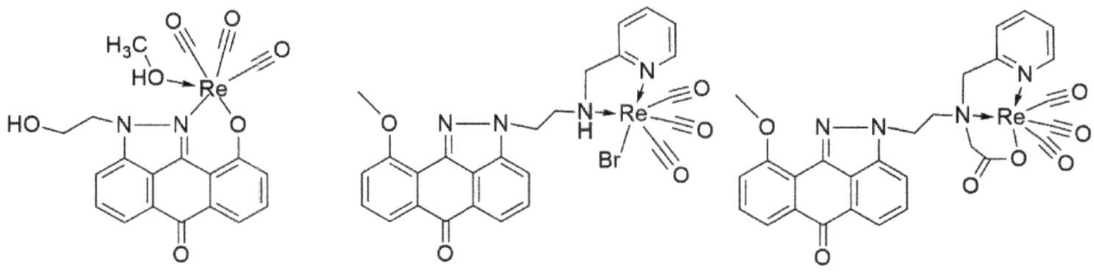

ReL1 ReL2 ReL3

Figure 1. Structures of rhenium anthrapyrazole complexes.

2. Results and Discussion

2.1. Synthesis and Spectroscopic Characterization of Rhenium Complexes

Synthesis of the ligands **L1**, **L2** and **L3** followed the route depicted in Scheme 1. The intermediate compound **3** was synthesized according to methods reported in the literature for similar anthraquinones [24].

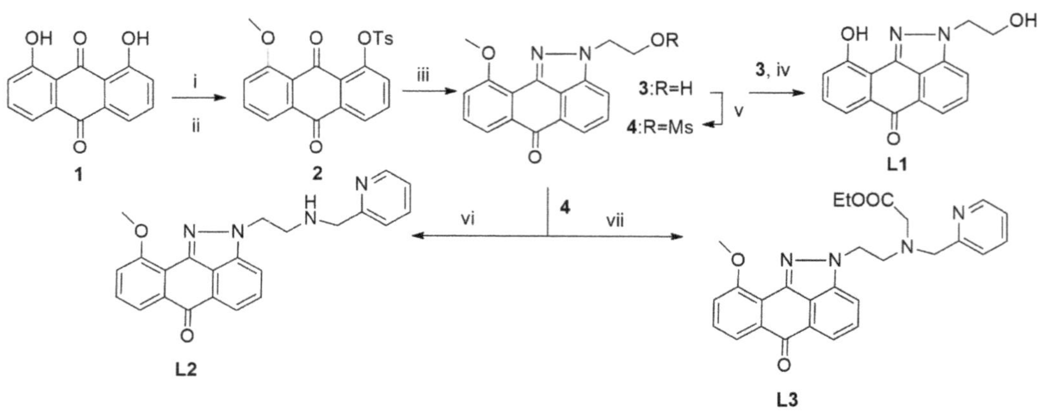

Scheme 1. (i) Me_2SO_4, 60 °C; (ii) p-toluenesulfonyl chloride (TsCl), 60 °C (46%); (iii) 2-hydroxyethylhydrazine, 130 °C (25%); (iv) BBr_3, 25 °C (65%); (v) methylsulfonyl chloride (MsCl), 25 °C (85%); (vi) 2-picolylamine, 65 °C (39%); (vii) ethyl 2-((pyridin-2-ylmethyl)amino)acetate, 80 °C (67%).

The bidentate ligand **L1** was prepared after the demethylation of **3** with BBr_3. **L1** offers a bidentate N,O-donor system for direct complexation of the phenolic O and pyrazole N with tricarbonylrhenium. The complex was prepared by reaction of **L1** with equimolar amounts of the precursor $[Re(CO)_5Br]$ in refluxing methanol for 24 h. The product was isolated by column chromatography as a red solid in average yield. The NMR analysis of the complex and the free ligand shows analogous proton and carbon signals, where in the ^{1}H-NMR spectrum of the complex, the phenolic proton signal of **L1** at 9.87 ppm is absent, and in the ^{13}C-NMR spectrum, the carbon of the C=N pyrazole ring is downfield-shifted to 160.30 ppm versus 154.43 ppm of **L1**. Furthermore, three CO carbons of the tricarbonylrhenium core appear at 197.89, 197.35 and 196.83 ppm. These NMR signals indicate the N,O-coordination of the metal core. The IR spectrum of the complex exhibits the characteristic bands of the ligand and a CO stretch at 2025, 1930 and 1900 cm^{-1} of the asymmetric tricarbonylrhenium core. High-resolution mass analysis revealed a signal that corresponded to negative molecular ions at *m/z* 579.0101 (60%) and 581.0139 (100%) for a $[M-H]^-$ pattern that matches the calculated values for M = $C_{20}H_{15}N_2O_7Re$ and corresponds to the formula *fac*-$[Re(CO)_3(\kappa^2$-**L1**$)(MeOH)]$, as shown in Figure 1.

Ligands **L2** and **L3** were synthesized after the mesylation of the intermediate compound **3** and reaction of the mesylate compound **4** with the bidentate chelator 2-picolylamine to form **L2** or the tridentate chelator ethyl 2-((pyridin-2-ylmethyl)amino)acetate to form **L3**. These ligands were designed as part of the pendant approach, where metal complexation takes place with donor atoms that do not belong to the anthrapyrazole moiety.

L2 offers a bidentate N,N′-donor system from 2-picolylamine moiety for complexation with tricarbonylrhenium. The complex was prepared by reaction of **L2** with equimolar amounts of the precursor [Re(CO)$_5$Br] in refluxing methanol for 3 h. The product precipitated from the reaction mixture as a yellow solid in average yield. The NMR analysis of the complex shows the presence of two diastereomers due to the pseudooctahedral rhenium coordination and the prochiral secondary amine nitrogen donor. The ratio of the isomers is approximately 60:40, based on NMR and HPLC integration of the signals. In the ^{1}H-NMR spectrum of the complex, differences in the shifts of the two isomers were detected, some of which are shown in Figure 2. In particular, both isomers exhibit the characteristic pattern of (N,N′) coordination due to picolylamine fragment, where the 2 protons of H-15 are split in two dd signals; in addition, the N-H can be observed at 5.64 and 6.73 ppm, respectively.

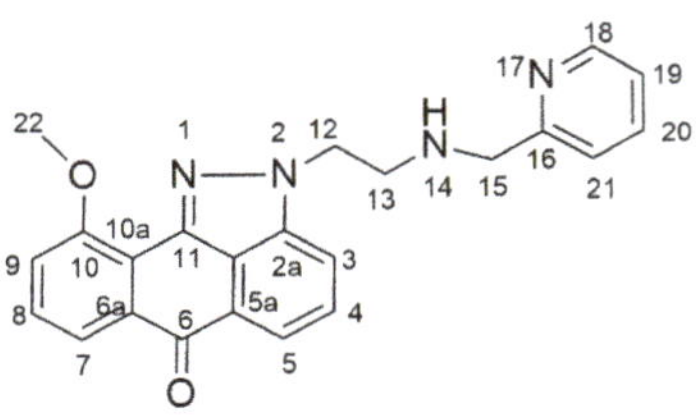

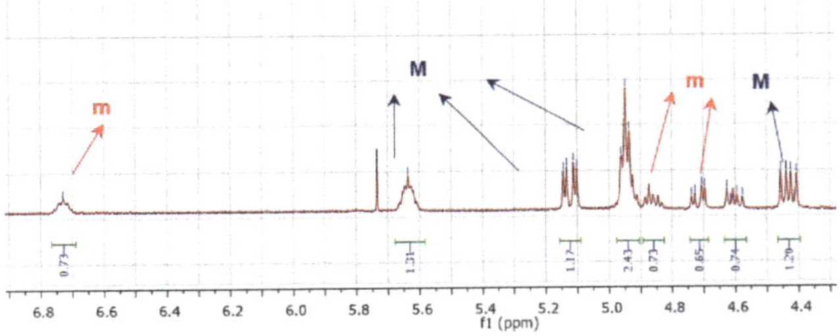

	L2	ReL2 (Major)	ReL2 (minor)
H-14	-	5.64 (m)	6.73 (m)
H-15	3.98 (s)	5.12 (dd, J = 15.9, 5.2 Hz), 4.43 (dd, J = 15.9, 8.8 Hz)	4.72 (dd, J = 15.6, 5.3 Hz), 4.60 (dd, J = 15.7, 8.8 Hz)
H-18	8.48 (d, J = 4.6 Hz)	8.78 (d, J = 5.1 Hz)	8.79 (d, J = 5.1 Hz)
H-19	7.12 (t, J = 6.1 Hz)	7.56 (t, J = 7.7 Hz)	7.48 (m)
H-20	7.58 (t, J = 7.6 Hz)	8.10 (td, J = 7.8, 1.5 Hz)	8.06 (td, J = 7.8, 1.5 Hz)
H-21	7.25 (d, J = 8.1 Hz)	7.79 (m)	7.68 (d, J = 7.9 Hz)

Figure 2. Representative section of the ^{1}H NMR spectrum of **ReL2** and table with proton shifts δ (in ppm) of **L2** ($CDCl_3$) and **ReL2** (d_6-DMSO); isomer designation M: major (blue), m: minor (red).

In the ^{13}C-NMR spectrum, the CO signals of the tricarbonylrhenium core exhibit shifts at 197.89, 197.35 and 196.83 ppm. The IR spectrum of the complex exhibits the characteristic bands of the ligand and in addition the CO stretch at 2021, 1909 and 1867 cm^{-1} of the asymmetric tricarbonylrhenium core. The NMR and IR signals are in agreement with a *fac*-[Re(CO)$_3$(N,N′)Br] coordination mode. Furthermore, the high-resolution mass analysis revealed a signals that correspond to positive molecular ions [M − Br + DMSO]$^+$ at *m/z* 731.0995 (60%), 733.1125 (100%). Diastereomeric tricarbonylrhenium complexes have been synthesized previously by our group and others in the literature [25–28].

L3 offers a tridentate N,N′,O-donor system of 2-((pyridin-2-ylmethyl)amino)acetate moiety for complexation with tricarbonylrhenium. The complex was prepared by reaction of **L3** with equimolar amounts of the precursor *fac*-[Re(CO)$_3$(MeOH)$_3$](OTf) in refluxing methanol for 24 h. The product was isolated by precipitation from methanol as a yellow solid in average yield. The NMR analysis of the complex and the free ligand shows analogous proton and carbon signals, where in the ^{1}H-NMR spectrum of the complex the characteristic pattern of this N,N′,O-coordination is evident as the methylene protons next to pyridine appear two doublets at 4.85 ppm (J = 15.7 Hz) and 3.59 ppm (J = 16.6 Hz), respectively, and the acetate protons appear as two doublets at 4.28 ppm (J = 13.8 Hz) and 4.16 ppm (J = 13.3 Hz), respectively. Furthermore, in the ^{13}C-NMR spectrum, the CO carbons of the tricarbonylrhenium core appear at 198.00 and 197.34 ppm. These NMR signals indicate the [N,N′,O] coordination of the metal core. The IR spectrum of the complex exhibits the characteristic bands of the ligand, with the exception of the carboxylate ester stretch at 1737 cm^{-1} of the ligand and in addition the CO stretch at 2023, 1912 and 1870 cm^{-1} of the asymmetric tricarbonylrhenium core. High-resolution mass analysis revealed a signal that corresponds to positive molecular ions at m/z 711.1006 (60%) and 713.1030 (100%) for an $[M + H]^+$ pattern that matches the calculated values for M = $C_{28}H_{21}N_4O_7Re$ and corresponds to the formula *fac*-[Re(CO)$_3$(κ^3-**L3**)], as shown in Figure 1.

Complex *fac*-[Re(CO)$_3$Br(**L4**)] (Figure 3) was synthesized from **L4** (synthesis is reported in Supplementary File, ESI: 1. Synthesis of Ligand **L4**) under the same conditions employed for **ReL2** and was used as a model compound to corroborate the coordination mode of **ReL2**, which also contains a picolylamine chelator and shares the same coordination mode of *fac*-[Re(CO)$_3$(κ^2-N,N′)Br]. The complex was crystallized, and its structure was solved by X-ray crystallography. The IR and NMR characterization of **ReL4** shows similar spectroscopic data with **ReL2** such as the stretching frequencies of the carbon monoxide coordinated to Re(I), which is 2021, 1913, 1875 cm^{-1} for **ReL4** vs. 2021, 1909, 1867 of **ReL2**. In the ^{13}C-NMR spectrum, the CO signals of the tricarbonylrhenium core of **ReL4** exhibit shifts at 196.40, 195.47, 191.60 ppm, similar to those of **ReL2**, indicating that the coordination of the two complexes is the same.

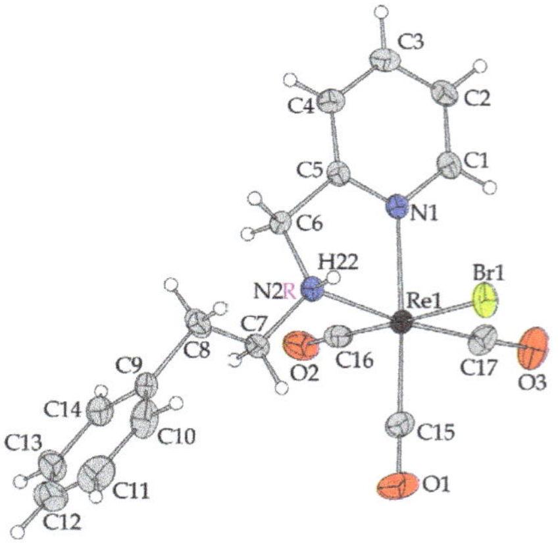

Figure 3. ORTEP of *fac*-[Re(CO)$_3$Br(**L4**)] R-conformer.

2.2. *X-ray-Structure*

A plot of the molecular structure of complex *fac*-[Re(CO)$_3$Br(**L4**)] is depicted in Figure 3, and bond distances and angles are shown in Table 1. The compound crystallizes in the monoclinic space group $P2_1/c$, and four neutral independent complex molecules can be found in the unit cell. The complex is mononuclear, and the coordination geometry is distorted octahedral. **L4** behaves as a bidentate ligand coordinated to rhenium(I) via the aromatic nitrogen atom N(1) and the amine nitrogen atom N(2), forming a five–membered chelate ring. Additionally, three carbon monoxide molecules are coordinated to Re(I) via the carbon atoms C(15), C(16), C(17) in facial orientation. The total coordination number of 6 is completed with the coordination of the bromine anion Br(1). The complex presents chirality on N(2) and from the four symmetrically equivalent molecules present in the unit cell, two of them are R type enantiomers, while the rest two are S conformers. In the ESI, we give both the cif files of the enantiomers and the plots of these isomers for comparison

(S conformer Figure S1). The bond distances and angles of complex *fac*-[Re(CO)$_3$Br(**L4**)] are similar to those found in other structures of rhenium(I)-tricarbonyl halide complexes coordinated to bidentate (N,N) donor ligands and forming five-membered ring, such as N,N′-bipyridine, pyridylimine and pyridylamine ligands [29–32].

Table 1. Selected geometric parameters (Å, °).

Re1—Br1	2.6219 (7)
Re1—N1	2.171 (4)
Re1—N2	2.223 (4)
Re1—C15	1.919 (6)
Re1—C16	1.894 (7)
Re1—C17	1.897 (6)
O1—C15	1.140 (8)
O2—C16	1.127 (7)
O3—C17	1.156 (7)
N1—C1	1.342 (7)
N1—C5	1.346 (7)
N2—C6	1.469 (7)
N2—C7	1.505 (7)
Br1—Re1—N1	85.82 (12)
Br1—Re1—N2	84.56 (12)
N1—Re1—N2	75.39 (16)
Br1—Re1—C15	92.5 (2)
N1—Re1—C15	174.8 (2)
N2—Re1—C15	99.5 (2)
Br1—Re1—C16	178.08 (18)
N1—Re1—C16	93.1 (2)
N2—Re1—C16	93.6 (2)
C15—Re1—C16	88.5 (3)
Br1—Re1—C17	91.2 (2)
N1—Re1—C17	98.5 (2)
N2—Re1—C17	172.8 (2)
C15—Re1—C17	86.5 (3)
C16—Re1—C17	90.5 (3)

In order to compare the distortion of the coordination octahedron of fac-[Re(CO)$_3$Br(**L4**)] with the reported analog [Re(CO)$_3$Br(**L**)] complexes, we used Octadist software version 3.1.0 [33] to compute the values of ζ and Σ (=deviations of the metal ion complex from an ideal octahedral structure) and Θ (=distortion from a perfect octahedral (O_h) to a trigonal prismatic (D_{3h}) geometry). The parameter ζ is the average of the sum of the deviation of six unique metal–ligand bond lengths around the central metal atom (d_i) from the average value (d_{mean}). The parameter Σ is the sum of the deviation of 12 unique cis ligand–metal–ligand angles (ϕ_i) from 90°. The parameter Θ is defined as the degree of trigonal distortion of the coordination geometry from an octahedron towards a trigonal prism. The Θ parameter is the sum of the deviation of 24 unique torsional angles between the ligand atoms on opposite triangular faces of the octahedron viewed along the pseudo-threefold axis (θ_i) from 60°. Comparison results are included in Table S1. From these results, it is concluded that [Re(CO)$_3$Br(**L4**)] presents nearly the same coordination parameters and distortion with the analog octahedral complexes already published. The pyridine rings of neighboring complexes have a centroid-to-centroid distance of 3.986 Å indicating weak π–π interactions present in the crystal structure of [Re(CO)$_3$Br(**L4**)], which give extra stability to the complex structure.

2.3. DNA-Binding Studies

The study of the interaction of anthrapyrazoles and their rhenium complexes with DNA is of great interest due to their ability to act as DNA intercalators. Interactions between a compound and CT DNA may cause changes to the absorption bands upon addition of

CT DNA in various ratios (*r*) values (=[compound]/[DNA]). The UV–vis spectra of the anthrapyrazoles and their complexes in DMSO in the presence of increasing amounts of CT DNA are shown in Figure 4 and Figure S2. The UV–vis spectra of the compounds exhibited similar changes of the intraligand absorption band after the addition of a CT DNA solution, such as hypochromism up to 20%, as well as, in few cases, a red-shift (Table 2). The DNA-binding constants (K_b) of the complexes calculated by the Wolfe–Shimer equation [34] (Equation (1)) and plots [DNA]/($\varepsilon_A - \varepsilon_f$) versus [DNA] (Figure S3) are similar to that of the corresponding free anthrapyrazoles (Table 2), except **ReL3** complex, which exhibited much higher K_b than **L3**, suggesting that its coordination to Re(I) results in a significant increase in the affinity for CT DNA. The K_b values suggest a strong binding of the anthrapyrazole ligands **L1**, **L2** and **L3** to CT DNA, which are similar to that of the classical intercalator EB (=1.23 (±0.07) × 10^5 M^{-1}) [35]. The K_b value of complex **ReL3** (=1.09 (±0.15) × 10^6 M^{-1}) is the highest DNA-binding constant among the herein examined compounds. However, the results obtained from the UV–vis spectroscopic titration studies (hypochromism) do not provide sufficient information to elucidate the type of interaction between the anthrapyrazoles and their rhenium complexes with DNA, and additional experiments are required to clarify the binding mode [24].

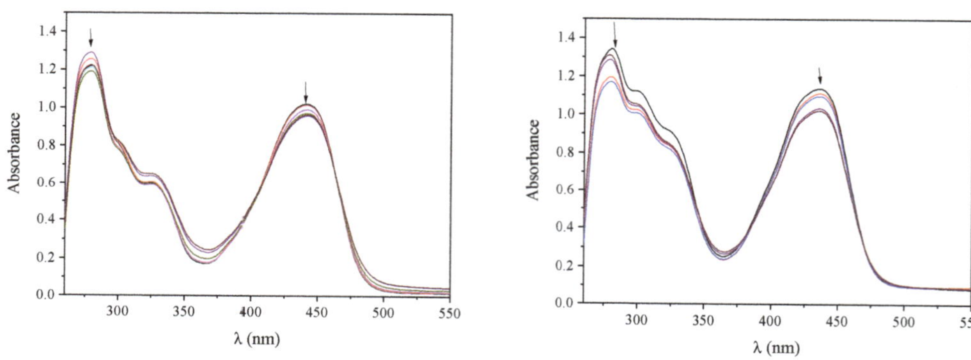

Figure 4. UV–vis spectra of DMSO solution (1 × 10^{-4} M) of **L3** and **ReL3** in the presence of increasing amounts of CT DNA. The arrows show the changes upon addition of CT DNA.

Table 2. Spectral features of the interaction of compounds **L1–L3** and **ReL1–ReL3** with CT DNA. UV band (λ_{max}, in nm), percentage of the observed hyper-/hypo-chromism ($\Delta A/A_0$, %), red-/blue-shift ($\Delta\lambda_{max}$, in nm) and DNA-binding constants (K_b, M^{-1}).

Compound	λ_{max} (nm) ($\Delta A/A_0$ (%) [a], $\Delta\lambda_{max}$ (nm) [b])	K_b (M^{-1})
L1	276 (−6 [a], 0 [b]); 301 (sh) [c] (−5, 0); 323 (sh) (−7, 0); 439 (−7, 0)	1.33 (±0.01) × 10^5
L2	278 (−10, 0); 326 (sh) (−3.5, 0); 441 (−4, +2)	7.74 (±0.03) × 10^4
L3	277 (−7, +2); 325 (sh) (−7, 0); 441 (−6, 0)	7.01 (±0.01) × 10^4
ReL1	276 (−14, +3); 304 (−12, +2); 395 (−6.5, 0); 437 (sh) (−9, 0); 530 (−6, 0)	1.24 (±0.04) × 10^5
ReL2	278 (−20, +1); 300 (sh) (−10, 0); 435 (−5, +2)	6.87 (±0.09) × 10^4
ReL3	278 (−12, +2); 300 (sh) (−10, 0); 322 (sh) (−10, +2); 436 (−10, 0)	1.09 (±0.15) × 10^6

[a] "+" denotes hyperchromism, "−" denotes hypochromism; [b] "+" denotes red-shift, "−" denotes blue-shift; [c] "sh" = shoulder.

Viscosity measurements were performed by the addition of increasing amounts of the anthrapyrazoles and their rhenium complexes on a CT DNA solution (10^{-4} M). The relative DNA viscosity (η/η_0) is sensitive to DNA length (L/L_0), and their relation is expressed by the equation $L/L_0 = (\eta/\eta_0)^{1/3}$ [36]. This study provides important information on the DNA-interaction mode of a tested compound by monitoring changes in DNA viscosity in the presence of a compound. During intercalation, the DNA bases are separated to host the intercalator, which leads to the elongation of DNA and increased DNA viscosity. In the case of partial and/or non-classic intercalation (e.g., groove-binding or electrostatic interaction), the compounds do not enter in-between the DNA bases, and a bend or a kink in the DNA helix may occur, which does not significantly affect the DNA length, and the DNA viscosity remains practically unchanged or may even show a slight decrease. In the viscosity measurement, all the compounds resulted in a relative increase in the DNA viscosity (Figure 5). Such results indicate the insertion of the compounds between the DNA bases due to an intercalative interaction.

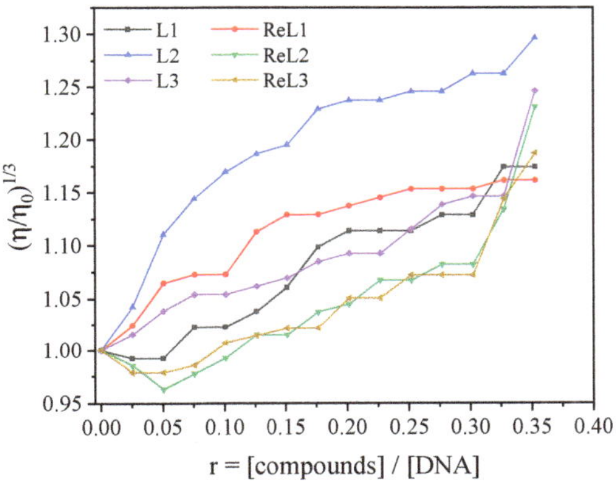

Figure 5. Relative viscosity $(\eta/\eta_0)^{1/3}$ of CT DNA (0.1 mM) in buffer solution (150 mM NaCl and 15 mM trisodium citrate at pH 7.0) in the presence of the compounds (anthrapyrazoles **L1–L3** and complexes **ReL1–ReL3**) at increasing amounts (r = [compound]/[DNA]).

The EB-displacing ability of the compounds by interaction with EB–DNA is considered to be useful in verifying the intercalation of a compound with DNA. EB is a typical intercalator, where the planar EB–phenanthridinium ring is inserted between adjacent base pairs on the double helix. The EB–DNA complex emits intense fluorescence at 592 nm (with $\lambda_{exitation}$ = 540 nm), which may be quenched by a DNA-intercalating compound which competes with EB for DNA intercalation and is therefore used as a fluorescence dye [37].

The ligands and the rhenium complexes did not exhibit significant fluorescence in the presence of CT DNA and EB at 540 nm excitation; therefore, the complexes can be used as EB competitors in this study. The fluorescence emission spectra of pre-treated EB–CT DNA were obtained for [EB] = 20 μM, [DNA] = 26 μM and for increasing amounts of the compounds (up to r = 0.04) (shown in Figures 6 and S4). The addition of increasing amounts of the ligands and rhenium complexes resulted in a substantial decrease in the intensity of the emission band of the DNA–EB complex at 592 nm (the fluorescence intensity in the highest competitor concentration was up to 43.5% of the initial EB–DNA fluorescence one, Table 3). Therefore, the tested compounds exhibited EB-displacing ability by competing with EB in binding to DNA (Figure 6B), proving thus, indirectly, their interaction with CT DNA via intercalation [38]. As seen from the Stern–Volmer plots of EB–DNA fluorescence studies in the presence of the competitors (Figure S5), the quenching of EB–DNA by the compounds is in agreement (R = 0.99) with the linear Stern–Volmer equation (Equation (2)), which proves the displacement of EB from EB–DNA by the compounds [34]. The obtained values of K_{SV} (Table 3) may show tight binding of the complexes to DNA. Since the fluorescence lifetime of EB–DNA (τ_0) is 23 ns [39], the K_q values were calculated with Equation (3). All quenching constants are higher than 10^{10} $M^{-1}s^{-1}$, indicating the presence of a static quenching mechanism which reveals the formation of a new adduct between the studied complexes and DNA, indirectly confirming intercalation as the most possible mode of interaction [37].

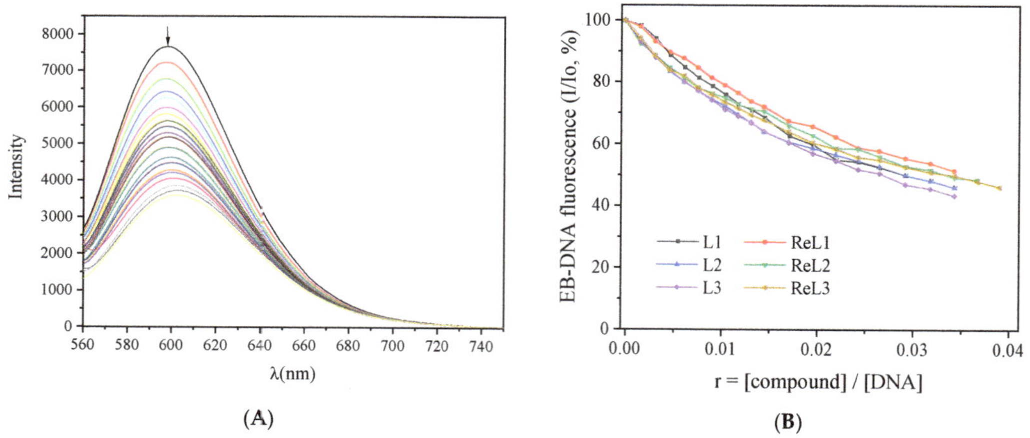

Figure 6. (**A**) Fluorescence emission spectra ($\lambda_{exitation}$ = 540 nm) of EB–DNA ([EB] = 20 μM, [DNA] = 26 μM) in buffer solution in increasing amounts of complex **ReL3** (up to the value of r = 0.4). The arrow shows the changes of intensity upon addition of **ReL3**. (**B**) Plot of EB–DNA relative fluorescence intensity at λ_{em} = 592 nm (I/I_0, in %) (in buffer solution of 150 mM NaCl and 15 mM trisodium citrate at pH = 7.0) vs. r (r = [compound]/[DNA]) in the presence of the compounds (up to 46.1% of the initial EB–DNA fluorescence emission intensity for **L1**, 46.0% for **L2**, 43.5% for **L3**, 51.5% for **ReL1**, 48.6% for **ReL2**, and 46.1% for **ReL3**).

Table 3. Percentage of EB–DNA fluorescence quenching (ΔI/Io, %), the Stern–Volmer (K_{SV}, in M^{-1}) and EB–DNA quenching constants (K_q, in $M^{-1}s^{-1}$) for compounds **L1-L3** and **ReL1–ReL3**.

Compound	$\Delta I/I_0$ (%)	K_{sv} (M^{-1})	K_q ($M^{-1}s^{-1}$)
L1	53.9	5.65 (±0.09) × 10^5	2.46 (±0.04) × 10^{13}
L2	54.0	5.27 (±0.08) × 10^5	2.29 (±0.03) × 10^{13}
L3	56.5	5.93 (±0.04) × 10^5	2.58 (±0.02) × 10^{13}
ReL1	48.5	4.47 (±0.05) × 10^5	1.94 (±0.02) × 10^{13}
ReL2	51.4	4.52 (±0.07) × 10^5	1.96 (±0.03) × 10^{13}
ReL3	53.9	4.60 (±0.07) × 10^5	2.00 (±0.03) × 10^{13}

2.4. In Vitro Cell Studies

The cytotoxicity of the anthrapyrazole ligands **L1**, **L2** and **L3**, as well as that of the respective rhenium complexes **ReL1**, **ReL2** and **ReL3**, was assessed in vitro in colorectal adenocarcinoma cells CT26. The anthrapyrazole ligands **L1**, **L2**, **L3** and their rhenium complexes **ReL1**, **ReL2** and **ReL3** were incubated at 10^{-5} M concentration with the cells. Compounds **L2** and **ReL2** affected cell viability, exhibiting in average 82.55 ± 1.19 and 98.61 ± 0.85% decrease in cell proliferation and 83.92 ± 1.66 and 79.46 ± 1.35% cell death (Figure 7). The other four compounds tested (**L1**, **L3**, **ReL1**, **ReL3**) resulted in no significant cytotoxicity or cell death after incubation with the cells in this concentration. The most cytotoxic compounds **L2** and **ReL2** were tested at lower concentrations up to 10^{-7} M, and the IC_{50} was calculated to be 0.36 μM for **L2** and 0.64 μM for **ReL2** (Figure S6). The cytotoxicity of these compounds is in the same range as for other standard cytotoxic agents from the literature, such as doxorubicin and cisplatin [40,41]. The fact that ligands **L1** and **L3** are not active in this concentration range is clearly attributed to their structure. By comparison to anthrapyrazoles and similar compounds in the literature, it is evident that the pyrazole nitrogen substitution is important for its cytotoxicity [24,42–44], while both losoxantrone and mitoxantrone clinical agents contain -$NCH_2CH_2NHCH_2CH_2OH$ moieties.

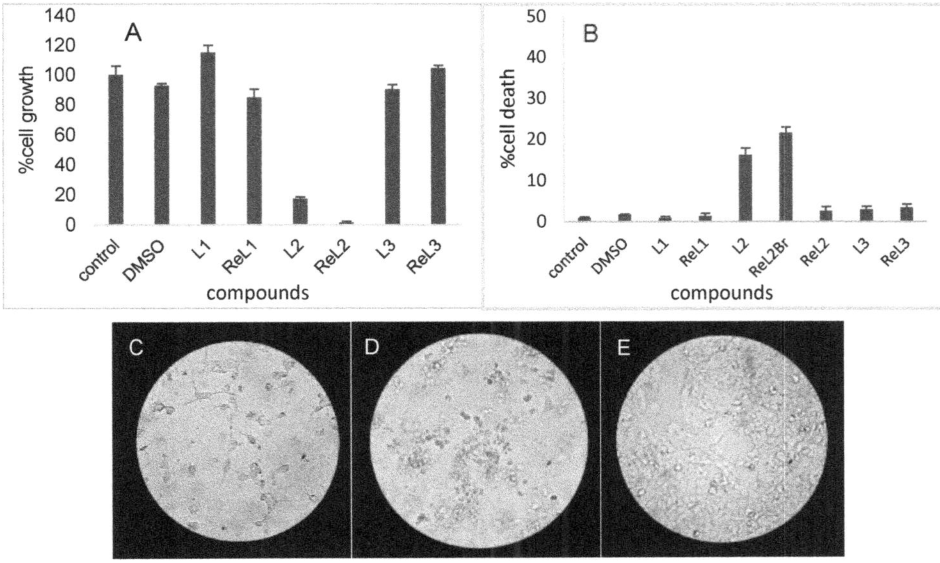

Figure 7. Assessment of cell growth (**A**) and the proportion of dead cells (**B**) in cultures of CT 26 incubated with 10^{-5} M of each compound. The morphology of CT 26 treated with 10^{-5} M of either L2 (**D**) and ReL2 (**C**) along with the untreated control culture (**E**) is shown.

2.5. Radiotracer Studies

Technetium-99m (^{99m}Tc) is a γ-emitting radionuclide with excellent physical properties ($t_{1/2}$: 6.01 h, 0.142 MeV). In addition, its low-cost commercial availability from ^{99}Mo/^{99m}Tc generators makes it an efficient choice in nuclear medicine for SPECT imaging. Technetium and rhenium are considered congener metals, and usually, they form isostructural complexes with similar biological properties [45]. The first in vivo evaluation of the potential ability of **ReL2** to be distributed to the tumor was conducted by preparing an analogous γ-emitting radiotracer, *fac*-[^{99m}Tc][Tc(CO)$_3$(**L2**)(H$_2$O)]$^+$ (**^{99m}TcL2**). Furthermore, in an effort to explain the low biological activity of the complex **ReL3**, we prepared the analogous radiotracer **^{99m}TcL3**.

^{99m}TcL2 and **^{99m}TcL3** were synthesized by reaction of the aqua ion *fac*-[^{99m}Tc][Tc(CO)$_3$(H$_2$O)$_3$]$^+$ with 10^{-3} M of the ligand **L2** or **L3** for 30 min at 55–65 °C, with high radiochemical purity of >95%. **^{99m}TcL2**, t_R = 19.2 min was identified by comparative RP-HPLC studies, using complexes [Re(CO)$_3$(**L2**)Br] (**ReL2**), t_R = 21.1/21.3 min and [Re(CO)$_3$(**L2**)(MeOH)]$^+$ (**ReL2′**) (which was formed from **ReL2** by precipitation of bromide with silver salts) t_R = 19.1 min as references (Figure S7). **^{99m}TcL3**, t_R = 21.3 min and **ReL3**, t_R = 21.4 min were identified. The lipophilicity value of the HPLC-purified **^{99m}TcL2** was found to be 2.45 ± 0.06, and that of **^{99m}TcL3** was found to be 2.27 ± 0.12, which indicates that the complexes could penetrate cell membranes and be distributed across various tissues and organs. The stability of **^{99m}TcL2** was tested after incubation with 1 mM histidine and rat plasma, and the percentage of intact tracer was found to be 95% at 4 h in histidine and 82% at 4 h in rat plasma, where it exhibited 54% protein binding. Respectively, **^{99m}TcL3** was 93% stable in histidine and 95% stable in rat plasma (with low protein binding of 16.4%) at 4 h.

The cellular uptake of tracer **^{99m}TcL2** and **^{99m}TcL3** was performed in CT26 cells over 240 min. **^{99m}TcL2** exhibited a time-dependent increase in cell uptake reaching 5.42 ± 0.19% cells at 240 min after incubation. **^{99m}TcL3** exhibited low cell uptake. In detail, Table 4 shows the results obtained:

Table 4. Percentage (%) of ^{99m}**TcL2** and ^{99m}**TcL3** uptake in CT26 cells.

	15 min	60 min	120 min	240 min
^{99m}**TcL2**	1.29 ± 0.09%	2.48 ± 0.25%	3.94 ± 0.68%	5.42 ± 0.19%
^{99m}**TcL3**	0.49 ± 0.06%	0.53 ± 0.04%	0.55 ± 0.08%	0.75 ± 0.11%

Biodistribution studies were performed in CT26 tumor-bearing mice at time points of 30 and 120 min post intravenous injection of ^{99m}**TcL2** (Table 5). The radiotracer ^{99m}**TcL2** exhibited hepatobiliary and renal elimination, while the percentage of radioactivity in the blood was 4.16 ± 1.31% ID/g at 120 min. The tumor uptake of ^{99m}**TcL2** was 2.42 ± 0.04 and 3.10 ± 1.21% ID/g at 30 and 120 min, respectively, which shows retention of the tracer in the tumor up to 2 h.

Table 5. Distribution of ^{99m}**TcL2** in CT26 tumor-bearing mice (%ID/g).

Organ	%ID/g	
	30 min	120 min
Blood	6.48 ± 2.02	4.16 ± 1.31
Tumor (CT26)	2.42 ± 0.04	3.10 ± 1.21
Heart	8.91 ± 3.77	5.90 ± 1.16
Liver	29.49 ± 9.43	24.51 ± 7.12
Lungs	15.36 ± 6.43	7.73 ± 2.29
Muscle	2.17 ± 0.25	2.20 ± 0.51
Kidneys	22.09 ± 8.91	17.04 ± 3.53
Spleen	6.78 ± 1.33	3.87 ± 0.87
Intestine	11.38 ± 3.43	19.84 ± 8.10
Stomach	8.58 ± 6.60	14.26 ± 7.53

3. Materials and Methods

3.1. General

All chemicals were reagent-grade. For the chromatographic purifications, Silica gel 60 (0.040–0.063 mm) from Merck (Darmstadt, Germany) was used. The precursors $[Re(CO)_5Br]$ and $[Re(CO)_5(OTf)]$ were prepared according to literature procedures [46,47]. For labeling with ^{99m}Tc, a vial containing 5.5 mg of $NaBH_4$, 4 mg of Na_2CO_3, and 15 mg of Na-K tartrate was purged with CO gas prior to addition of $Na^{99m}TcO_4$, as described in the literature [48]. Solvents used for high-performance liquid chromatography (HPLC) were HPLC-grade, and solvents used for mass spectroscopy (MS) were MS-grade. To prepare the HPLC mobile phase, the solvents were filtered through membrane filters (0.22 μm, Millipore, Milford, MA, USA) and degassed. UV–visible (UV–vis) spectra were recorded on a Hitachi U-2001 (Hitachi, Tokyo, Japan) dual beam spectrophotometer. C, H and N elemental analysis were performed on a Perkin-Elmer 240B elemental analyzer (Perkin Elmer, Waltham, MA, USA). Fluorescence spectra were recorded in solution on a Hitachi F-7000 (Hitachi, Tokyo, Japan) fluorescence spectrophotometer. Viscosity experiments were carried out using an ALPHA L Fungilab (Barcelona, Spain) rotational viscometer equipped with an 18 mL LCP spindle and the measurements were performed at 100 rpm. The ESI-HRMS spectra were recorded on an Agilent Q-TOF Mass Spectrometer, G6540B model with Dual AJS ESI-MS (Santa Clara, CA, USA). IR spectra were recorded on a Spectrum BX spectrophotometer (Perkin Elmer, Waltham, MA, USA) in the region 4000–500 cm^{-1}. NMR spectra were recorded on a DD2 500 MHz spectrometer (Agilent, Santa Clara, CA, USA), respectively. The HPLC system used comprised an Agilent HP 1100 series pump (HP, Waldbronn, Germany), connected to a Gabi gamma detector (Raytest, Straubenhardt, Germany) and an HP 1100 multiple wavelength detector. RP-HPLC analyses of the rhenium and technetium-99m complexes were performed using an Agilent Eclipse XDB C18 column (25 cm × 4.6 mm, 5 μm) by applying a binary gradient method of Solvent A: H_2O—0.1% TFA and Solvent B: Methanol.

The flow rate was set at 1 mL/min, and the composition was as follows: (min, A%, B%); (0, 100, 0); (15, 25, 75); (20, 5, 95); (25, 5, 95); (27, 100, 0); (30, 100, 0).

Calf thymus (CT) DNA (Merck, Darmstadt, Germany) was diluted in buffer (consisting of 15 mM trisodium citrate and 150 mM NaCl at pH 7.0) followed by continuous stirring for three days for the preparation of the DNA stock solution, which was kept at 4 °C for up to a week. The ratio of UV absorbance of the CT DNA stock solution at 260 and 280 nm (A_{260}/A_{280}) was measured to be 1.85, indicating that it was free of protein contamination [49]. The DNA concentration was determined by the UV absorbance at 260 nm after 1:20 dilution using ε = 6600 $M^{-1}cm^{-1}$ [50].

CAUTION! ^{99m}Tc is a gamma (γ)-emitter with nuclear properties of $t_{1/2}$ = 6 h and γ-energy, 140 keV). Its handling was performed according to Greek legislation and the EU regulations (2013/59/Euratom). ^{99m}Tc was obtained as sodium pertechnetate in sterile saline from a commercial ^{99}Mo/^{99m}Tc generator (AHEPA General Hospital, Thessaloniki, Greece).

3.2. Syntheses

1-Tosyl-8-methoxy-9,10-anthraquinone (**2**): 1,8-dihydoxy-anthraquinone (240 mg, 1 mmol) and potassium carbonate (138 mg, 1 mmol) were dissolved in dry acetone (50 mL). To this solution, dimethyl sulfate (130 µL, 1.3 mmol) was added dropwise, and the mixture was refluxed for 24 h. Then, sulfuric acid 0.1 M (5 mL) was added, and the mixture was concentrated to dryness under vacuum, followed by the addition of distilled water (300 mL). The precipitate that contained 1-hydroxy-8-methoxy-9,10-anthraquinone was isolated through filtration under vacuum, and the crude product was purified by silica gel (15 g) column chromatography with petroleum ether:ethyl acetate, 90:10–80:20, to afford a yellow solid. Yield: 120 mg, 47.2%. R_f: (SiO_2, petroleum ether:ethyl acetate, 9:1) 0.62. ^{1}H NMR (500 MHz, $CDCl_3$, ppm) δ: 12.99 (s, 1H), 7.97 (d, *J* = 7.7 Hz, 1H), 7.79–7.72 (m, 2H), 7.62 (t, *J* = 7.9 Hz, 1H), 7.37 (d, *J* = 8.5 Hz, 1H), 7.29 (d, *J* = 8.4 Hz, 1H), 4.08 (s, 3H). Then, 1-hydroxy-8-methoxy-9,10-anthraquinone (3 g, 11.8 mmol), potassium carbonate (1.63 g, 11.8 mmol) and p-toluenesulfonyl chloride (4.6 g, 24 mmol) were mixed in dry acetone (300 mL). The mixture was refluxed for 72 h. A brown solid formed, which was filtered under vacuum and dried. Yield: 2.2 g, 45.7%. R_f: (SiO_2, petroleum ether:ethyl acetate, 7:3) 0.5. m.p. 190–193 °C [24]. ^{1}H NMR (500 MHz, $CDCl_3$, ppm) δ: 8.17 (dd, *J* = 7.7, 1.2 Hz, 1H), 7.87 (d, *J* = 8.3 Hz, 2H), 7.83 (dd, *J* = 7.6, 0.7 Hz, 1H), 7.66 (t, *J* = 8.0 Hz, 2H), 7.57 (dd, *J* = 8.1, 1.1 Hz, 1H), 7.30 (t, *J* = 8.9 Hz, 3H), 4.02 (s, 3H), 2.38 (s, 3H). ^{13}C NMR (126 MHz, $CDCl_3$, ppm) δ: 197.22, 187.33, 182.69, 180.65, 159.56, 146.91, 145.36, 134.65, 134.59, 134.47, 133.29, 132.40, 129.98, 129.59, 129.16, 128.91, 125.76, 123.23, 119.17, 118.26, 56.65, 21.68. ESI-HRMS (*m/z*): Calc. for M = $C_{22}H_{16}SO_6$: 409.0740 $[M + H]^+$, 431.0560 $[M + Na]^+$; Found: 409.0726 $[M + H]^+$, 431.0566 $[M + Na]^+$.

2-(2-Hydroxyethyl)-10-methoxy-anthra [1,9-cd]pyrazol-6(2H)-one (**3**): Compound **2** (200 mg, 0.49 mmol) and 2-hydroxyethylhydrazine (168 µL, 2.47 mmol) were dissolved in anhydrous N,N-dimethylformamide (DMF) (10 mL). The mixture was heated at 130 °C, under nitrogen, for 4 h. The reaction mixture was partitioned between ethyl acetate (50 mL) and water (50 mL), and the organic phase was collected and concentrated to dryness. The crude product was purified by silica gel column chromatography (60 g) with ethyl acetate:petroleum ether, starting with 70:30 up to 90:10, to afford a yellow solid. Yield: 36 mg, 25%. R_f: (SiO_2, ethyl acetate:petroleum ether, 9:1) 0.01. m.p. 245–248 °C [8]. ^{1}H NMR (500 MHz, $CDCl_3$, ppm) δ: 8.11 (d, *J* = 7.8 Hz, 1H), 8.03 (d, *J* = 7.1 Hz, 1H), 7.76 (d, *J* = 8.2 Hz, 1H), 7.66 (t, *J* = 7.7 Hz, 1H), 7.51 (t, *J* = 8.0 Hz, 1H), 7.28 (d, *J* = 5.4 Hz, 1H), 4.68 (t, *J* = 4.8 Hz, 2H), 4.27 (t, *J* = 4.8 Hz, 2H), 4.12 (s, 3H). ^{13}C NMR (126 MHz, d_6-DMSO, ppm) δ: 183.01, 156.40, 139.46, 136.47, 134.31, 129.09, 128.28, 125.28, 123.38, 121.07, 120.97, 120.44, 117.40, 116.83, 60.92, 56.62, 52.71. ESI-HRMS (*m/z*): Calc. for M=$C_{17}H_{14}N_2O_3$: 295.1077 $[M + H]^+$, 317.0897 $[M + Na]^+$; Found: 295.1083 $[M + H]^+$, 317.0897 $[M + Na]^+$.

2-(2-Hydroxyethyl)-10-hydroxy-anthra [1,9-cd]pyrazol-6(2H)-one (***L1***): Compound **3** (260 mg, 0.88 mmol) was dissolved in anhydrous dichloromethane (50 mL) and was stirred at −10 °C. To this solution, boron tribromide 1 M (5.85 mL, 5.2 mmol) in dichloromethane was added dropwise, under nitrogen, for 1 h. The mixture was then stirred at rt (25 °C) for 24 h. Then, distilled water was added (50 mL) dropwise at −10 °C, for 1 h. The reaction mixture was partitioned between aqueous $NaHCO_3$ 1 M (50 mL) and dichloromethane (50 mL) and the organic phase was collected and concentrated to dryness. The crude product was washed three times with dichloromethane (3 × 10 mL), and a brown solid formed, which was filtered and dried. Yield: 160 mg, 65%. RP-HPLC retention time (t_R): 18.2 min. R_f: (SiO_2, ethyl acetate) 0.38. m.p. 190–193 °C. IR (cm^{-1}, KBr): 3406, 3076, 2934, 1636, 1593, 1459, 1396, 1351, 1282, 1262. ^1H NMR (500 MHz, d_6-DMSO, ppm) δ: 9.87 (s, 1H), 8.11 (d, *J* = 8.2 Hz, 1H), 7.94 (d, *J* = 7.1 Hz, 1H), 7.85 (d, *J* = 7.6 Hz, 1H), 7.75 (t, *J* = 7.6 Hz, 1H), 7.44 (t, *J* = 7.9 Hz, 1H), 7.35 (d, *J* = 7.9 Hz, 1H), 4.97 (br, 1H), 4.66 (t, *J* = 5.3 Hz, 2H), 3.92 (t, *J* = 5.3 Hz, 2H). ^{13}C NMR (126 MHz, d_6-DMSO, ppm) δ: 183.21, 154.43, 139.40, 137.17, 134.26, 129.19, 128.71, 125.56, 122.95, 121.11, 120.49, 120.31, 118.67, 117.37, 60.87, 52.68. ESI-HRMS (*m/z*): Calc. for $M{=}C_{16}H_{12}N_2O_3$: 279.0775 $[M-H]^-$, 248.0580 $[M-CH_3OH]^-$; Found: 279.0658 $[M-H]^-$, 248.0489 $[M-CH_3OH]^-$.

*2-(2-Methylsulfonylethyl)-10-methoxy-anthra [1,9-cd]pyrazol-6(2H)-one (**4**):* Compound **3** (250 mg, 0.85 mmol) and triethylamine (840 µL, 5.9 mmol) were dissolved in anhydrous dichloromethane (50 mL). To this solution, methanesulfonyl chloride (490 µL, 5.9 mmol) in dry dichloromethane (5 mL) was added dropwise at −10 °C, under nitrogen, for 30 min. The mixture was stirred at 25 °C, under nitrogen, for 4 h. The reaction mixture was concentrated to dryness under vacuum and was extracted between distilled water (50 mL) and ethyl acetate (50 mL). The organic phase was concentrated to dryness under vacuum, to afford a yellow solid. Yield: 317 mg, 85%. R_f: (SiO_2, ethyl acetate) 0.26. m.p. 155–158 °C. ^1H NMR (500 MHz, $CDCl_3$, ppm) δ: 8.14 (d, *J* = 7.8 Hz, 1H), 8.08 (d, *J* = 7.1 Hz, 1H), 7.81 (d, *J* = 8.2 Hz, 1H), 7.70 (t, *J* = 7.7 Hz, 1H), 7.53 (t, *J* = 8.0 Hz, 1H), 7.32 (d, *J* = 8.2 Hz, 1H), 4.92 (t, *J* = 5.1 Hz, 2H), 4.81 (t, *J* = 5.1 Hz, 2H), 4.14 (s, 3H), 2.80 (s, 3H). ^{13}C NMR (126 MHz, $CDCl_3$, ppm) δ (ppm): 182.93, 156.40, 138.95, 138.01, 134.79, 129.80, 129.51, 126.26, 123.38, 121.77, 121.46, 119.48, 115.87, 115.15, 68.16, 56.45, 48.92, 37.42. ESI-HRMS (*m/z*): Calc. for $M{=}C_{18}H_{16}N_2SO_5$: 373.0853 $[M + H]^+$, 395.0672 $[M + Na]^+$. Found: 373.0850 $[M + H]^+$, 395.0666 $[M + Na]^+$.

*2-(2-Picolylaminolethyl)-10-methoxy-anthra [1,9-cd]pyrazol-6(2H)-one (**L2**):* Compound **4** (290 mg, 0.75 mmol) and 2-(aminomethyl)pyridine (1.48 mL, 15 mmol) were dissolved in anhydrous methanol (70 mL), and the mixture was refluxed for 24 h. The reaction mixture was concentrated to dryness under vacuum and extracted in ethyl acetate–water (3 × 50 mL). The organic phase was concentrated to dryness under vacuum. The crude product was purified by silica gel (20 g) column chromatography with dichloromethane:methanol, 95:5, to afford a brown oil. Yield: 110 mg, 39%. t_R: 16.4 min. R_f: (SiO_2, dichloromethane:methanol, 9:1) 0.18. IR (cm^{-1}, KBr): 3451, 1639, 1578, 1455, 1281, 1270. ^1H NMR (500 MHz, $CDCl_3$, ppm) δ: 8.48 (d, *J* = 4.6 Hz, 1H), 8.12 (d, *J* = 7.9 Hz, 1H), 8.04 (d, *J* = 7.2 Hz, 1H), 7.76 (d, *J* = 8.2 Hz, 1H), 7.63 (t, *J* = 7.7 Hz, 1H), 7.58 (t, *J* = 7.6 Hz, 1H), 7.48 (td, *J* = 8.1, 1.1 Hz, 1H), 7.28 (d, *J* = 8.2 Hz, 1H), 7.25 (d, *J* = 8.1 Hz, 1H), 7.12 (t, *J* = 6.2 Hz, 1H), 4.74 (t, *J* = 6.0 Hz, 2H), 4.09 (s, 3H), 3.98 (s, 2H), 3.34 (t, *J* = 5.8 Hz, 2H), 2.37 (s, 1H). ^{13}C NMR (126 MHz, $CDCl_3$, ppm) δ: 183.65, 159.02, 156.30, 149.24, 138.76, 137.78, 136.48, 134.81, 128.62, 128.06, 126.16, 124.00, 122.13, 122.06, 121.67, 121.06, 120.62, 115.71, 114.97, 56.46, 54.72, 49.86, 48.98. ESI-HRMS (*m/z*): Calc. for $C_{23}H_{20}N_4O_2$: 385.1659 $[M + H]^+$ 407.1478 $[M + Na]^+$. Found: 385.1640 $[M + H]^+$, 407.1466 $[M + Na]^+$.

Ethyl 2-((2-(10-methoxy-6-oxodibenzo[cd,g]indazol-2(6H)-yl)ethyl)(pyridin-2-ylmethyl)amino) acetate (**L3**): First, ethyl 2-((pyridin-2-ylmethyl)amino)acetate was prepared by a modification of the procedure published elsewhere [51]. Specifically, ethyl bromoacetate (1.67 g, 10 mmol), 2-(aminomethyl)pyridine (1.48 mL, 15 mmol) and potassium carbonate (4 g, 30 mmol) were dissolved in dry acetonitrile (200 mL). The mixture was refluxed for 1 h. The reaction mixture was filtered and concentrated to dryness under vacuum. The crude product was purified by silica gel (60 g) flash chromatography with dichloromethane:methanol: aqueous ammonia solution 10%, 90:10:0.1, to afford a brown oil. Yield: 1.3 g, 67%. R_f: (SiO_2, dichloromethane:methanol: aqueous ammonia solution

10%, 9:1:0.1) 0.69. ^{1}H NMR (500 MHz, $CDCl_3$) δ (ppm): 8.56 (d, J = 4.2 Hz, 1H), 7.65 (td, J = 7.7, 1.8 Hz, 1H), 7.34 (t, J = 7.8 Hz, 1H), 7.19–7.14 (m, 1H), 4.18 (dq, J = 14.6, 7.1 Hz, 2H), 3.96 (s, 2H), 3.48 (s, 2H), 1.27 (t, J = 7.1 Hz, 3H).

Consequently, compound **4** (390 mg, 1.01 mmol) and ethyl 2-((pyridin-2-ylmethyl) amino)acetate (1.3 g, 6.7 mmol) were dissolved in dry methanol (100 mL). The mixture was refluxed, under nitrogen, for 72 h. The reaction mixture was concentrated to dryness under vacuum and extracted in ethyl acetate–water (3 × 50 mL). The organic phase was concentrated to dryness under vacuum. The crude product was purified by silica gel (60 g) column chromatography with dichloromethane:methanol, 98:2, to afford a brown oil. Yield: 160 mg, 34%. t_R: 17.3 min. R_f: (SiO_2, dichloromethane:methanol, 9:1) 0.55. IR (cm^{-1}, KBr): 3451, 2936, 1737, 1647, 1455, 1290, 1270. ^{1}H NMR (500 MHz, $CDCl_3$, ppm) δ: 8.44 (d, *J* = 4.8 Hz, 1H), 8.12 (d, *J* = 7.8 Hz, 1H), 8.03 (d, *J* = 7.1 Hz, 1H), 7.69 (dd, *J* = 8.2, 2.0 Hz, 1H), 7.59 (t, *J* = 7.7 Hz, 1H), 7.48 (t, *J* = 8.0 Hz, 1H), 7.36 (m, 1H), 7.28 (d, *J* = 8.2 Hz, 1H), 7.08 (d, *J* = 7.8 Hz, 1H), 7.03 (m, 1H), 4.66 (t, *J* = 6.4 Hz, 2H), 4.13 (dd, *J* = 14.0, 6.8 Hz, 2H), 4.10 (s, 3H), 4.00 (d, *J* = 2.3 Hz, 2H), 3.51 (d, *J* = 2.7 Hz, 2H), 3.39 (t, *J* = 6.4 Hz, 2H), 1.22 (t, *J* = 7.1 Hz, 3H). ^{13}C NMR (126 MHz, d_6-DMSO, ppm) δ: 183.71, 171.18, 158.69, 156.22, 148.53, 138.63, 137.42, 136.65, 134.75, 128.55, 127.88, 126.01, 123.97, 123.06, 122.14, 121.65, 121.12, 120.52, 115.67, 115.26, 60.57, 56.48, 55.56, 53.92, 51.53, 48.81, 14.22. ESI-HRMS (*m/z*): Calc. for M=$C_{27}H_{26}N_4O_4$: 471.2027 $[M + H]^+$, 493.1846 $[M + Na]^+$. Found: 471.2032 $[M + H]^+$, 493.1846 $[M + Na]^+$.

*Synthesis of **ReL1**:* [$Re(CO)_5Br$] (81.2 mg, 0.20 mmol) and **L1** (56 mg, 0.20 mmol) were dissolved in methanol (40 mL), and the mixture was refluxed for 24 h. The mixture was concentrated to dryness under vacuum, followed by the addition of distilled water (10 mL). After 24 h, a precipitate was formed at room temperature and was isolated through filtration under vacuum. The crude product was purified by silica gel (60 g) column chromatography with dichloromethane:methanol, 98:2–90:10, to afford a red solid. Yield: 45 mg, 38.5%. t_R: 18.8 min. R_f: (SiO_2, dichloromethane:methanol, 9:1) 0.67. m.p. 249–252 °C. Calc. for $C_{20}H_{15}N_2O_7Re$: C, 41.31; H, 2.60; N, 4.82; Found: C 41.55; H 2.91; N 4.71. IR (cm^{-1}, KBr): 3432 (O-H), 2936(C-H), 2025(CO), 1930(CO), 1900(CO), 1636 (C=O), 1566(C=N), 1459 (arom. C-C, C-N), 1282(C-O). ^{1}H NMR (500 MHz, d_6-DMSO, ppm) δ: 8.10 (d, *J* = 8.3 Hz, 1H), 7.95 (d, *J* = 7.0 Hz, 1H), 7.87 (dd, *J* = 8.2,7.2 Hz, 1H), 7.54 (dd, *J* = 7.5, 1.0 Hz, 1H), 7.40 (t, *J* = 8.2, 7.6 Hz, 1H), 7.18 (dd, *J* = 7.3, 1.0 Hz, 1H), 5.07 (t, *J* = 5.4 Hz, 1H), 4.89–4.84 (m, 1H), 4.05–3.87 (m, 2H), 5.03 (t, *J* = 5.3 Hz, 1H), 4.90–4.85 (m, 2H), 4.03–3.96 (m, 1H), 3.96–3.89 (m, 1H). ^{13}C NMR (126 MHz, d_6-DMSO, ppm) δ: 197.88, 197.36, 196.86, 182.96, 160.30, 140.80, 138.65, 133.86, 131.39, 131.13, 126.59, 126.39, 121.40, 120.48, 117.96, 116.92, 116.49, 60.30, 52.36. ESI-HRMS (*m/z*): Calc. for M = $C_{20}H_{15}N_2O_7Re$: 579.0318 (60%), 581.0311 (100%) $[M-H]^-$. Found: 579.0101 (60%), 581.0139 (100%) $[M-H]^-$.

*Synthesis of **ReL2**:* [$Re(CO)_5Br$] (65 mg, 0.16 mmol) and **L2** (60 mg, 0.16 mmol) were dissolved in methanol (35 mL), and the mixture was refluxed for 3 h. The yellow solid formed was filtered under vacuum, recrystallized from methanol–water and dried. Yield: 55 mg, 47%. t_R: 21.1 min, 21.3 min. R_f: (SiO_2, dichloromethane:methanol, 9:1) 0.89. m.p. 336–339 °C. Calc. for $C_{26}H_{20}N_4O_5ReBr$: C, 42.51; H, 2.74; N, 7.63; Found: C 42.82; H 2.99; N 7.68. IR (cm^{-1}, KBr): 3448, 3168 (N-H), 2935 (C-H), 2021 (CO), 1909 (CO), 1867 (CO), 1654 (C=O), 1636 (C=N), 1438 (arom. C-C, C-N), 1284 (C-O), 1267 (C-O). ^{1}H NMR (500 MHz, d_6-DMSO, ppm) *Major/δ:* 8.78 (d, *J* = 5.1 Hz, 1H), 8.17 (d, *J* = 8.2 Hz, 1H), 8.10 (td, *J* = 7.8, 1.5 Hz, 1H), 7.96 (dd, *J* = 7.1, 1.7 Hz, 1H),7.92 (dd, *J* = 7.8, 1.0 Hz, 1H), 7.79 (m, 2H), 7.56 (t, *J* = 7.7 Hz, 1H), 7.55 (t, *J* = 8.0 Hz, 1H), 7.45 (d, *J* = 7.6 Hz, 1H), 5.64 (s, 1H), 5.12 (dd, *J* = 15.9, 5.2 Hz, 1H), 4.94 (dd, *J* = 12.2, 5.8 Hz, 2H), 4.43 (dd, *J* = 15.9, 8.8 Hz, 1H), 3.80 (s, 3H), 3.74 (m, 2H); *minor/δ:* 8.79 (d, *J* = 5.1 Hz), 8.18 (d, *J* = 8.2 Hz, 1H), 8.06 (td, *J* = 7.8, 1.5 Hz), 7.96 (dd, *J* = 7.1, 1.7 Hz, 1H), 7.93 (dd, *J* = 7.8, 1.0 Hz, 1H), 7.79 (t, *J* = 7.7 Hz, 1H), 7.68 (d, *J* = 7.9 Hz), 7.55 (t, *J* = 8.0 Hz, 1H), 7.49 (d, *J* = 7.6 Hz, 1H), 7.48 (m, 1H), 6.73 (s, 1H), 4.97–4.83 (m, 2H), 4.72 (dd, *J* = 15.6, 5.3 Hz, 1H), 4.60 (dd, *J* = 15.7, 8.8 Hz), 3.91 (s), 3.74 (m, 2H). ^{13}C NMR (126 MHz, d_6-DMSO, ppm) *Major/δ:* 197.87, 196.58, 192.21, 182.97, 160.69, 156.57, 153.12, 140.38, 138.80, 137.46, 134.33, 129.57, 128.99, 125.90, 125.53, 123.52, 123.26,

120.94, 120.45, 116.90, 109.98, 58.64, 57.44, 56.62, 47.26; *minor/δ:* 197.87, 196.58, 192.21, 182.97, 161.80, 156.51, 153.05, 140.34, 138.92, 137.26, 134.34, 129.63, 128.91, 125.79, 125.54, 123.31, 123.26, 121.05, 120.62, 117.06, 116.77, 58.80, 57.68, 56.69, 47.26. ESI-HRMS (*m/z*): Calc. for M=$C_{26}H_{20}N_4O_5ReBr$: 653.0946 (60%), 655.0783 (100%) $[M-Br]^+$, 731.0995 (60%), 733.1125 (100%) $[M-Br + DMSO]^+$. Found: 653.0747 (60%), 655.0960 (100%) $[M-Br]^+$, 731.1110 (60%), 733.1138 (100%) $[M-Br + DMSO]^+$.

*Synthesis of **ReL3**:* $[Re(CO)_5(OTf)]$ (22 mg, 0.05 mmol) was dissolved in methanol (7 mL), and the mixture was refluxed for 1 h for the formation of $[Re(CO)_3(MeOH)_3]^+$. **L3** (24 mg, 0.05 mmol) was added, and the mixture was refluxed for 24 h. The reaction mixture was concentrated to ~3 mL, and after cooling, a yellow solid was formed, which was filtered under vacuum and dried. Yield: 21 mg, 60%. t_R: 21.4 min. R_f: (SiO_2, dichloromethane:methanol, 9:1) 0.78. m.p. 327–330 °C. Calc. for $C_{28}H_{21}N_4O_7Re$: C, 47.25; H, 2.97; N, 7.87; Found: C, 47.48; H 3.15; N 7.56. IR (cm^{-1}, KBr): 3449, 2941 (C-H), 2023 (CO), 1912 (CO), 1370 (CO), 1651 (C=O), 1457 (arom. C-C, C-N), 1296 (C-O), 1267 (C-O). ^1H NMR (500 MHz, d_6-DMSO, ppm) δ: 8.78 (d, *J* = 5.4 Hz, 1H), 8.30 (d, *J* = 8.2 Hz, 1H), 8.16 (t, *J* = 7.7 Hz, 1H), 7.98 (d, *J* = 7.1 Hz, 1H), 7.94 (d, *J* = 7.7 Hz, 1H), 7.81 (t, *J* = 7.8 Hz, 1H), 7.72 (d, *J* = 7.9 Hz, 1H), 7.59 (t, *J* = 6.6 Hz, 1H), 7.56 (t, J = 7.9 Hz, 1H), 7.51 (d, *J* = 8.1 Hz, 1H), 5.13–5.03 (m, 2H), 5.04 (d, *J* = 15.5 Hz, 1H), 4.85 (d, *J* = 15.7 Hz, 1H), 4.28 (dt, *J* = 13.8, 7.0 Hz, 1H), 4.16 (dt, *J* = 13.3, 6.7 Hz, 1H), 4.03 (d, *J* = 16.6 Hz, 1H), 4.02 (s, 3H), 3.59 (d, *J* = 16.6 Hz, 1H). ^{13}C NMR (126 MHz, d_6-DMSO, ppm) δ: 198.00, 197.34, 182.96, 178.92, 159.68, 156.54, 152.55, 141.18, 138.80, 137.28, 134.42, 129.62, 128.93, 126.45, 125.56, 124.27, 123.60, 121.16, 120.96, 120.58, 117.20, 117.16, 67.99, 67.45, 61.32, 56.68, 45.69. ESI-HRMS (*m/z*): Calc. for M=$C_{28}H_{21}N_4O_7Re$: 711.0999 (60%), 713.0994 (100%) $[M + H]^+$. Found: 711.1006 (60%), 713.1050 (100%) $[M + H]^+$.

*Synthesis of tricarbonylrheniumbromo(2-phenyl-N-(pyridin-2-ylmethyl)ethanamine), **ReL4**:* $[Re(CO)_5Br]$ (40 mg, 0.1 mmol) and **L4** (21 mg, 0.1 mmol) (synthesis of **L4** is reported in ESI)) were dissolved in methanol (3 mL), and the mixture was refluxed for 3 h. The complex crystallized from methanol 1 mL at 4 °C, after 2 days. Yield: 25 mg (45%). Calc. for $C_{17}H_{16}BrN_2O_3Re$: C, 36.30; H, 2.87; N, 4.98; Found: C, 36.59; H, 3.15; N, 4.77. IR (cm^{-1}, KBr): 3448, 3209 (N-H), 2926 (C-H), 2021 (CO), 1913 (CO), 1875 (CO), 1489 (arom. C-C, C-N), 1438 (arom. C-C, C-N). ^1H NMR (500 MHz, $CDCl_3$, ppm) δ: 8.82 (d, *J* = 5.2 Hz, 1H), 7.84 (t, *J* = 7.8 Hz, 1H), 7.42 (d, *J* = 7.8 Hz, 1H), 7.36–7.23 (m, 6H), 4.71 (dd, *J* = 14.7, 3.9 Hz, 1H), 4.18–4.09 (m, 1H), 3.91–3.89 (m, 1H), 3.70–3.57 (m, 1H), 3.43–3.34 (m, 1H), 3.17–3.01 (m, 2H).; ^{13}C NMR (126 MHz, $CDCl_3$, ppm) δ: 196.40 (CO), 195.47 (CO), 191.60 (CO), 158.43, 153.26, 139.02, 136.65, 129.03, 128.80, 127.20, 125.28, 121.96, 60.16, 59.23, 35.11 ESI-HRMS (*m/z*): Calc. for M=$C_{17}H_{16}BrN_2O_3Re$: 481.0685 (60%), 483.0689 (100%) $[M-Br]^+$; 559.0807 (60%), 561.0795 (100%) $[M-Br + DMSO]^+$; Found: 481.0724 (60%), 483.0746 (100%) $[M-Br]^+$; 559.0864 (60%), 561.0893 (100%) $[M-Br + DMSO]^+$.

3.3. X-ray Crystallography of fac-[Re(CO)$_3$Br(**L4**)]

X-ray quality crystals of compound *fac*-[Re(CO)$_3$Br(**L4**)] were grown in the mother liquor. A crystal suitable for X-ray diffraction with dimensions 0.14 × 0.09 × 0.07 mm was mounted at rt on a Bruker Kappa APEX2 diffractometer equipped with a triumph monochromator using Mo Kα (λ = 0.71073 Å, source operating at 50 kV and 30 mA) radiation. Unit cell dimensions were determined and refined using the angular settings of 173 high-intensity reflections (>10σ(I)) in the range 11 < 2θ < 36°. Intensity data were recorded using φ and ω–scans. The crystal remained intact during the data collection. The frames collected were integrated using the Bruker SAINT Software package V7.60A [52], and a narrow-frame algorithm. Data were corrected for absorption using the numerical method (SADABS) based on crystal dimensions [53]. The structure was solved using the SUPERFLIP package [54], incorporated in CRYSTALS. Data refinement (full-matrix least-squares methods on F^2), and all subsequent calculations were carried out using the CRYSTALS version 14.61 build 6236 program package [55–57]. All non–hydrogen atoms were refined anisotropically.

Hydrogen atoms riding on parent carbon atoms were located from difference Fourier maps and refined at idealized positions riding on the parent atoms with isotropic displacement parameters Uiso(H) = 1.2Ueq(C) or 1.5Ueq(methylene and -NH hydrogens) and at distances C--H 0.95 Å and N-H 0.85 Å. All methylene and NH hydrogen atoms were allowed to rotate but not to tip. Illustrations with 50% ellipsoids probability were drawn by CAMERON [58,59]. Crystallographic data for the complex are presented in Table 6.

Table 6. Experimental details.

Crystal Data	
Chemical formula	$C_{17}H_{16}BrN_2O_3Re$
M_r	562.43
Crystal system, space group	Monoclinic, $P2_1/c$
Temperature (K)	295
a, b, c (Å)	11.5730 (8), 20.8663 (11), 7.7637 (5)
β (°)	99.867 (2)
V (Å^3)	1847.1 (2)
Z	4
Radiation type	Mo $K\alpha$
μ (mm^{-1})	8.76
Crystal size (mm)	0.14 × 0.09 × 0.07
Data collection	
Diffractometer	Bruker Kappa Apex2
Absorption correction	Numerical Analytical Absorption [56]
T_{min}, T_{max}	0.45, 0.54
No. of measured, independent and observed [$I > 2.0\sigma(I)$] reflections	17060, 3533, 3106
R_{int}	0.029
$(\sin \theta/\lambda)_{max}$ (Å^{-1})	0.613
Refinement	
$R[F^2 > 2\sigma(F^2)]$, $wR(F^2)$, S	0.036, 0.057, 1.00
No. of reflections	3106
No. of parameters	217
H-atom treatment	H-atom parameters constrained
$\Delta\rho_{max}$, $\Delta\rho_{min}$ (e Å^{-3})	1.54, −1.55

Further details on the crystallographic studies as well as atomic displacement parameters are given as Supporting Information in the form of cif files.

Crystallographic data were submitted to the Cambridge Crystallographic Data Center, No. 2374416. Copies of the data are available free of charge upon application to CCDC, 12 Union Road, Cambridge, CB2 1EZ, UK. Telephone: +(44)-1223-336033; E-mail: deposit@ccdc.ac.uk, or via https://www.ccdc.cam.ac.uk/structures/ (accessed on 30 July 2024).

3.4. DNA-Binding Studies

3.4.1. Study by UV–Vis Spectroscopy

The interaction of the compounds with CT DNA as well as their possible binding modes were investigated by UV–vis spectroscopy and the respective binding constants (K_b) were calculated. The UV–vis spectra of CT DNA were recorded under a constant DNA concentration in the presence of each compound at various mixture ratios (r) [compound]/[DNA]. To obtain the binding constant K_b (in M^{-1}), the changes in the absorbance of each compound at the corresponding λ_{max} of their UV–vis spectra were recorded at increasing CT DNA concentrations (different r values), and it was calculated by the ratio of slope to the y intercept in plots $[DNA]/(\varepsilon_A - \varepsilon_f)$ vs. [DNA], according to the Wolfe–Shimer equation [34]:

$$\frac{[DNA]}{\left(\varepsilon_A - \varepsilon_f\right)} = \frac{[DNA]}{\left(\varepsilon_b - \varepsilon_f\right)} + \frac{1}{K_b\left(\varepsilon_b - \varepsilon_f\right)} \quad (1)$$

where [DNA] is the concentration of DNA in base pairs, $\varepsilon_A = A_{obsd}/[compound]$, ε_f = the extinction coefficient of the unbound compound, and ε_b = the extinction coefficient of the compound in the fully bound form.

3.4.2. Viscometry

The viscosity of DNA ([DNA] = 0.1 mM) was measured in buffer solution (150 mM NaCl and 15 mM trisodium citrate at pH 7.0) and in increasing amounts of the tested compounds. The measurements were performed at room temperature. The data are presented as $(\eta/\eta_0)^{1/3}$ vs. r, where η is the viscosity of DNA in the presence of the compound, and η_0 is the viscosity of DNA without the compound.

3.4.3. Competition Studies with Ethidium Bromide (EB) via Fluorescence Spectroscopy

The ability of the compounds to displace EB from its DNA–EB complex was investigated by fluorescence emission spectroscopy. The DNA–EB adduct was prepared by mixing 20 μM EB and 26 μM CT DNA in buffer (150mM NaCl and 15mM trisodium citrate at pH 7.0). The intercalating effect of the compounds was studied by adding incremental amounts of the compound into a solution of the DNA–EB adduct. The effect of the addition of each compound to DNA–EB was obtained by measuring the changes of fluorescence emission spectra at the excitation wavelength of 540 nm. The compounds did not exhibit significant fluorescence at rt with or without DNA under the same conditions; therefore, the observed quenching is attributed to the displacement of EB from its EB–DNA adduct. The Stern–Volmer constant (K_{SV}) was employed to evaluate the quenching efficiency of the compounds. The K_{SV} value (in M^{-1}) of the compounds was calculated as the slope of the plot I_0/I vs. [Q], according to the linear Stern–Volmer equation (Equation (2)) [37]:

$$\frac{Io}{I} = 1 + K_q\tau_0[Q] = 1 + K_{SV}[Q] \quad (2)$$

where I is the emission intensity in the presence of the compound, I_0 is the emission intensity without the presence of the quencher (i.e., the compound under study), K_q = the quenching constant of the EB–DNA system, and τ_o = the average lifetime of EB–DNA without the quencher. Taking τ_o = 23 ns as the fluorescence lifetime of the EB–DNA system [38], the EB–DNA quenching constants (K_q, in $M^{-1}s^{-1}$) of the compounds were determined from Equation (3) [37]:

$$K_{SV} = K_q \cdot \tau_o \quad (3)$$

3.5. In Vitro Cell Studies

3.5.1. Cell Cultures

Murine Balb/c colorectal carcinoma CT26 cells were a gift from Prof. C. Chlichlia and were grown in a culture containing DMEM medium that was supplemented with 10% (*v/v*) FBS in the presence of penicillin and streptomycin (1%) at 37 °C and in a humidified atmosphere containing 5% (*v/v*) CO_2. The medium was renewed every 2 days (<90% confluency in the plates) to allow logarithmic cellular growth in culture. The cells were detached by using trypsin-EDTA solution (25%) (GIBCO Laboratories, Grand Island, NY, USA). The compounds were dissolved in DMSO and then added at the appropriate concentrations in the cell cultures.

3.5.2. Assessment of Proliferation Capacity of CT26 Cells Exposed to Anthrapyrazole Derivatives

The CT26 cells were plated in 24-well plates at a density of 1×10^5 cells/mL. Following the attachment of cells to the plate, the compounds were added in the cultures at various concentrations (10^{-5}–10^{-7} M). Then, the cells were permitted to grow for an additional 48 h. Subsequently, the cells were detached via trypsinization, and the cell density (number of cells/mL) was measured using a Neubauer chamber. The cell proliferation rate of CT26 cultures was expressed as a percentage (%) of cell growth compared to the control-untreated cell cultures. Furthermore, viability and cellular death were also assessed by using the trypan-blue dye exclusion method.

3.6. ^{99m}*Tc Radiochemistry and In Vitro Radiotracer Studies*

Synthesis of ^{99m}***TcL2*** *and* ^{99m}***TcL3****:* A fresh solution of *fac*-[$^{99m}Tc(CO)_3(H_2O)_3]^+$ (400 μL, 370–480 MBq) at pH 7 was transferred to a vial containing a methanolic solution of the ligand **L2** or **L3** (50 μL of 10^{-2} M) and an aqueous solution of ascorbic acid (50 μL of 10^{-1} M). The vial was sealed, flushed with N_2 for 5 min and heated at 55–65 °C for 60 min. The reaction mixtures were analyzed by HPLC. The ^{99m}Tc complexes were purified by HPLC, and after evaporation of the solvents, they were reconstituted in 1% Tween 80 saline solution prior to further study.

Lipophilicity: The lipophilicity of the radiocomplexes was determined by the shake-flask method. ^{99m}**TcL2** or ^{99m}**TcL3** (20 μL) were mixed with 2 mL of 1-octanol and 1.98 mL of phosphate buffer (PBS, 0.1 M, pH 7.4) in a centrifuge tube. The mixture was vortexed at rt for 1 min and centrifuged at 3500 rpm for 5 min. Aliquots (50 μL) of both 1-octanol and PBS phases were withdrawn and counted in a gamma counter. The experiment was conducted in triplicates. The distribution coefficient (D) was calculated by dividing the radioactivity of the organic phase with that of the aqueous phase, and the results are expressed as $\log D_{7.4}$.

Stability studies: Histidine challenge: The purified ^{99m}**TcL2** or ^{99m}**TcL3** (50 μL, approx. 11–15 MBq) was mixed with a solution of L-histidine (1 mM) in 0.1 M PBS, pH 7.4 (0.45 mL) and incubated at 37 °C for 4 h. The mixtures were analyzed by HPLC at 1 and 4 h.

Rat plasma stability: ^{99m}**TcL2** or ^{99m}**TcL3** (100–120 μL, 25–30 MBq) was mixed with undiluted rat plasma (0.5 mL) at 37 °C for 4 h. Samples were withdrawn at 1 and 4 h, which were mixed with three times the volume of acetonitrile, to precipitate the proteins. The mixture was centrifuged at 5000 rpm for 10 min and the supernatant solution was separated from the solids. The radioactivity of the solution and the solid was measured in a γ-counter, and the solution was analyzed by HPLC.

Cellular uptake of ^{99m}***TcL2 and*** ^{99m}***TcL3****:* The CT26 cells were seeded at a density of 1×10^6 cells/mL in 24-well plates and were allowed to attach for 6 h. Aliquots of ^{99m}**TcL2** or ^{99m}**TcL3** (20–40 μL, 0.9 MBq) were added to each plate and the cells were incubated for 15, 60, 120, and 240 min at 37 °C in an atmosphere containing 5% (*v/v*) CO_2. The culture medium was withdrawn and the cells were detached using trypsin-EDTA (200 μL 0.25% *w/v*). After 2 min incubation, DMEM was added in the culture. The solution containing the cells was transferred to a tube followed by centrifugation, (5 min, 2000 rpm) and the cells were washed with 1 × PBS twice (150 and 120 μL). The radioactivity of the cells and supernatant was counted in γ–counter to evaluate the cellular uptake. The experiment was conducted in triplicates for each time point.

3.7. *Biodistribution Studies of* ^{99m}***TcL2*** *in Mice*

The experiment was approved by the Aristotle University Committee for Animal Experimentation (License No 114251/528), and was performed according to the European guidelines 2010/63/EU and Greek legislation (PD 56) for animal experimentation. Balb/c mice, 10–12 weeks old, with a median weight of 20–25 g, were housed in suitable animal facilities (Laboratory of Development-Breeding of Animal Models and Biomedical Research, Faculty of Health Sciences, Aristotle University, EU License No EL 54 BIOexp-10) with food and water ad libitum and constant conditions of temperature, humidity and regular light cycles of 12/12 h light/dark. The 3R alternatives (Replacement, Refinement, Reduction) were considered in all animal experiments, while the mice were not subjected to pain or discomfort during the experimentation.

Balb/c were implanted subcutaneously with 5×10^6 CT26 cells on the hind right flank. One week after inoculation, when the tumor size was between 0.4–1 cm, the animals were injected in the tail vein with ~370 kBq of the HPLC-purified tracer **^{99m}TcL2** in 0.1 mL saline each. Animals were sacrificed at 30 and 120 min post-injection (p.i.) by cervical dislocation, which was followed by blood withdrawal and myocardial excision. Organs and tissues of interest were excised and weighed, and their radioactivity was measured by a γ scintillator. The radioactivity of the samples was decay-corrected by the use of a standard solution corresponding to 1% of the injected dose. The radioactivity of the tissues and organs is expressed as a percentage of the injected dose per gram tissue (% ID/g). Values are quoted as the mean% ID = standard deviation (SD) of the four mice per group. Blood volume and muscle mass were estimated at 7 and 43% of body weight, respectively.

4. Conclusions

In this work, three new anthrapyrazole ligands and their respective tricarbonylrhenium complexes were synthesized and characterized. All compounds possess the ability to intercalate with DNA. Complex **ReL1** acts as a metallointercalator with strong DNA-binding affinity. However, it was not cytotoxic at the tested concentration range and cell line, which is attributed to the ligand's N-substituent. The anthrapyrazole pendant complexes **ReL2** and **ReL3** show strong DNA-binding affinities, with **ReL3** exhibiting the highest DNA-binding constant among the tested compounds. Tumor cell cytotoxicity was high for ligand **L2** and complex **ReL2** with submicromolar IC_{50} values of 0.36 and 0.64 μM, respectively. The low cytotoxicity of **ReL3** can be attributed also to its low cell uptake, based on the radiotracer studies with its analogous **^{99m}TcL3**. As complex **ReL2** was the most cytotoxic, its γ-emitting analogue **^{99m}TcL2** was evaluated for its tumor cell uptake and biodistribution properties in tumor-bearing mice. It was observed that this tracer exhibits high tumor uptake in vitro as well as accumulation in the tumor in vivo. **ReL2** is a cytotoxic DNA-intercalator with suitable pharmacokinetic properties to be distributed in tumors in vivo and may be considered for further studies. Also, future design should focus on compounds with suitable substituents which is critical for cytotoxicity.

Supplementary Materials: The following supporting information can be downloaded at: https://www.mdpi.com/article/10.3390/inorganics12090254/s1, 1. Synthesis of Ligand **L4**; 2. NMR spectra of **L1-L4** and **ReL1-ReL4**; 3. IR spectra; Figure S1: ORTEP diagram of the S enantiomer of **ReL4**; Table S1: Computed octahedral distortion parameters; Figure S2: UV-vis spectra of DMSO solution of **L1**, **L2**, **ReL1** and **ReL2** in the presence of increasing amounts of CT DNA; Figure S3: Plot of ([DNA])/(ε_A-ε_f)) versus [DNA] for compounds **L1–L3** and **ReL1–ReL3**. Figure S4: Fluorescence emission spectra of compounds **L1-L3** and **ReL1-ReL2**; Figure S5: Stern–Volmer quenching plot of EB–DNA fluorescence for compounds **L1–L3** and **ReL1–ReL3**; Figure S6: IC_{50} curve of **L2** and **ReL2**. Figure S7: HPLC analysis of **L2**, **ReL2**, **ReL2′** and **^{99m}TcL2**.

Author Contributions: Conceptualization, D.P. and G.P. (Georgios Paparidis); methodology, G.P. (George Psomas), I.S.V., A.H. and C.G.; resources, D.S., I.S.V., G.P. (George Psomas) and D.P.; data curation, G.P. (Georgios Paparidis) and M.A.; writing—original draft preparation, G.P. (Georgios Paparidis), G.P. (George Psomas), I.S.V. and A.H.; writing—review and editing, D.P.; supervision, D.P., I.S.V. and G.P. (George Psomas). All authors have read and agreed to the published version of the manuscript.

Funding: This research received no external funding.

Institutional Review Board Statement: The experiment was approved by the Aristotle University Committee for Animal Experimentation (License No 114251/528), and was performed according to the European guidelines 2010/63/EU and Greek legislation (PD 56) for animal experimentation.

Data Availability Statement: The data of this work are provided within the manuscript and in the Supplementary Materials.

Acknowledgments: The authors are grateful to I. Iakovou, Head of the AHEPA Hospital Nuclear Medicine Laboratory, for donating ^{99m}Tc to perform the study.

Conflicts of Interest: The authors declare no conflicts of interest.

References

1. Muhammad, N.; Guo, Z. Metal-Based Anticancer Chemotherapeutic Agents. *Curr. Opin. Chem. Biol.* **2014**, *19*, 144–153. [CrossRef]
2. Ye, R.; Tan, C.; Chen, B.; Li, R.; Mao, Z. Zinc-Containing Metalloenzymes: Inhibition by Metal-Based Anticancer Agents. *Front. Chem.* **2020**, *8*, 402. [CrossRef]
3. Simpson, P.V.; Desai, N.M.; Casari, I.; Massi, M.; Falasca, M. Metal-Based Antitumor Compounds: Beyond Cisplatin. *Future Med. Chem.* **2019**, *11*, 119–135. [CrossRef]
4. Wernitznig, D.; Kiakos, K.; Del Favero, G.; Harrer, N.; Machat, H.; Osswald, A.; Jakupec, M.A.; Wernitznig, A.; Sommergruber, W.; Keppler, B.K. First-in-Class Ruthenium Anticancer Drug (KP1339/IT-139) Induces an Immunogenic Cell Death Signature in Colorectal Spheroids in Vitro†. *Metallomics* **2019**, *11*, 1044–1048. [CrossRef]
5. King, A.P.; Wilson, J.J. Endoplasmic Reticulum Stress: An Arising Target for Metal-Based Anticancer Agents. *Chem. Soc. Rev.* **2020**, *49*, 8113–8136. [CrossRef]
6. Anthony, E.J.; Bolitho, E.M.; Bridgewater, H.E.; Carter, O.W.L.; Donnelly, J.M.; Imberti, C.; Lant, E.C.; Lermyte, F.; Needham, R.J.; Palau, M.; et al. Metallodrugs Are Unique: Opportunities and Challenges of Discovery and Development. *Chem. Sci.* **2020**, *11*, 12888–12917. [CrossRef]
7. Ortega, E.; Vigueras, G.; Ballester, F.J.; Ruiz, J. Targeting Translation: A Promising Strategy for Anticancer Metallodrugs. *Coord. Chem. Rev.* **2021**, *446*, 214129. [CrossRef]
8. Ghosh, S. Cisplatin: The First Metal Based Anticancer Drug. *Bioorg. Chem.* **2019**, *88*, 102925. [CrossRef]
9. Clède, S.; Lambert, F.; Saint-Fort, R.; Plamont, M.A.; Bertrand, H.; Vessières, A.; Policar, C. Influence of the Side-Chain Length on the Cellular Uptake and the Cytotoxicity of Rhenium Triscarbonyl Derivatives: A Bimodal Infrared and Luminescence Quantitative Study. *Chem. A Eur. J.* **2014**, *20*, 8714–8722. [CrossRef] [PubMed]
10. Leonidova, A.; Pierroz, V.; Adams, L.A.; Barlow, N.; Ferrari, S.; Gra, B. Enhanced Cytotoxicity through "Click" Conjugation of a Luminescent Re (I) Complex to a Cell-Penetrating Lipopeptide. *ACS Med. Chem. Lett.* **2014**, *5*, 809–814. [CrossRef] [PubMed]
11. Leonidova, A.; Pierroz, V.; Rubbiani, R.; Heier, J.; Ferrari, S.; Gasser, G. Towards Cancer Cell-Specific Phototoxic Organometallic Rhenium (I) Complexes. *Dalton Trans.* **2014**, *43*, 4287–4294. [CrossRef] [PubMed]
12. Knopf, K.M.; Murphy, B.L.; Macmillan, S.N.; Baskin, J.M.; Barr, M.P.; Boros, E.; Wilson, J.J. In Vitro Anticancer Activity and in Vivo Biodistribution of Rhenium (I) Tricarbonyl Aqua Complexes. *J. Am. Chem. Soc.* **2017**, *139*, 14302–14314. [CrossRef] [PubMed]
13. Schindler, K.; Zobi, F. Anticancer and Antibiotic Rhenium Tri- and Dicarbonyl Complexes: Current Research and Future Perspectives. *Molecules* **2022**, *27*, 539. [CrossRef] [PubMed]
14. Rescifina, A.; Zagni, C.; Varrica, M.G.; Pistarà, V.; Corsaro, A. Recent Advances in Small Organic Molecules as DNA Intercalating Agents: Synthesis, Activity, and Modeling. *Eur. J. Med. Chem.* **2014**, *74*, 95–115. [CrossRef] [PubMed]
15. Liu, H.-K.; Sadler, P.J. Metal Complexes as DNA Intercalators. *Acc. Chem. Res.* **2011**, *44*, 349–359. [CrossRef]
16. Lee, L.C.-C.; Leung, K.-K.; Lo, K.K.-W. Recent Development of Luminescent Rhenium (i) Tricarbonyl Polypyridine Complexes as Cellular Imaging Reagents, Anticancer Drugs, and Antibacterial Agents. *Dalton Trans.* **2017**, *46*, 16357–16380. [CrossRef]
17. Pagoni, C.C.; Xylouri, V.S.; Kaiafas, G.C.; Lazou, M.; Bompola, G.; Tsoukas, E.; Papadopoulou, L.C.; Psomas, G.; Papagiannopoulou, D. Organometallic Rhenium Tricarbonyl—Enrofloxacin and—Levofloxacin Complexes: Synthesis, Albumin—Binding, DNA*i*—Interaction and Cell Viability Studies. *JBIC J. Biol. Inorg. Chem.* **2019**, *24*, 609–619. [CrossRef]
18. Nasiri Sovari, S.; Kolly, I.; Schindler, K.; Djuric, A.; Srdic-Rajic, T.; Crochet, A.; Pavic, A.; Zobi, F. Synthesis, Characterization, and in Vivo Evaluation of the Anticancer Activity of a Series of 5- and 6-(Halomethyl)-2,2′-Bipyridine Rhenium Tricarbonyl Complexes. *Dalton Trans.* **2023**, *52*, 6934–6944. [CrossRef]
19. Frederick, C.A.; Williams, L.D.; Ughetto, G.; Van der Marel, G.A.; Van Boom, J.H.; Rich, A.; Wang, A.H.J. Structural Comparison of Anticancer Drug-DNA Complexes: Adriamycin and Daunomycin. *Biochemistry* **1990**, *29*, 2538–2549. [CrossRef]
20. Mattioli, R.; Ilari, A.; Colotti, B.; Mosca, L.; Fazi, F.; Colotti, G. Doxorubicin and Other Anthracyclines in Cancers: Activity, Chemoresistance and Its Overcoming. *Mol. Asp. Med.* **2023**, *93*, 101205. [CrossRef]
21. Imstepf, S.; Pierroz, V.; Rubbiani, R.; Felber, M.; Fox, T.; Gasser, G.; Alberto, R. Organometallic Rhenium Complexes Divert Doxorubicin to the Mitochondria. *Angew. Chem. Int. Ed.* **2016**, *55*, 2792–2795. [CrossRef]
22. Cardinale, D.; Iacopo, F.; Cipolla, C.M. Cardiotoxicity of Anthracyclines. *Front. Cardiovasc. Med.* **2020**, *7*, 26. [CrossRef]
23. Gogas, H.; Mansi, J.L. New Drugs. The Anthrapyrazoles. *Cancer Treat. Rev.* **1996**, *21*, 541–552. [CrossRef] [PubMed]
24. Tan, J.H.; Zhang, Q.X.; Huang, Z.S.; Chen, Y.; Wang, X.D.; Gu, L.Q.; Wu, J.Y. Synthesis, DNA Binding and Cytotoxicity of New Pyrazole Emodin Derivatives. *Eur. J. Med. Chem.* **2006**, *41*, 1041–1047. [CrossRef] [PubMed]
25. Haiyang, H.; Morley, J.E.; Twamley, B.; Groeneman, R.H.; Bucar, D.K.; MacGillivray, L.R.; Benny, P.D. Investigation of the Coordination Interactions of S-(Pyridin-2-Ylmethyl)-L-Cysteine Ligands with $M(CO)_3^+$ (M = Re, ^{99m}Tc). *Inorg. Chem.* **2009**, *48*, 10625–10634.

26. Makris, G.; Radford, L.; Gallazzi, F.; Jurisson, S.; Hennkens, H.; Papagiannopoulou, D. Synthesis and Evaluation of fac-$[^{99m}Tc/Re(CO)_3]^+$ Complexes with a New (N,S,N) Bifunctional Chelating Agent: The First Example of a fac-$[Re(CO)_3$(N,S,N-Sst2-ANT)] Complex Bearing a Somatostatin Receptor Antagonist Peptide. *J. Organomet. Chem.* **2016**, *805*, 100–107. [CrossRef]
27. He, H.; Lipowska, M.; Xu, X.; Taylor, A.T.; Marzilli, L.G. Rhenium Analogues of Promising Renal Imaging Agents with a $\{^{99m}Tc(CO)_3\}^+$ Core Bound to Cysteine-Derived Dipeptides, Including Lanthionine. *Inorg. Chem.* **2007**, *46*, 590–596. [CrossRef]
28. Suárez-Ortiz, G.A.; Hernández-Correa, R.; Morales-Moreno, M.D.; Toscano, R.A.; Ramirez-Apan, M.T.; Hernandez-Garcia, A.; Amézquita-Valencia, M.; Araiza-Olivera, D. Diastereomeric Separation of Chiral Fac-Tricarbonyl (Iminopyridine) Rhenium (I) Complexes and Their Cytotoxicity Studies: Approach toward an Action Mechanism against Glioblastoma. *J. Med. Chem.* **2022**, *65*, 9281–9294. [CrossRef]
29. Song, X.; Lim, M.H.; Mohamed, D.K.B.; Wong, S.M.; Zhao, J.; Hor, T.S.A. Re (I) Carbonyl Complexes Containing Pyridyl-Imine and Amine Ligands: Synthesis, Characterization and Their Catalytic Olefin Epoxidation Activities. *J. Organomet. Chem.* **2016**, *814*, 1–7. [CrossRef]
30. Saund, S.S.; Siegler, M.A.; Thoi, V.S. Electrochemical Degradation of a Dicationic Rhenium Complex via Hoffman-Type Elimination. *Inorg. Chem.* **2021**, *60*, 13011–13020. [CrossRef]
31. Liu, W.; Heinze, K. Rhenium (I) and Platinum (II) Complexes with Diimine Ligands Bearing Acidic Phenol Substituents: Hydrogen-Bonding, Acid–Base Chemistry and Optical Properties. *Dalton Trans.* **2010**, *39*, 9554–9564. [CrossRef] [PubMed]
32. Wang, W.; Spingler, B.; Alberto, R. Reactivity of 2-Pyridine–Aldehyde and 2-Acetyl–Pyridine Coordinated to $[Re(CO)_3]^+$ with Alcohols and Amines: Metal Mediated Schiff Base Formation and Dimerization. *Inorganica Chim. Acta* **2003**, *355*, 386–393. [CrossRef]
33. Ketkaew, R.; Tantirungrotechai, Y.; Harding, P.; Chastanet, G.; Guionneau, P.; Marchivie, M.; Harding, D.J. OctaDist: A Tool for Calculating Distortion Parameters in Spin Crossover and Coordination Complexes. *Dalton Trans.* **2021**, *50*, 1086–1096. [CrossRef]
34. Wolfe, A.; Shimer, G.; Meehan, T. Polycyclic Aromatic Hydrocarbons Physically Intercalate into Duplex Regions of Denatured DNA. *Biochemistry* **1987**, *26*, 6392–6396. [CrossRef] [PubMed]
35. Zhao, G.; Lin, H.; Zhu, S.; Sun, H.; Chen, Y. Dinuclear Palladium (II) Complexes Containing Two Monofunctional [Pd(En)(Pyridine)Cl]$^+$ Units Bridged by Se or S. Synthesis, Characterization, Cytotoxicity and Kinetic Studies of DNA-Binding. *J. Inorg. Biochem.* **1998**, *70*, 219–226. [CrossRef]
36. Luis García-Giménez, J.; González-Álvarez, M.; Liu-González, M.; Macías, B.; Borrás, J.; Alzuet, G. Toward the Development of Metal-Based Synthetic Nucleases: DNA Binding and Oxidative DNA Cleavage of a Mixed Copper (II) Complex with N-(9H-Purin-6-Yl) Benzenesulfonamide and 1,10-Phenantroline. Antitumor Activity in Human Caco-2 Cells and Jurkat T Lymphocy. *J. Inorg. Biochem.* **2009**, *103*, 923–934. [CrossRef]
37. Lakowicz, J.R. *Principles of Fluorescence Spectroscopy*, 3rd ed.; Plenum Press: New York, NY, USA, 2006.
38. Wilson, W.D.; Ratmeyer, L.; Zhao, M.; Strekowski, L.; Boykin, D. The Search for Structure-Specific Nucleic Acid-Interactive Drugs: Effects of Compound Structure on RNA versus DNA Interaction Strength. *Biochemistry* **1993**, *32*, 4098–4104. [CrossRef] [PubMed]
39. Heller, D.P.; Greenstock, C.L. Fluorescence Lifetime Analysis of DNA Intercalated Ethidium Bromide and Quenching by Free Dye. *Biophys. Chem.* **1994**, *50*, 305–312. [CrossRef]
40. Sharma, A.; Özayral, S.; Caserto, J.S.; ten Cate, R.; Anders, N.M.; Barnett, J.D.; Kandala, S.K.; Henderson, E.; Stewart, J.; Liapi, E.; et al. Increased Uptake of Doxorubicin by Cells Undergoing Heat Stress Does Not Explain Its Synergistic Cytotoxicity with Hyperthermia. *Int. J. Hyperth.* **2019**, *36*, 712–720. [CrossRef]
41. Shimolina, L.E.; Gulin, A.A.; Paez-Perez, M.; López-Duarte, I.; Druzhkova, I.N.; Lukina, M.M.; Gubina, M.V.; Brooks, N.J.; Zagaynova, E.V.; Kuimova, M.K.; et al. Mapping Cisplatin-Induced Viscosity Alterations in Cancer Cells Using Molecular Rotor and Fluorescence Lifetime Imaging Microscopy. *J. Biomed. Opt.* **2020**, *25*, 126004. [CrossRef]
42. Showalter, H.D.; Johnson, J.L.; Hoftiezer, J.M.; Turner, W.R.; Werbel, L.M.; Leopold, W.R.; Shillis, J.L.; Jackson, R.C.; Elslager, E.F. Anthrapyrazole Anticancer Agents. Synthesis and Structure-Activity Relationships against Murine Leukemias. *J. Med. Chem.* **1987**, *30*, 121–131. [CrossRef] [PubMed]
43. Kostakis, I.K.; Magiatis, P.; Pouli, N.; Marakos, P.; Skaltsounis, A.L.; Pratsinis, H.; Léonce, S.; Pierré, A. Design, Synthesis, and Antiproliferative Activity of Some New Pyrazole-Fused Amino Derivatives of the Pyranoxanthenone, Pyranothioxanthenone, and Pyranoacridone Ring Systems: A New Class of Cytotoxic Agents. *J. Med. Chem.* **2002**, *45*, 2599–2609. [CrossRef] [PubMed]
44. Zee-Cheng, R.K.Y.; Podrebarac, E.G.; Menon, C.S.; Cheng, C.C. Structural Modification Study of Bis (Substituted Aminoalkylamino) Anthraquinones. An Evaluation of the Relationship of the [2-[(2-Hydroxyethyl)Amino]Ethyl] Amino Side Chain with Antineoplastic Activity. *J. Med. Chem.* **1979**, *22*, 501–505. [CrossRef]
45. Papagiannopoulou, D. Technetium-99m Radiochemistry for Pharmaceutical Applications. *J. Label. Comp. Radiopharm.* **2017**, *60*, 502–520. [CrossRef] [PubMed]
46. Schmidt, S.; Trogler, W.; Basolo, F. Pentacarbonyl Rhenium Halides. *Inorg. Synth.* **1985**, *23*, 41.
47. Nitschke, J.; Schmidt, S.P.; Trogler, W.C. Properties of (Trifluoromethanesulfonato) Pentacarbonylmanganese (I) and Rhenium (I). Reactions in Superacid Solvents. *Inorg. Chem.* **1985**, *24*, 1972–1978. [CrossRef]
48. Alberto, R.; Schibli, R.; Egli, A.; Schubiger, A.P.; Abram, U.; Kaden, T.A. A Novel Organometallic Aqua Complex of Technetium for the Labeling of Biomolecules: Synthesis of $[^{99m}Tc(OH_2)_3(CO)_3]^+$ from $[^{99m}TcO_4]^-$ in Aqueous Solution and Its Reaction with a Bifunctional Ligand. *J. Am. Chem. Soc.* **1998**, *120*, 7987–7988. [CrossRef]

49. Marmur, J. A Procedure for the Isolation of Deoxyribonucleic Acid from Micro-Organisms. *J. Mol. Biol.* **1961**, *3*, 208-IN1. [CrossRef]
50. Reichmann, M.E.; Rice, S.; Thomas, C.; Doty, P. A Further Examination of the Molecular Weight and Size of Desoxypentose Nucleic Acid. *J. Am. Chem. Soc.* **1954**, *76*, 3047–3053. [CrossRef]
51. Policar, C.; Lambert, F.; Cesario, M.; Morgenstern-Badarau, I. An Inorganic Helix [Mn(IPG)(MeOH)]*n*[PF6]*n*[‡]: Structural and Magnetic Properties of a Syn-Anti Carboxylate-Bridged Manganese (II) Chain Involving a Tetradentate Ligand. *Eur. J. Inorg. Chem.* **1999**, *1999*, 2201–2207. [CrossRef]
52. McInally, C. *Apex2*; Bruker Analytical X-Ray Systems, Inc.: Madison, WI, USA, 2006; p. M86-E01078.
53. Sheldrick, G.M. *SADABS: Area-Detector Absorption Correction*; Siemens Industrial Automation, Inc.: Madison, WI, USA, 1996.
54. Palatinus, L.; Chapuis, G. SUPERFLIP—A computer program for the solution of crystal structures by charge flipping in arbitrary dimensions. *J. Appl. Crystallogr.* **2007**, *40*, 786–790. [CrossRef]
55. Betteridge, P.W.; Carruthers, J.R.; Cooper, R.I.; Prout, K.; Watkin, D.J. CRYSTALS version 12: Software for guided crystal structure analysis. *J. Appl. Crystallogr.* **2003**, *36*, 1487. [CrossRef]
56. Watkin, D. The Control of Difficult Refinements. *Acta Crystallogr. Sect. A* **1994**, *50*, 411–437. [CrossRef]
57. Prince, E. *Mathematical Techniques in Crystallography and Materials Science*; Springer: Berlin/Heidelberg, Germany, 1982.
58. Watkin, D.J.; Prout, C.K.; Pearce, L.J. *CAMERON, Chemical Crystallography Laboratory*; University of Oxford: Oxford, UK, 1996.
59. De Meulenaer, J.; Tompa, H. The Absorption Correction in Crystal Structure Analysis. *Acta Crystallogr.* **1965**, *19*, 1014–1018. [CrossRef]

inorganics

Article

Accomplishment of α-Chymotrypsin on Photodynamic Effect of Octa-Substituted Zn(II)- and Ga(III)-Phthalocyanines against Melanoma Cells

Vanya Mantareva [1,*], Diana Braikova [1], Neli Vilhelmova-Ilieva [2], Ivan Angelov [1] and Ivan Iliev [3,4]

1 Institute of Organic Chemistry with Centre of Phytochemistry, Bulgarian Academy of Sciences, 1113 Sofia, Bulgaria
2 The Stephan Angeloff Institute of Microbiology, Bulgarian Academy of Sciences, Bl. 26, 1113 Sofia, Bulgaria
3 Institute of Experimental Morphology, Pathology and Anthropology with Museum, Bulgarian Academy of Sciences, 1113 Sofia, Bulgaria
4 Department of Biotechnology, University of Chemical Technology and Metallurgy, 8 Kliment Ohridski Blvd., 1756 Sofia, Bulgaria
* Correspondence: mantareva@yahoo.com; Tel.: +359-9606-181

Abstract: Octa-methylpyridiloxy-substituted Zn(II)- and Ga(III)-phthalocyanines (**ZnPc1** and **GaPc1**) were studied on human pigmented melanoma (SH4) and keratinocyte (HaCaT) cell lines. The efficacy of **ZnPc1** and **GaPc1** against melanoma cells was compared to the results in the presence of a proteaseα-chymotrypsin (ChT). The synthesis and characterization of compounds were carried out using well-known approaches. The formation of physical conjugates due to the addition of ChT was studied via absorption and fluorescence. The proteolytic activity of ChT was verified with casein as a substrate. The photosafety of compounds was proven on embryonal cells (BALB 3T3) under solar exposure (LED 360–1100 nm). The photodynamic activity of **GaPc1** and **ZnPc1** was studied for a concentration range of irradiation (LED 660 nm). The reduction of the proteolytic activity of ChT was observed only for the irradiation of **ZnPc1** or **GaPc1**. **GaPc1** and ChT and their conjugates, except **ZnPc1** (PIF ~6), were evaluated as photo-safe to solar light (PIF < 2). The efficiency of **GaPc1** was shown to be much higher than that of **ZnPc1** in their individual applications. The phototherapeutic index of **GaPc1** (PI = 1.71) on SH4 cells was higher for the conjugate. α-Chymotrypsin and phthalocyanine have the advantages of reducing high toxicity and increasing the phototherapeutic index.

Keywords: zinc and gallium complexes; phthalocyanines; α-Chymotrypsin; pigmented melanoma; keratinocytes; photodynamic therapy

1. Introduction

Melanoma continues to be a serious threat to human health. Despite the scientific advancements in chemical science and the current pharmaceutical novelties, existing melanoma treatments still have many unfavorable side effects and low efficiency [1]. Photodynamic therapy (PDT) has emerged as a tolerable clinical procedure for localized tumors [2]. The method is based on the production of cytotoxic oxygen species in the spot of irradiation on the tumor area, avoiding the cytotoxic effect on surrounding healthy cells and the whole body. PDT involves the administration of a photosensitizer (PS), its accumulation in the malignant cells, and irradiation with light of specific wavelengths and properties to activate the PS in a triplet excited state molecule that produces a variety of reactive oxygen species (ROS) that induce cell death [3]. The potential of PDT against tumors with no other alternative treatment options is well admitted [4]. Bioactive antibody molecules, drug-delivery biomolecules, and other tumor-specific biomolecules are well known for targeting PDT [5].

Phthalocyanine complexes (MPcs) are heterocyclic macromolecules that are identifiable as second-generation PSs for PDT because of their unique photo-properties that can be modified by their easy-to-tailor structure, which facilitates the achievement of compatible properties for PDT [6]. Despite their promising photochemistry, the MPc-photosensitizers feature several complications for clinical applications, such as undesirable dark toxicity, slow body clearance, and a lack of water solubility, which necessitate further studies for improvement [7].

Proteolytic enzymes take part in the chemical reaction of hydrolysis of peptide bonds and in bio-synthesis and are used for cell detachment in practice. Serine proteases, among which is α-chymotrypsin (ChT), have been evaluated for their promising potential as anticancer agents [8]. A recent study reported that nano-particles with a protease can be powerful medical tools to slow chemoresistance [8]. It was found that therapy that includes proteases may improve drug bioavailability, reduce circulation time, and increase the selective uptake in tumors [9]. These necessities arise due to the need for lower drug doses for body protection, a lack of side effects, and a better quality of life [10]. The human body produces proteins and enzymes that may serve as natural drug vehicles in circulation, as well as for specific carriers of anticancer drugs [11]. A promising approach based on the different mechanisms of action of different molecular species for targeted PDT was reported [12]. Another option is the possibility of obtaining quenched PS-conjugates to restore fluorescence emission and activate the generation of ROS after protease digestion [13]. The role of protease activity was exploited in drug delivery for the site-specific release of chemo-drugs at the tumor and their rapid activation by the target protease [14]. The results of preclinical studies with protease-sensitive prodrugs or fluorescence imaging agents suggested the prospect for the next clinical estimation of site-selective prodrugs or activators [15]. In addition to the wide-ranging usage of enzymes in proteins and cell research, they are well-accepted as catalysts in clinical diagnostics and routine therapeutic practice [16].

Enzyme-activated PDT drugs were evaluated without dependence on the enzyme function because of the main mechanism of singlet oxygen generation that directly destroys malignant cells [17]. Generally, proteases are part of a wide variety of diseases, and their importance to sickness pathology makes them excellent therapeutic targets. An initial idea of protease-sensitive prodrugs, also referred to as PDT action, has the key goal of increasing PS uptake [18]. The conjugation was shown to lower the fluorescence quantum yield, triplet state quantum yield, and ability for photosensitivity [19]. This study reported that enzymes reduced the singlet oxygen quantum yield of pheophorbides (from 0.52 to 0.17 and 0.04). Another study reported the development of a strategy termed protease-mediated PDT (PM-PDT) with an activation step from proteases on PS and a positive effect on PDT due to proteases' catalytic activity [20]. Some examples are serine endoproteinases, trypsin, and chymotrypsin, which were studied and found to have promising therapeutic efficacy against cancer [21]. Among the proteolytic enzymes, only ChT in combination with trypsin has been approved as a medication for clinical usage since the early 1960s [22]. Proteases are plentiful and sufficient for several diseases, so these inhibitory therapies are often used in cure regimes together with other cancer therapies. Chen and co-authors [23] offered a first illustration of synergistically destroying the extracellular matrix (ECM) by combining digestive enzymes with the generation of reactive oxygen species. This approach of combining ChT as an enzyme with oxygen species generated during PDT showed a tendency to improve the uptake and penetration of drugs into tumors. The acidity of tumors, plus the sufficient singlet oxygen due to PDT, can enhance their efficacy because of the enzymes that catalyze protein decomposition. Thus, it has resulted in effective damage to the tumor matrix and increased the therapeutic effect. The incorporation of proteases permitted a better permeation of the drug in cancer tissues and the ability for more efficient singlet oxygen saturation than for the free-state molecule [24].

The present study aims to investigate the influence of the protease α-Chymotrypsin on the PDT efficacy of two phthalocyanine complexes (**ZnPc1** and **GaPc1**) against human pigmented melanoma (SH-4) in comparison to normal keratinocyte cells (HaCaT). The effect of the macromolecule of ChT on phthalocyanines' photophysical properties was studied in a buffer solution (pH 7.8). The photosafety of compounds was analyzed on a fibroblast cell line (BALB/3T3) under artificial solar light exposure. The proteolytic activity of ChT was tested in a prior in vitro study, and the inhibition ability of the complexes (**ZnPc1** and **GaPc1**) was proven after irradiation with LED 660 nm and UV 254 nm spectra. The PDT efficiency of both phthalocyanines applied alone and with the addition of the protease chymotrypsin was studied in a concentration-dependent and comparison-based manner for a set of light parameters.

2. Results

2.1. Synthesis

Zn(II)-phthalocyanine (**ZnPc1**) with eight peripheral methylpyridiloxy substitution groups was successfully synthesized following a slightly modified well-known procedure summarized in Scheme 1. The previous synthetic procedure of cyclotetramerization was applied [25,26]. The synthetic procedure starts from commercial dinitrile **1**. It includes a nucleophilic displacement reaction to obtain dinitrile **2**. Direct cyclotetramerization and a multiple-step procedure were used to obtain **ZnPc1**. A metal-free phthalocyanine is a favorable starting compound for phthalocyanine complexes and facilitates the production of high-purity complexes. The metalation was conducted with zinc acetate as a metal salt. The complex of gallium (**GaPc1**) was possible to obtain only at a high temperature (~220 °C) in the presence of a catalyst (DBU) starting from a monomer [27]. Both metal complexes (**ZnPc1** and **GaPc1**) were isolated from the reaction mixture by precipitation in hexane and washed with several solvents. The purification was carried out by column chromatography on silica gel with an eluent mixture of dichloromethane (DCM) and methanol (DCM:MeOH, 9:1). All compounds were characterized by analytical methods.

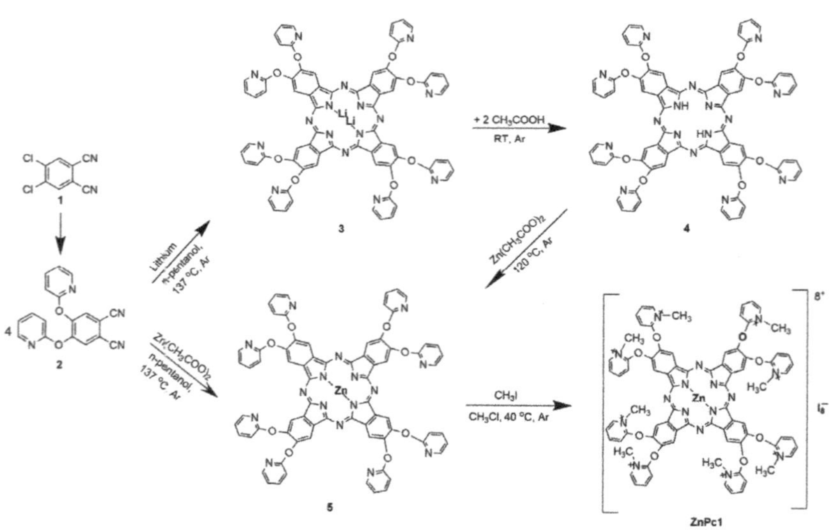

Scheme 1. Synthesis of octa-substituted Zn(II)-phthalocyanine (**ZnPc1**). The same pathway was applied for Ga(III)-phthalocyanine (**GaPc1**).

2.2. Spectrophotometric Study

Spectrophotometric studies of cationic complexes (**ZnPc1** and **GaPc1**) were performed in a comparison-based manner in phosphate buffer (PB, pH 7.8) and in DMSO solutions at room temperature Figure 1a). The absorption spectra in the visible region presented the typical Q-bands of the monomeric compound for **GaPc1** and the molecular associates for **ZnPc1** in water solutions. The maxima were recorded at 681 nm (DMSO) and 678 nm (PB) for **GaPc1**, assigned by π–π* transitions. The complex **ZnPc1** showed a Q-band at 676 nm (DMSO) and the split and a low-intensity Q-band (640 nm and 672 nm) in an aqueous solution due to aggregation. This observation is associated with the formation of J-aggregates of phthalocyanine molecules. A wide B-band was recorded between 328 nm and 395 nm for **ZnPc1** (DMSO), and there was a low-intensity vibrionic band in the visible region with a peak at 615 nm (**ZnPc1**) and 612 nm (**GaPc1**). The recorded Q-band maxima of both compounds showed that eight substituents to the Pc-ring molecule slightly shifted the Q-band to the red spectrum region. Both are relatively similar to the peaks of similar peripheral tetra-substituted Ga(III) or Zn(II) phthalocyanine complexes [28–30].

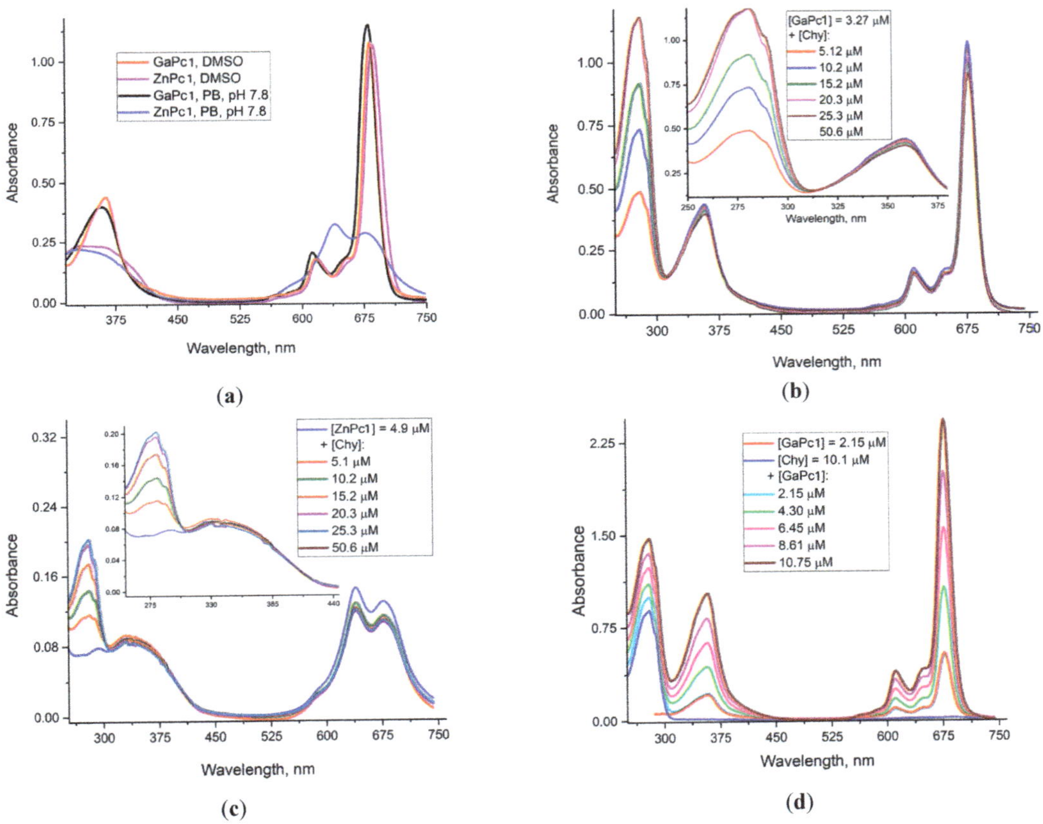

Figure 1. Absorption spectra of Zn(II)- and Ga(III)-phthalocyanines (**ZnPc1** and **GaPc1**) in dimethylsulfoxide (DMSO) and phosphate buffer, pH 7.8 (PB) (**a**); Changes in absorption spectra by addition of chymotrypsin (ChT) with isosbestic points in inset spectra (**b**) and the same measurements for **ZnPc1** (**c**), and the absorption spectra for the titration of ChT with **GaPc1** (**d**).

The absorption spectra were recorded during the titration of the solutions of the complexes (**ZnPc1** and **GaPc1**) in PB, pH 7.8 with ChT for a concentration range of 5.12–50.6 μM ChT (Figure 1b,c). The Q-band of the aggregated **ZnPc1** was not changed by an increase of ChT, suggesting that the compound was not changed. The spectra show two isosbestic points at 325 nm and 375 nm (**GaPc1**) and at 304 nm and 406 nm (**ZnPc1**), which are most likely due to the dilution for the spectra in the visible region and the increase in the UV-vis region due to additional aromatic amino acids (due to ChT). The addition of **GaPc1** in a concentration range (2.15–10.75 μM) to the ChT solution showed an increase in the maxima with no spectral changes (Figure 1d).

The fluorescence spectroscopy properties of **GaPc1** and **ZnPc1** were evaluated for both compounds as fluorescent molecules individually and with the addition of ChT to predict possible interactions (Figure 2a,b). The wide excitation band with a maximum of 280 nm was recorded for the ChT emission of 350 nm (Figure 2a, A). The high-intensity emission maximum of ChT was registered at 350 nm for excitation at 280 nm (Figure 2a, B). Both fluorescence bands were shown to increase the intensity with the addition of ChT (0.07–2.1 μM) because of the rise of the ChT concentration in the solution. The studies on the fluorescence emission of **GaPc1** at an excitation wavelength of 610 nm showed the overlapping of the fluorescence spectra (em: 693 nm) for concentrations over 10.2 μM due to the inner filter effect (not shown). The emission spectra of the mixtures containing both ChT and **GaPc1** at the excitation of 280 nm, which is typical for tryptophan (Trp) molecular residues, slightly increased the intensity of the emission band of **GaPc1** at 693 nm (Figure 2b). However, at the excitation of 365 nm with the addition of ChT (0.033–0.33 μM), the fluorescence intensity of **GaPc1** decreased due to dilution. Following the emission of ChT (347 nm), the fluorescence intensity increased with the splitting of the band in the same solutions. The experiment of **ZnPc1** titration with ChT was carried out for the excitation of 280 nm, and the recorded emission spectra of ChT showed an increase in the intensity of the emission band at 347 nm because of the ChT concentration without changes in the band shape. Octa-substituted **GaPc1** and **ZnPc1** (DMSO) were evaluated and found to have similar fluorescence quantum yields, which were lower than peripheral tetra-substituted complexes with similar substitution groups (0.21 and 0.23). This may be explained through the physical quenching due to the eight peripheral substitution groups to the ring molecule. In addition, the presence of ChT seems to lower the fluorescence of phthalocyanine because of the quenching by the ChT macromolecule. The physical interactions between **GaPc1** and ChT were shown by the quenching of **GaPc1** fluorescence. The excitation of ChT at 280 nm and also of **GaPc1** at 610 nm after the addition of ChT resulted in different spectra. This phenomenon may be explained by the energy transfer between ChT (exc: 280 nm; em: 347 nm), of which the emission may be absorbed by **GaPc1** and the fluorescence intensity at 693 nm may increase. The higher concentrations of **GaPc1** make changes to the ChT macromolecule, as seen by the emission spectra.

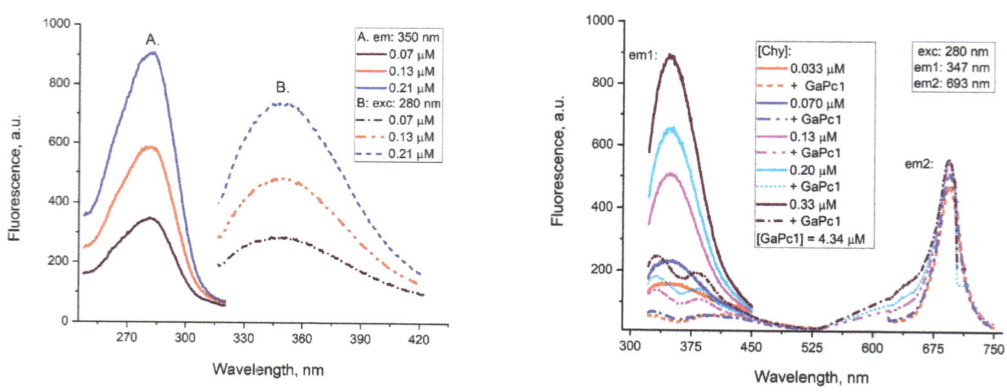

(a) (b)

Figure 2. Fluorescence excitation (A) and emission (B) spectra of α-chymotrypsin (ChT) at em/exc: 350 nm/280 nm (**a**) and the emission spectra of the conjugate (**GaPc1**+ChT) for a range of concentrations for ChT and **GaPc1**, at exc: 280 nm (**b**).

2.3. Proteolytic Activity

The proteolytic activity of serine protease ChT is an important property that needs to be purposely investigated [31]. Experiments were carried out before in some in vitro studies using casein as a substrate by absorption at 275 nm. In this experiment, it is important to know if the applied ChT is active in a range given by the supplier. Further experiments were carried out for the inhibition of ChT proteolytic activity with the addition of complexes (**ZnPc1** and **GaPc1**) kept in the dark and after irradiation with two light sources: UV-C (254 nm) and LED 660 nm (Figure 3). The choice of the light source was predicted by the absorption spectra of the target molecules, namely an enzyme (ChT) with absorbance mostly in the deep UV region and phthalocyanine complexes (**ZnPc1** and **GaPc1**) with intensive absorption bands with maximums around 675 nm (LED 660 nm has a spectrum with the maximum of the wavelength in half of the width of the Q-band).

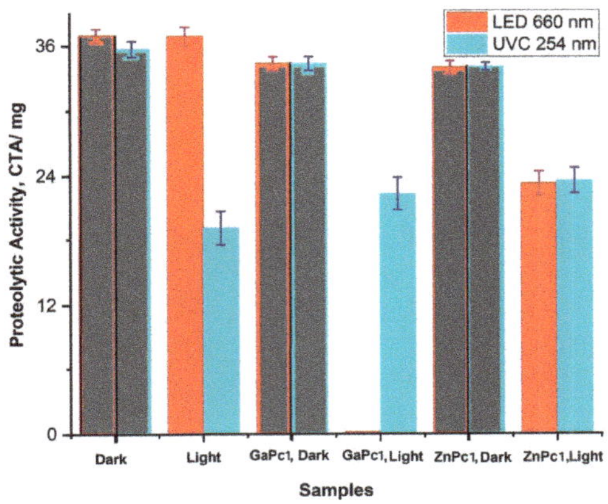

Figure 3. Proteolytic activity of α-chymotrypsin (ChT) studied by photosensitization with for Ga(III)- or Zn(II)-phthalocyanine complexes (**ZnPc1** and **GaPc1**) in the dark and by irradiation light-emitting diode (LED) at 660 nm or with UV-lamp (254 nm).

The inhibition of ChT activity because of photosensitization with complexes (**ZnPc1** and **GaPc1**) was observed. The assay was carried out in parallel for three control experimental groups: (1) only ChT, (2) added complexes (**ZnPc1** and **GaPc1**) in the dark, and (3) ChT irradiated with UV-C or LED 660 nm. The dark controls showed no loss of enzymic activity for the experiments that were carried out in atmospheric oxygen. The red-light exposure (LED 660 nm) showed no influence on inhibition of proteolytic activity. The typical photodynamic inhibition of ChT activity was observed for **GaPc1** during irradiation with LED 660 nm (full inhibition). Similar results were obtained for **GaPc1** and **ZnPc1** in the dark samples. Comparable inhibition was determined for **ZnPc1** at LED660 nm and under UVC exposure. The different inhibition ability of complexes (**ZnPc1** and **GaPc1**) refers to the difference in their photosensitization capability. However, both compounds showed a lower inhibition at UV-C irradiation than the control samples with UV-C alone (without phthalocyanine).

2.4. Photo-Safety Validation

The photosafety of octa-substituted phthalocyanines (**GaPc1** and **ZnPc1**) and the conjugates with an enzyme (**GaPc1**+ChT and **ZnPc1**+ChT) was studied on the model mouse embryo fibroblast normal cell line BALB 3T3. The results are presented by CC_{50} values ± SD (μM) at irradiation with a spectrum from a light-emitting diode (LED) known as a solar simulator with a light dose of 10 J/cm^2 (for details, see the Experimental section). The collected data were used to calculate parameters such as the photo-irritation factor (PIF), which for PIF < 2 means the lack of toxicity for **GaPc1** and prevalent phototoxicity with PIF > 5 for **ZnPc1**. Relatively high values of CC_{50}, such as 57.74 µM and 58.39 µM, and a factor PIF = 1.023 were obtained for **GaPc1**, suggesting that it is a very promising compound that is non-cytotoxic under solar light exposure. An increase in the PIF value means the lessening of the toxicity with the addition of ChT to **GaPc1** (0.626 vs. 1.023 for **GaPc1**) and **ZnPc1** (1.521 vs. 6.104), which is in advance of the high photosafety of compounds (Table S3 in Supplementary Materials).

The photosafety at solar light exposure for **GaPc1** was identical as for the sample of cells kept in the dark (Figure 4). The lack of cytotoxicity was observed for ChT in concentrations below 0.1 mg/mL (not shown). An increase in cytotoxicity on BALB 3T3 cells was determined by the addition of ChT (0.05%), with a big concentration gap between the samples with irradiation and in the dark (Figure 4a). The toxicity of the conjugate **ZnPc1**+ChT was almost similar to that of **ZnPc1** alone (Figure 4b). High phototoxicity was observed for concentrations over 1 µM **ZnPc1** and the addition of ChT (0.05%) increase the photosafety. The photosafety of ChT showed no influence on phototoxicity with PIF = 0.726 (Table S3).

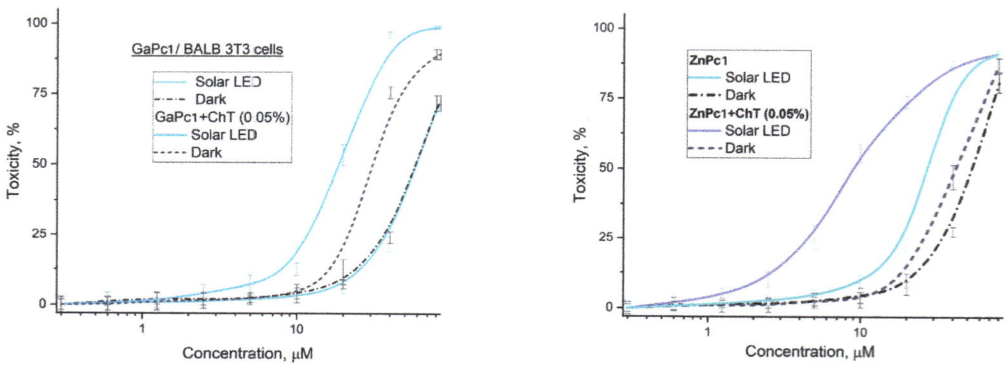

(a) (b)

Figure 4. Dark and photo-cytotoxicity presented as a percentage of cell death: (**a**) Ga(III)-phthalocyanine (**GaPc1**) and in the presence of α-chymotrypsin (**GaPc1**+ChT), and (**b**) Zn(II)-phthalocyanine (**ZnPc1**) and in the presence of α-chymotrypsin (**ZnPc1**+ChT) studied on embryonal cell line (BALB 3T3) at exposure with a solar light-emitting diode (LED 360–1100 nm, 10 J/cm^2). $p < 0.05$ for each point was considered significant.

The high cytotoxicity **ZnPc1** with PIF = 6.104 was observed to be lower with the addition of ChT (PIF = 1.152). The results showed that for concentrations between 0.001 µM and 80 µM, **GaPc1** has proper photosafety to solar light exposure (360–1100 nm). In addition, the obtained values suggested the positive low-toxicity effect of a proteolytic serine enzyme ChT as a biomolecule with an advance in photosafety on the cells.

2.5. Photodynamic Therapeutic Efficacy

The photodynamic efficacy of octa-substituted phthalocyanine complexes (**GaPc1** and **ZnPc1**) was studied on two skin-originated cell lines, namely a human pigmented melanoma cell line (SH-4) and keratinocyte cell line (HaCaT). The activity of compounds was evaluated after their individual application, as well as for their fresh mixtures with a proteolytic enzyme (ChT). The results are presented as concentration–cell toxicity curves. The dark and phototoxicity of **ZnPc1** and **GaPc1** were studied on SH-4 versus HaCaT (Figure 5). The toxicity curves for compounds applied individually showed no difference in the inactivation efficiency of both cell lines with better activity for **GaPc1** (Figure 5b). The difference in both toxicities and for both cell lines was observed with the addition of ChT for the conjugates **GaPc1**+ChT and **ZnPc1**+ChT (Figure 5a,b). The phototoxic effect was higher for **GaPc1** than for **ZnPc1** in the presence of ChT, with a difference in the effect between both cell lines. The cytotoxicity of both **GaPc1** and **ZnPc1** was shown to diminish in the presence of a serine protease (ChT).

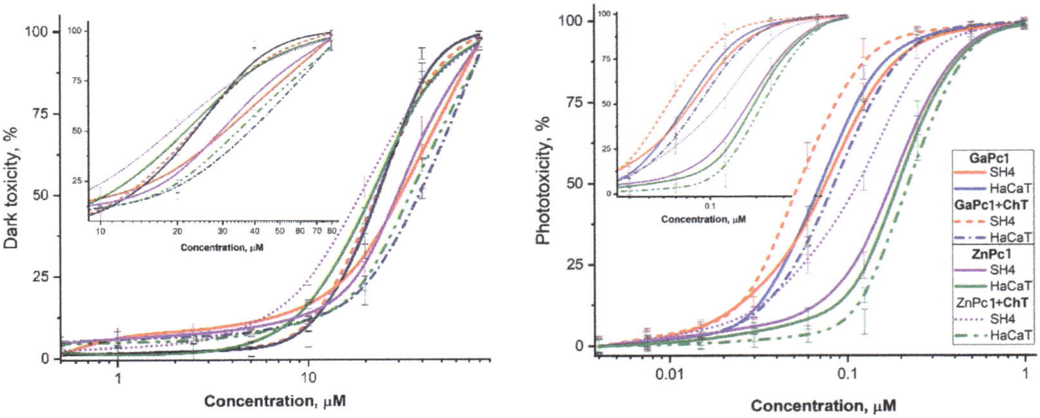

(**a**) (**b**)

Figure 5. Cytotoxicity of Ga(III)- and Zn(II)-phthalocyanines (**GaPc1** and **ZnPc1**) in the presence of chymotrypsin (ChT), (**a**) without light (dark toxicity) and (**b**) at irradiation with a light-emitting diode (LED) at 660 nm with a dose of 50 J/cm^2 (inset: the part of the toxicity curves). $p < 0.05$ for each point was considered significant.

In vitro PDT was carried out with **ZnPc1** or **GaPc1** with the enzyme ChT, and the results suggest the advantage of reducing their strong dark and photo-cytotoxicity effects on normal cell lines. This observation is likely due to the competition for the generated reactive species between ChT and the cells' components (lipids, proteins, and other biomolecules). The phototherapeutic index (PI) of the compounds in the presence of ChT is presented in Table S4. The phototherapeutic index (PI) was observed to have ratios of 1.3 and 1.27 for the normal cells treated with **ZnPc1** or **GaPc1** with the addition of ChT, suggesting the lower toxicity of the compounds on normal keratinocyte cells. The index PI was determined, with higher values for tumor cells for **GaPc1** (PI: 461 vs. 269 on SH4 cells) and similar values for **ZnPc1** (154 vs. 156).

3. Discussion

Octa-substituted phthalocyanines and gallium and zinc phthalocyanine complexes with eight methylpyridiloxy groups (**ZnPc1** and **GaPc1**) were synthesized by a well-developed synthetic procedure [26,27]. **ZnPc1** was prepared by including an additional step, as shown in Scheme 1. Both complexes were studied and compared for their efficiency in the method of photodynamic therapy (PDT). The water-soluble **GaPc1** exists as a monomolecular and photoactive in-phosphate buffer (PB) solution (Figure 1a). The interaction between the proteolytic enzyme α-chymotrypsin (ChT) and **ZnPc1** or **GaPc1** was assessed using molecular spectroscopy techniques. The changes in the absorption spectra of the resulting mixtures were recorded (Figure 1b,c). The increase in the absorption of the Q-band at 681 nm (**GaPc1**) indicated a rise in the **GaPc1** concentration. It may assume an electrostatic contact between the negative charges of ChT and the complex with eight positive charges (**GaPc1** or **ZnPc1**). This may lead to a high local concentration of molecules, which further creates favorable conditions for the stacking interaction between both molecules may lead to physical conjugates. The spectra of **ZnPc1** in buffer solution (PB) are typical for aggregated molecules (Figure 1d). However, the addition of ChT showed only a slight reduction in absorbance due to dilutions without monomerization in the PB solution. The increase in the emission band of ChT and, respectively, tryptophan residue (Trp) fluorescence was recorded for the solutions in buffer, pH 7.8 (Figure 2a). The intensity of the fluorescence of **GaPc1** was lessened by an increase in ChT concentration at a constant concentration of **GaPc1**, as was observed for both excitation wavelengths (exc: 365 and 610 nm). The intrinsic fluorescence of ChT, which has the characteristics of a globular protein containing eight Trp residues, mainly depends on the Trp residue because other amino acids such as phenylalanine (Phe) have a very low quantum yield, and the fluorescence of tyrosine (Tyr) is almost completely quenched by ionization, or by an amino or a carboxyl group, or a Trp. However, the presence of **GaPc1** in solution was shown to change the emission band (~350 nm) of Trp residue, resulting in a low intensity and the splitting of the emission band (Figure 2b). The fluorescence quenching of the ChT may be explained by the more polar location of Trp residues. One of the attributes of enzymes is their short lifetimes, which are used because of their sensibility to some changes in the media. Farhadian et al. [31] reported variations in the fluorescence intensity of ChT with dynamic quenching during the binding of spermine, a polyamine that is involved in cellular metabolism and is found in all eukaryotic cells. The phenomenon of phthalocyanines quenching enzymic activity was also observed for cholinesterase and amylase, as they are natural physiologically active biomolecules [32,33].

The present study of a photosensitizer (**GaPc1** or **ZnPc1**) and the red LED 660 nm treatment of an enzyme such as ChT showed a significant proteolytic inhibition of its activity. Irradiation with UV-C light was also observed to induce the lower activity of ChT by the Trp residues (Figure 3). The incubation of ChT with phthalocyanine showed a lack of inhibition of the proteolytic activity, which suggests that the compound is not connected to the active center. As is known, the serine proteases widely circulating in the human body have many important functions such as blood coagulation, tissue morphogenesis, cell death, swelling, and wound healing, and assimilation [34]. The changes in proteins because of the photosensitization reactions resulted in the degradation of histidine (His), methionine (Met), cysteine (Cys), Trp, and Tyr [35]. The photooxidation kinetics of His, Met, and Cys residues were found to be insensitive to the pH changes of media within the range of 6–8, which indicated that the photooxidation rates of proteins must be lower than the corresponding ones for the mixture of the individual amino acids. The sensitized photooxidation of ChT using perinaphthenone, rose bengal, and eosine with a high susceptibility of enzymes to photosensitization in water was reported.

A high photosafety due to the sunlight spectra of exposure was observed for **GaPc1** alone and in the presence of ChT on the tested model embryonal cell line BALB 3T3 (Figure 4a). It was determined that there was an increase in toxicity for **ZnPc1** used individually, which was much higher for **ZnPc1** and ChT (0.05%) for the range of concentrations (Figure 4b). The studies suggested an increase in toxicity due to solar light (LED 350–1100 nm). The phototoxicity to solar light and the calculated indexes PIF < 1 or ~1 suggest high photosafety, except for **ZnPc1** (Table S3).

The phototherapeutic effect of **GaPc1** and **ZnPc1** studied on two cell lines (SH-4 and HaCaT) showed relatively high efficiency for **GaPc1** compared to **ZnPc1** (Figure 5a,b). The gallium complex has higher activity than the zinc one, which may be related to the ionic radius of the gallium ion and the deformation of the phthalocyanine ring. The cytotoxic effect was lessened, suggesting a lower toxicity because of the addition of ChT. The positive effect of ChT is of importance, especially for the dark toxicity of phthalocyanines (Figure 5a). The phototoxicity of compounds (**ZnPc1** and **GaPc1**) was observed with a slight concentration gap between the curves by the addition of ChT in non-toxic concentrations (Figure 5b). These changes in the photo- and dark toxicity were seen for both cell lines (melanoma tumor and keratinocyte cells). The calculated index (PI) as an indicator of PDT efficiency was seen to increase by a double value for **GaPc1** in the presence of ChT (Table S4).

The photosensitizers **ZnPc1** and **GaPc1** showed differences in anticancer PDT activity on the melanoma cell line (SH-4) as well as with the addition of ChT (Figure 5b). The lowering of the PDT activity due to ChT may be a result of the quenching effect on the generated ROS from the enzyme (ChT) and probably the enzymic inhibition of the tumor mass growth. Thus, it might also influence the diminishment of the harsh toxicity of both photoactive compounds. A difference in the cells' toxicity was observed with the addition of ChT in a PDT study of **ZnPc1** and **GaPc1**. This could possibly be explained by the PAR1 overexpression typical for melanoma cells and probably the inhibition of catalytic activity of the ChT due to photosensitization with **GaPc1** or **ZnPc1** and LED 660 nm. The option of light screening due to the large molecular structure of ChT may also lead to limitations in the light absorption and lower phototoxicity.

A recent study on pancreatic enzymes showed that they weaken tumor growth by slicing and decomposing the binding proteins of cellular structures [36]. The knowledge that the tumor matrix of 3D collagen promotes the migration of tumor cells may be used for protease applications on tumors. In vitro studies in which the tumor microenvironment-associated structures are not present have also shown antitumor potential of pancreatic enzymes [37]. In this study, it was reported that proteases, like trypsin or α-chymotrypsin, cleave extracellular precursors, resulting in the generation of the active form of numerous proteins. For example, trypsin can cleave pro-insulin to generate active insulin through a proteolytic mechanism. The membrane receptors, known as proteinase-activated-receptors (PARs), have been described as potential targets for pancreatic proteolytic enzymes [38]. In the present study, ChT can cleave PAR1, which is overexpressed in many kinds of malignancies, including pigmented melanoma. Chymotrypsin could also make an inhibitory cleavage on PAR1, removing its tethered activating ligand and making the receptor insensitive to the action of other proteinases. The local character of the method plus the positive effect of the proteolytic enzymes such as ChT for anticancer therapy may have several constructive aspects for minimizing the undesirable effects of the PDT procedure. The promising effects were seen after combining the application of a proteolytic enzyme and PDT, then only the PDT. This study presents a promising start for future research on the dual macromolecular assembly of enzymes and photosensitizers in the direction of advanced PDT.

4. Materials and Methods

4.1. Phthalocyanines and Chemicals

Two octa-methylpyridiloxy-substituted phthalocyanine complexes of zinc and gallium (**ZnPc1** and **GaPc1**) were synthesized following a well-established procedure [25,26]. The dinitrile 4,5-dichloro-1,2-dicyanobenzene was dried at 80 °C in a glass oven before the reaction. A large-scale amount of 4,5-bis(pyridiloxy)-1,2-dicyanobenzene was prepared in an excess of dry potassium carbonate in dimethylformamide (DMF). The cyclotetramerization was carried out with $Zn(OAc)_2$ or $GaCl_3$ with addition of a catalyst 1,8-diazabicyclo[5.4.0]undec-7-ene (DBU) in dry 1-pentanol at reflux temperature for both phthalocyanine complexes before quaternization. Synthesis of **GaPc1** was carried out as was previously described [26]. Bovine α-chymotrypsin (ChT) was the salt-free enzyme supplied by a commercial source (FOT, Bulgaria). Casein acc. to HAMMARSTEN is a product of Sigma-Aldrich and Merck (FOT and Labimex, Bulgaria). The solids and other chemicals were used as supplied from different commercial vendors. The used solvents were additionally purified.

4.1.1. Synthesis of 2,3,9,10,16,17,23,24-Octakis-[(2-N-pyridiloxy) Phthalocyaninato] Zinc(II) Complex (**6**)

A mixture of 4,5-bis(pyridiloxy)-1,2-dicyanobenzene, 2 (400 mg, 1.27 mmol), DBU (1.7 mL, 1 mmol) and anhydrous $Zn(OAc)_2$ (0.179 g, 1 mmol) in n-pentanol (4 mL) was stirred at reflux (137 °C) for 6 h under argon. After cooling, the reaction mixture was poured into n-hexane (25 mL). The greenish precipitate was collected and washed with n-hexane. The crude product was purified by column chromatography (SiO_2) using CH_2Cl_2-MeOH (95:5). Yield: 0.67 g (48%). IR [ν_{max}/cm^{-1}]: 3052 (Ar-CH), 2930, 1657, 1624, 1583 (C=C),

1267, 1238, 1123, 1115 (C–O–C), 901, 865, 704. ^{1}H-NMR ($CDCl_3$): δ, ppm 7.84–7.21 (12H, m, Pc-H and Pyridyl-H), 6.80–5.90 (28H, m, Pc-H and benzene-H). Calc. for $C_{72}H_{40}N_{16}O_8Zn$; MS (DCI ng., NH_3, 20 mA/s): *m/z* (% intensity): 1322 (100 %, M^+), 1245 (7.5%, M^+-C_5H_3N), 1229 (17%, M^+-C_5H_3NO).

4.1.2. Synthesis of 2,3,9,10,16,17,23,24-Octakis-{[2-(N-methyl) Pyridiloxy] Phthalocyaninato- Zinc(II) Octaiodide (**ZnPc1**)

Phthalocyanine 4 (100 mg, 0.1mmol) was dissolved in dry DMF (1 mL) and methyl iodide (5 mL, 80 mmol) was added drop wisely. The reaction continued under argon for 16 h at 40 °C. The product was precipitated in hot acetone and collected by centrifugation. The green solid was washed with several solvents and dried over P_4O_{10}. Yield: 0.86 g (81%). UV/Vis (DMSO), λ_{max} nm (log ε): 351 (4.46), 615 (4.11), 678 (4.35). IR [ν_{max}/cm^{-1}]: 3014 (Ar-CH), 1655, 1633, 1585, 1540 (C=C), 1497, 1467, 1445, 1401, 1264, 1030 (C–O–C), 1169, 1135, 1023, 954; 867, 827, 746 (CH). ^{1}H-NMR (300 MHz, DMSO-d_6): δ, ppm 9.55 (s, 8H), 9.35 (s, 8H), 8.9 (d, *J* = 7.5 Hz, 8H), 8.65 (dd, *J* = 7.5, 1.5 Hz, 8H), 7.99–7.24 (m, 8H), 4.39 (s, 24H, CH_3). Calc. 1442.85 for $C_{80}H_{64}N_{16}O_8Zn$. ESI-MS (positive ion mode) *m/z*: 181.36 $[M]^{8+}$.

4.2. Proteolytic Activity Measurements

The proteolytic activity was determined by an enzyme assay in proteolytic units (CTA) using casein as a substrate. The method was first published by Johnson et al. [39] and further modified [40]. This method was chosen because of the specificity in the absorption spectra of the studied phthalocyanines. All assays were made in triplicate and the average of the measurements was taken to evaluate the activity units. One CTA-unit of enzyme caseinolytic activity corresponds to the enzyme amount that releases 1 μeq/min of tyrosin in the substrate (37 °C). The reaction mixtures of the samples consisted of 0.016 mg/mL α-chymotrypsin in 0.06 M Tris buffer, pH 7.4 and either 5 μM **GaPc1** or **ZnPc1** (control groups). The experimental samples of 2.5 mL volumes were irradiated with a UV-C lamp for 3 min and 15 min by stirring in air-open cuvettes at room temperature (T = 20 °C). The light-emitting diode (LED) 660 nm was applied with the doses (10 J/cm^2 and 50 J/cm^2) and did not alter the proteolytic activity of ChT.

4.3. Photo-Physicochemical Study

The absorption spectra were measured using a spectrophotometer Perkin Elmer Precisely Lambda 25 UV/Vis with quartz cuvettes with 1.0 cm at room temperature. The stock solutions of **GaPc1** and **ZnPc1** (~2 mM) in DMSO were used. The dilutions were made with phosphate buffer (PB) with concentration 0.01 M, pH 7.8. The absorption spectra were recorded for the full spectral range 190–750 nm. Absorbance of ChT was observed in the UV-C region with maxima (192 nm and 254 nm in PB (pH 7.8) for different concentrations. The absorption spectra of the mixture **GaPc1** and ChT were registered at the titration of a solutions of **GaPc1** (3.27 μM or 4.34 μM) or **ZnPc1** (4.89 μM) in PB with ChT (5.12–50.6 μM) or the opposite titration, which involved the addition of **GaPc1** (2.15–10.75 μM) to a solution of ChT (10.1 μM). Fluorescence studies were carried out using Perkin-Elmer LS 55 (Switzerland). The emission spectra were recorded for wavelengths of excitation at 280 nm of tryptophan (ChT) and at 610 nm for phthalocyanines following the previous study [41].

4.4. Cell Cultures

Three cell lines were used in this study: a human skin melanoma cell line SH-4 (ATCC® CRL-7724™), a mouse embryonal fibroblast BALB/c 3T3 clone A31 (ATCC® CCL-163TM), which are products of the American Type Cultures Collection (ATCC, Virginia, USA), and a human keratinocyte cell line HaCaT (CLS, cat. № 300493) from CLS Cell Lines Service GmbH (CLS, Eppelheim, Germany). The cultivation of the cells was conducted in 25 cm^2 and 75 cm^2 tissue culture flasks in DMEM-high glucose (4.5 g/L), 10% FBS, 2 mM glutamine and antibiotics (penicillin 100 U/mL and streptomycin 100 μg/mL) at 37 °C, 5%

CO_2 and 90% relative humidity. Suspensions of cells were plated in a 96-well microtiter plate (1×10^4 cells/100 µL/well) and were cultured for 24 h.

4.5. Light Sources

Two different light sources were used in the present studies. Both light emitted diodes (LED) have different spectra and powers of exposure specific to the devices used in the photosafety experiment and the PDT study. The so-called solar simulator has spectra that cover the sunlight. This is a commercially available LED Helios-iO from SERIC Ltd. (Tokyo, Japan). The fluence rate was constant as measured with a power-meter PM 100D with a sensor S120VC (Thorlabs Inc., North Newton, KS, USA). The working dimension between 0.05–50 mW was within the spectrum 360–1100 nm. The intensity was achieved for a distance of 20 cm distance with a normal diffusion in the experimental zone of 1.16 W/m^2. A second light source is a LED at 660 nm was configured by ELO Ltd. (Bulgaria) with the fixed power density of 100 mW/cm^2, which was experimentally observed as an optimal fluence rate for in vitro PDT experiments.

4.6. Phototoxicity Study

The toxicity studies were performed after serial dilutions of **GaPc1** and **ZnPc1** from the stocks in DMSO. The incubations were made in the culture medium to the working concentrations between 0.0025–80 µM. The enzyme (ChT) used for the experiment was dissolved in phosphate buffered saline (PBS), pH 7.4, at a concentration of 50 mg/mL used as a stock solution. **GaPc1, ZnPc1**, and ChT were applied freshly prepared in the culture medium. In vitro tests were carried out on the cells in their exponential stage of grown. After trypsinization the cells were brought to the required cell density in each of the 96-well plates (1×10^4 cells/well). The NRU-assay was applied to register the results after treatments. The evaluation of the cells' viability in vitro was carried out. The used experimental protocols followed the known procedure. Briefly, it includes the standard conditions for the cell incubation for 24 h to reach a good adhesion. The cells were incubated after addition of double rise of the testing concentrations for the compound (**ZnPc1, GaPc1**, ChT and mixtures with ChT). The culture medium was replaced, and the cells were washed before the experiments. The study was performed in parallel plates at the same time. One plate was kept in the dark place to have the dark control in equal conditions as the light exposure for comparative assessment of photo and dark cytotoxicity. The second plate was irradiated with LED 660 nm with light dose of 50 J/cm^2 and 24 h post-irradiation, the medium was replaced with the medium containing the NR dye. Three hours later, the wells were washed with phosphate-buffered saline (PBS, pH 7.4), and a mixture of distilled water/ethanol/acetic acid in a ration 50:49:1 was supplemented. The optical density was measured on a TECAN microplate reader (λ = 570 nm). The cellular toxicity was calculated by Equation (1):

$$\text{Cytotoxicity (\%)} = \{1 - (OD_{570}(\text{treated sample})/OD_{570}(\text{negative control})\} \times 100 \quad (1)$$

Other parameters calculated based on the photo- and dark cytotoxicity studies are the phototherapeutic index (PI), toxicity in the dark (IC_{50}), after-irradiation photocytotoxicity (IC_{50}), and photo-irritation factor (PIF), calculated using the Equation (2):

$$PIF = IC_{50}(\text{Dark})/IC_{50}(\text{LED 360-1100 nm}) \quad (2)$$

where Dark is the value in the absence of light and Light is the same after solar LED 360–1100 nm irradiation.

Phototherapeutic index (PI) is defined as the ratio of dark to light IC_{50} values, and it is used to determine the light-induced effectiveness. The phototherapeutic index was defined as the dark IC_{50} value divided by the light IC_{50} value and calculated using Equation (3).

$$PI = IC_{50}(\text{Dark})/IC_{50}(\text{LED 660 nm}) \quad (3)$$

where Dark is the value in the absence of light and Light is after LED 660 nm irradiation. The ratio between the IC_{50} value (half maximal inhibitory concentration) of the resting and this value of the activated compound must be as high as possible.

4.7. Statistics

The experiments were carried out in triplicate. The data are presented as a mean value ± standard deviation (SD) and the difference between two values was compared by an unpaired Student's test. The values of $p < 0.05$ were considered significant.

5. Conclusions

Octa-methylpyridiloxy-substituted Ga(III)- and Zn(II)-phthalocyanines (**GaPc1** and **ZnPc1**) were prepared and studied for PDT activity on melanoma cells and keratinocyte cells. The results suggest a higher photoactivity for **GaPc1** than for **ZnPc1** on the melanoma tumor cell line (SH-4) by individual application which may be because of the size of the coordinated atom of gallium in the macrocycle. The phototoxicity was shown to be similar for both tested cell lines (SH-4 and HaCaT), and the addition of ChT resulted in the changes between the dark toxicity and the phototoxicity between both model cell lines. The addition of a serine protease α-chymotrypsin to the studied phthalocyanines (**ZnPc1** and **GaPc1**) suggested its potential positive influence on PDT, such as reducing dark toxicity and improving therapeutic response.

Supplementary Materials: The following supporting information can be downloaded at: https://www.mdpi.com/article/10.3390/inorganics12080204/s1, Methods, Instrumentation, Solubility of phthalocyanines, Singlet oxygen quantum yield study, Cytotoxicity of Ga(III)- and Zn(II)-phthalocyanines and α-chymotrypsin, Analyses (1H-NMR, MS) [42–45].

Author Contributions: Conceptualization, V.M.; methodology, V.M., D.B., N.V.-I., I.A. and I.I.; software, V.M., I.A. and I.I.; validation, V.M., N.V.-I., I.A. and I.I.; investigation, V.M., D.B., N.V.-I. and I.I.; data curation, V.M., D.B., N.V.-I. and I.I.; writing—original draft preparation, V.M.; writing—review and editing, V.M. and I.I.; supervision, V.M. and I.I.; funding acquisition, I.A. and I.I. All authors have read and agreed to the published version of the manuscript.

Funding: This research was partly funded by the Grant of European Union-Next Generation EU, through the National Recovery and Resilience Plan of the Republic of Bulgaria, project No. BG-RRP-2.004-0002, "BiOrgaMCT" and the Bulgarian National Science Fund (project КП-06-Н38/13/2019).

Institutional Review Board Statement: Not applicable.

Informed Consent Statement: Not applicable.

Data Availability Statement: The new data are available at request of the corresponding author.

Acknowledgments: Thanks to the European Union Next Generation EU, for the National Recovery and Resilience Plan of the Republic of Bulgaria (project No. BG-RRP-2.004-0002, "BiOrgaMCT") and the Bulgarian National Science Fund (project КП-06-Н38/13/2019) for supporting our research studies.

Conflicts of Interest: The authors declare no conflicts of interest. The founders had no role in the design of the study; in the analyses, or interpretation of data; in the writing of the manuscript; or in the decision to publish the results.

References

1. Lopes, J.; Rodrigues, C.M.P.; Gaspar, M.M.; Reis, C.P. How to Treat Melanoma? The Current Status of Innovative Nanotechnological Strategies and the Role of Minimally Invasive Approaches like PTT and PDT. *Pharmaceutics* **2022**, *14*, 1817. [CrossRef] [PubMed]
2. Dos Santos, A.F.; De Almeida, D.R.Q.; Terra, L.F.; Baptista, M.S.; Labriola, L. Photodynamic therapy in cancer treatment—An update review. *J. Cancer Metastasis Treat.* **2019**, *5*, 25. [CrossRef]
3. Bacellar, I.; Tsubone, T.; Pavani, C.; Baptista, M. Photodynamic Efficiency: From Molecular Photochemistry to Cell Death. *Int. J. Mol. Sci.* **2015**, *16*, 20523–20559. [CrossRef]

4. Gunaydin, G.; Gedik, M.E.; Ayan, S. Photodynamic Therapy for the Treatment and Diagnosis of Cancer–A Review of the Current Clinical Status. *Front. Chem.* **2021**, *9*, 686303. [CrossRef] [PubMed]
5. Crous, A.; Chizenga, E.; Hodgkinson, N.; Abrahamse, H. Targeted Photodynamic Therapy: A Novel Approach to Abolition of Human Cancer Stem Cells. *Int. J. Opt.* **2018**, *2018*, 1–9. [CrossRef]
6. Ömeroğlu, İ.; Durmuş, M. Water-soluble phthalocyanine photosensitizers for photodynamic therapy. *Turk. J. Chem.* **2023**, *47*, 837–863. [CrossRef]
7. Güzel, E.; Koçyigit, Ü.M.; Taslimi, P.; Erkan, S.; Taskin, O.S. Biologically active phthalocyanine metal complexes: Preparation, evaluation of α-glycosidase and anticholinesterase enzyme inhibition activities, and molecular docking studies. *J. Biochem. Mol. Toxicol.* **2021**, *35*, 22765. [CrossRef] [PubMed]
8. González-Titos, A.; Hernández-Camarero, P.; Barungi, S.; Marchal, J.A.; Kenyon, J.; Perán, M. Trypsinogen and chymotrypsinogen: Potent anti-tumor agents. *Expert Opin. Biol. Ther.* **2021**, *21*, 1609–1621. [CrossRef] [PubMed]
9. Blagosklonny, M.V. Selective protection of normal cells from chemotherapy, while killing drug-resistant cancer cells. *Oncotarget* **2023**, *14*, 193–206. [CrossRef]
10. Rudzińska, M.; Daglioglu, C.; Savvateeva, L.V.; Kaci, F.N.; Antoine, R.; Zamyatnin Jr, A.A. Current Status and Perspectives of Protease Inhibitors and Their Combination with Nanosized Drug Delivery Systems for Targeted Cancer Therapy. *Drug Des. Devel. Ther.* **2021**, *15*, 9–20. [CrossRef]
11. Gierlich, P.; Mata, A.I.; Donohoe, C.; Brito, R.M.M.; Senge, M.O.; Gomes-da-Silva, L.C. Ligand-Targeted Delivery of Photosensitizers for Cancer Treatment. *Molecules* **2020**, *25*, 5317. [CrossRef] [PubMed]
12. Alvarez, N.; Sevilla, A. Current Advances in Photodynamic Therapy (PDT) and the Future Potential of PDT-Combinatorial Cancer Therapies. *Int. J. Mol. Sci.* **2024**, *25*, 1023. [CrossRef] [PubMed]
13. Liu, B.; Bian, Y.; Yuan, M.; Zhu, Y.; Liu, S.; Ding, H.; Gai, S.; Yang, P.; Cheng, Z.; Lin, J. L-buthionine sulfoximine encapsulated hollow calcium peroxide as a chloroperoxidase nanocarrier for enhanced enzyme dynamic therapy. *Biomaterials* **2022**, *289*, 121746. [CrossRef] [PubMed]
14. Zuluaga, M.F.; Gabriel, D.; Lange, N. Enhanced prostate cancer targeting by modified protease sensitive photosensitizer prodrugs. *Mol. Pharm.* **2012**, *9*, 1570–1579. [CrossRef] [PubMed]
15. Choi, K.Y.; Swierczewska, M.; Lee, S.; Chen, X. Protease-activated drug development. *Theranostics* **2012**, *2*, 156–178. [CrossRef]
16. Qin, X.; Wu, C.; Niu, D.; Qin, L.; Wang, X.; Wang, Q.; Li, Y. Peroxisome inspired hybrid enzyme nanogels for chemodynamic and photodynamic therapy. *Nat. Commun.* **2021**, *12*, 5243–5315. [CrossRef]
17. López-Otín, C.; Bond, J.S. Proteases: Multifunctional enzymes in life and disease. *J. Biol. Chem.* **2008**, *283*, 30433–30437. [CrossRef] [PubMed]
18. Chen, J.; Stefflova, K.; Niedre, M.J.; Wilson, B.C.; Chance, B.; Glickson, J.D.; Zheng, G. Protease-triggered photosensitizing beacon based on singlet oxygen quenching and activation. *J. Am. Chem Soc.* **2004**, *126*, 11450–11451. [CrossRef] [PubMed]
19. Hackbarth, S.; Ermilov, E.A.; Röder, B. Interaction of Pheophorbide a molecule covalently linked to DAB dendrimers. *Opt. Commun.* **2005**, *248*, 295–306. [CrossRef]
20. Choi, Y.; Weissleder, R.; Tung, C.-H. Selective antitumor effect of novel protease-mediated photodynamic agent. *Cancer Res.* **2006**, *66*, 7225–7229. [CrossRef]
21. Leipner, J.; Saller, R. Systemic enzyme therapy in oncology: Effect and mode of action. *Drugs* **2000**, *59*, 769–780. [CrossRef] [PubMed]
22. Chandanwale, A.; Langade, D.; Sonawane, D.; Gavai, P. A Randomized, Clinical Trial to Evaluate Efficacy and Tolerability of Trypsin: Chymotrypsin as Compared to Serratiopeptidase and Trypsin: Bromelain: Rutoside in Wound Management. *Adv. Ther.* **2017**, *34*, 180–198. [CrossRef] [PubMed]
23. Chen, X.; Zhang, X.; Wu, Y.; Chen, Y.; Guo, Y.; Jana, D.; Wang, D.; Yuan, W.; Zhao, Y. Tumor extracellular matrix-targeted nanoscavengers reverse suppressive microenvironment for cocktail therapy. *Mater. Today* **2022**, *61*, 78–90. [CrossRef]
24. Jiao, J.; He, J.; Li, M.; Yang, J.; Yang, H.; Wang, X.; Yang, S. A porphyrin-based metallacage for enhanced photodynamic therapy. *Nanoscale* **2022**, *14*, 6373–6383. [CrossRef] [PubMed]
25. Wöhrle, D.; Eskes, M.; Shigehara, K.; Yamada, A. A Simple Synthesis of 4,5-Disubstituted 1,2-Dicyanobenzenes and 2,3,9,10,16,17,23,24- Octasubstituted Phthalocyanines. *Synthesis* **1993**, *1993*, 194–196. [CrossRef]
26. Mantareva, V.; Kussovski, V.; Angelov, I.; Wöhrle, D.; Dimitrov, R.; Popova, E.; Dimitrov, S. Non-aggregated Ga(III)-phthalocyanines in the photodynamic inactivation of planktonic and biofilm cultures of pathogenic microorganisms. *Photochem. Photobiol. Sci.* **2011**, *10*, 91–102. [CrossRef] [PubMed]
27. Mantareva, V.; Iliev, I.; Sulikovska, I.; Durmuş, M.; Genova, T. Collagen Hydrolysate Effects on Photodynamic Efficiency of Gallium (III) Phthalocyanine on Pigmented Melanoma Cells. *Gels* **2023**, *9*, 475. [CrossRef] [PubMed]
28. Angelov, I.; Mantareva, V.; Kussovski, V.; Wohrle, D.; Borisova, E.; Avramov, L. Improved antimicrobial therapy with cationic tetra- and octa-substituted phthalocyanines. In Proceedings of the 15th International School on Quantum Electronics: Laser Physics and Applications, Bourgas, Bulgaria, 15–19 September 2008; SPIE: Philadelphia, PA, USA, 2008; Volume 7027, pp. 378–385.
29. Mantareva, V.; Angelov, I.; Kussovski, V.; Wöhrle, D.; Dimitrov, S. Metallophthalocyanines as photodynamic sensitizers for treatment of pathogenic bacteria: Synthesis and singlet oxygen formation. *Comp. Rend. Acad. Bulg. Sci.* **2009**, *62*, 1521–1526.
30. Mack, J.; Kobayashi, N.; Stillman, M.J. Re-examination of the emission properties of alkoxy- and thioalkyl-substituted phthalocyanines. *J. Inorg. Biochem.* **2010**, *104*, 310–317. [CrossRef]

31. Farhadian, S.; Shareghi, B.; Saboury, A.A. Exploring the thermal stability and activity of α-chymotrypsin in the presence of spermine. *J. Biomol. Struct. Dyn.* **2016**, *35*, 435–448. [CrossRef]
32. Barut, B.; Demirbas, Ü. Synthesis, anti-cholinesterease, α-glucosidase inhibitory, antioxidant and DNA nuclease properties of non-peripheral triclosan substituted metal-free, copper (II), and nickel (II) phthalocyanines. *J. Organomet. Chem.* **2020**, *23*, 121423. [CrossRef]
33. Patel, S. A critical review on serine protease: Key immune manipulator and pathology mediator. *Allergol. Immunopathol.* **2017**, *45*, 579–591. [CrossRef] [PubMed]
34. Antalis, T.M.; Buzza, M.S. Extracellular: Plasma Membrane Proteases—Serine Proteases. *Encycl. Cell Biol.* **2016**, 650–660. [CrossRef]
35. Vizovisek, M.; Ristanovic, D.; Menghini, S.; Christiansen, M.G.; Schuerle, S. The Tumor Proteolytic Landscape: A Challenging Frontier in Cancer Diagnosis and Therapy. *Int. J. Mol. Sci.* **2021**, *22*, 2514. [CrossRef] [PubMed]
36. Biasutti, M.A.; Posadaz, A.; Garcıa, N.A. A comparative kinetic study on the singlet molecular oxygen-mediated photoxidation of a- and b-chymotrypsins. *J. Peptide Res.* **2003**, *62*, 11–18. [CrossRef] [PubMed]
37. Altrogge, L.M.; Monard, D. An assay for high-sensitivity detection of thrombin activity and determination of proteases activating or inactivating protease-activated receptors. *Anal. Biochem.* **2000**, *277*, 33–45. [CrossRef] [PubMed]
38. Jiang, P.; De Li, S.; Li, Z.G.; Zhu, Y.C.; Yi, X.J.; Li, S.M. The expression of protease-activated receptors in esophageal carcinoma cells: The relationship between changes in gene expression and cell proliferation, apoptosis in vitro and growing ability in vivo. *Cancer Cell Int.* **2018**, *18*, 81. [CrossRef]
39. Johnson, A.J.; Kline, D.L.; Alkjaersig, N. Assay methods and standard preparations for plasmin, plasminogen and urokinase in purified systems. *Thromb. Diath. Haemorrh.* **1969**, *21*, 259–272. [PubMed]
40. Gousterova, A.; Goshev, I.; Christov, P.; Tsvetkova, R.; Nedkov, P. Characterisation of Collagenolytic Enzymes Produced by Thermophylic Actinomycetes. *Biotechnol. Biotechnol. Equip.* **2003**, *17*, 81–8632. [CrossRef]
41. Hamad, O.A.; Kareem, R.O.; Omer, P.K. Recent Developments in Synthesize, Properties, Characterization, and Application of Phthalocyanine and Metal Phthalocyanine. *J. Chem. Rev.* **2024**, *6*, 39–75.
42. Ogunsipe, A.; Chen, J.-Y.; Nyokong, T. Photophysical and photochemical studies of zinc(II) phthalocyanine derivatives—Effects of substituents and solvents. *New J. Chem.* **2004**, *28*, 822–827. [CrossRef]
43. Gürel, E.; Pişkin, M.; Altun, S.; Odabaş, Z.; Durmuş, M. The novel mesityloxy substituted metallo-phthalocyanine dyes with long fluorescence lifetimes and high singlet oxygen quantum yields. *J. Photochem. Photobiol. A Chem.* **2016**, *315*, 42–51. [CrossRef]
44. Brannon, J.H.; Madge, D. Picosecond laser photophysics. Group 3A phthalocyanines. *J. Am. Chem. Soc.* **1980**, *102*, 62–65. [CrossRef]
45. Ogunsipe, A.; Nyokong, T. Photophysical and photochemical studies of sulphonated non-transition metal phthalocyanines in aqueous and non-aqueous media. *J. Photochem. Photobiol. A Chem.* **2005**, *173*, 211–220. [CrossRef]

inorganics

MDPI

Review

Noble Metal Complexes in Cancer Therapy: Unlocking Redox Potential for Next-Gen Treatments

Alina Stefanache [1], Alina Monica Miftode [1], Marcu Constantin [2,*], Roxana Elena Bogdan Goroftei [2,*], Iulia Olaru [2,†], Cristian Gutu [2,†], Alexandra Vornicu [3] and Ionut Iulian Lungu [1]

1 Faculty of Pharmacy, "Grigore T. Popa" University of Medicine and Pharmacy, 700115 Iasi, Romania; alina.stefanache@umfiasi.ro (A.S.)
2 Faculty of Medicine and Pharmacy, University "Dunarea de Jos", 47 Domneasca Str., 800008 Galati, Romania
3 Faculty of Medicine, "Grigore T. Popa" University of Medicine and Pharmacy, 700115 Iasi, Romania
* Correspondence: marcu.saar@yahoo.de (M.C.); roxana.goroftei@ugal.ro (R.E.B.G.)
† These authors contributed equally to this work.

Abstract: **(1) Context**: Cancer is still a major problem worldwide, and traditional therapies like radiation and chemotherapy often fail to alleviate symptoms because of side effects, systemic toxicity, and mechanisms of resistance. Beneficial anticancer effects that spare healthy tissues are made possible by the distinctive redox characteristics of noble metal complexes, especially those containing palladium, gold, silver, and platinum. **(2) Methods**: The redox processes, molecular targets, and therapeutic uses of noble metal complexes in cancer have been the subject of much study over the last 20 years; novel approaches to ligand design, functionalization of nanoparticles, and tumor-specific drug delivery systems are highlighted. **(3) Results**: Recent developments include Pt(IV) prodrugs and terpyridine-modified Pt complexes for enhanced selectivity and decreased toxicity; platinum complexes, like cisplatin, trigger reactive oxygen species (ROS) production and DNA damage. Functionalized gold nanoparticles (AuNPs) improve targeted delivery and theranostic capabilities, while gold complexes, particularly Au(I) and Au(III), inhibit redox-sensitive processes such as thioredoxin reductase (TrxR). **(4) Conclusions**: Ag(I)-based compounds and nanoparticles (AgNPs) induce DNA damage and mitochondrial dysfunction by taking advantage of oxidative stress. As redox-based anticancer medicines, noble metal complexes have the ability to transform by taking advantage of certain biochemical features to treat cancer more effectively and selectively.

Keywords: cancer; noble metals; redox; theranostics; nanoparticles

1. Introduction

Cancer remains a major global health challenge, with approximately 19.3 million new cases and 10 million deaths reported around the globe in 2020 alone. Even with improvements in early detection and treatment, the complexity and adaptability of cancer cells often renders traditional therapies, like chemotherapy and radiation, insufficient and in vain. Resistance mechanisms also present a current threat; besides off-target effects and systemic toxicity, these dangers require the development of novel therapeutic strategies that can specifically target cancer cells while sparing healthy tissues [1–4].

Metal-containing compounds have been crucial in the development of cancer therapies over time. The identification of cisplatin during the 1960s highlighted a change in a fresh period of treatment methods that capitalized on the distinct characteristics of metal compounds. Cisplatin's effectiveness prompted investigations into metal-based medications, focusing notably on noble metals such as platinum, gold, silver, and palladium. These metals have chemical and physical traits, such as electron density and different oxidation states, that allow them to bind with different ligands effectively. These qualities make them well-suited for interacting with biomolecules and influencing the environments within cells [5–8].

The significance of metal complexes redox activity is notably important (Figure 1) in the field of oncology because cancer cells often experience increased stress compared to healthy cells as a result of their rapid growth and metabolic irregularities. This susceptibility can be taken advantage of by metal complexes that can either enhance stress levels or interfere with redox sensitive pathways to trigger cell death. For instance, platinum complexes create ROS that harm DNA and proteins, while gold complexes impede TrxR, an enzyme that maintains redox balance, in cancer cells [9–11].

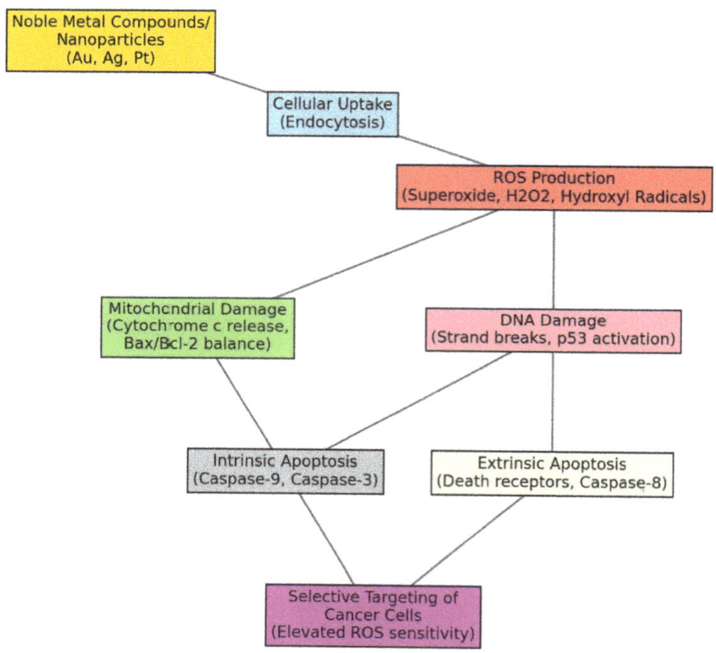

Figure 1. The pathways noble metal compounds follow to induce apoptosis in cancer cells.

This review explores the multifaceted roles of noble metal complexes in cancer therapy, focusing on their redox mechanisms, molecular targets, and clinical implications. By synthesizing insights from recent studies, we aim to provide a comprehensive understanding of their therapeutic potential and inspire further innovations in this promising field.

2. Results and Discussions

2.1. Platinum Complexes

Medications containing platinum, like cisplatin, have reinvented cancer therapy by utilizing their ability to bind to DNA and induce cellular apoptosis. This mechanism involves forming covalent bonds with the purine bases, resulting in intra-strand and inter-strand cross-links that disrupt DNA replication and transcription, leading to cell death (Figure 2). These interactions aid multiple apoptotic mechanisms, such as p53-mediated cell death and deactivation of mitochondrial activity. Besides DNA binding, Pt complexes are powerful redox agents, amplifying oxidative stress by generating ROS, which further damage cellular macromolecules and enhance cytotoxicity [12–14].

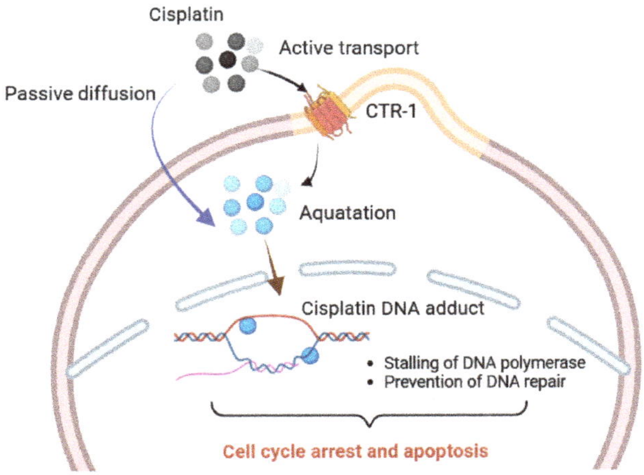

Figure 2. The steps of cisplatin-induced apoptosis in cancer cells explained by intracellular transport, aquatation in the cytoplasm, and further formation of adducts with DNA, inducing apoptosis.

Cummings et al. (2009) showed that cisplatin therapy increased ROS levels in ovarian cancer cells, which in turn led to mitochondrial membrane depolarization and activation of caspase. Similarly, other studies on oxaliplatin have highlighted its ability to produce oxidative DNA damage in colorectal cancer cells, while carboplatin exhibited a more minimized redox potency due to its slower reactivity [15]. Recent research has been focusing on designing new Pt complexes (Table 1) that exploit cancer-specific oxidative weaknesses. For instance, Pt(IV) prodrugs are reduced intracellularly to their active Pt(II) form, achieving targeted ROS generation in the hypoxic tumor environment [16–19].

Table 1. Pt-based anticancer medication currently for therapy.

Platinum Complex	Primary Target	Redox Mechanism	References
Cisplatin	DNA (guanine residues)	ROS generation, DNA damage	[12,20–22]
Oxaliplatin	DNA, ROS pathways	Oxidative stress, apoptosis	[23–27]
Pt(IV) prodrugs	Tumor-specific reduction	Redox-sensitive reduction, ROS activation	[28,29]

Pt complexes, depicted in Figure 3, are evolving at a fast pace to address the constant challenges of resistance and toxicity, which subdue their clinical efficiency, even though they are widely used. The design of these complexes is constantly updated with a focus on modifying ligands to improve pharmacokinetics, selectivity, and therapeutic effectiveness. For example, incorporating sizeable terpyridine groups has been shown to enhance the stability and cellular uptake of Pt complexes, while hydrophilic pairings improve their solubility, facilitating better in vivo biodistribution [30,31].

A study by Zhang et al. (2017) demonstrated that non-traditional ligands, like N-heterocyclic carbene (NHC) or terpyridine derivatives, offered improved redox activity and cytotoxic effects against cancer cell lines that showed resistance to cisplatin. These ligands stabilized the metallic center and improved interactions with specific biomolecules (DNA and proteins), thereby enhancing the therapeutic potential of Pt complexes. Complexes having these ligands exhibited selective toxicity, sparing normal cells and reducing side effects [32].

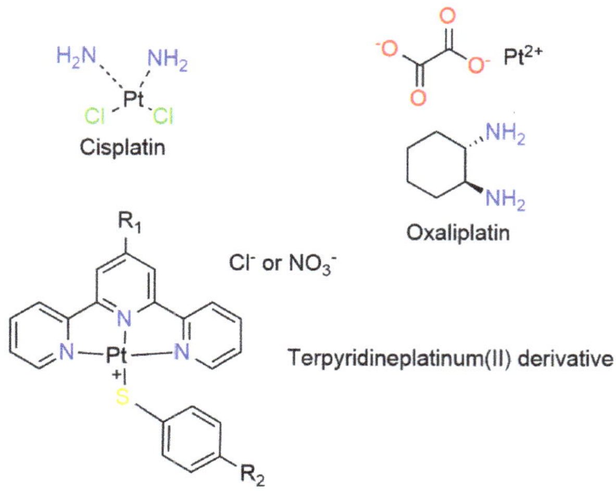

Figure 3. Pt(II) and (IV) complexes with potent redox effects on cancer cells.

Pt(II) complexes having terpyridine ligands showed particular and special cytotoxic activity thanks to their ability to intercalate DNA and inhibit enzymes like topoisomerase, which are crucial for DNA replication and repair. Lo et al. (2009) showed that complexes like these are effective at inhibiting mammalian topoisomerase II and TrxR, both of which play key roles in cancer cell proliferation and redox homeostasis. By targeting these enzymes, these Pt complexes disrupted redox balance in cancerous cells leading to apoptosis [33].

Other new and improved approaches to overcoming resistance and enhancing drug delivery are the development of Pt nanocarriers functionalized with targeting ligands. These nanocarriers increase the enhanced permeability and retention effect to accumulate selectively within tumor tissue. Modified with targeting groups such as peptides, antibodies, or small molecules, these structures ensure precise delivery of redox-active Pt medicine to cancerous cells while saving healthy tissues [34,35].

Recent research explored the potential use of Pt-terpyridine nanocarriers combined with NHC ligands, showcasing better and improved antiproliferative activity in vivo subjects with reduced systemic toxicity. This dual-ligand approach enhanced both the stability and specificity of the drug towards cancer cells. Additionally, the inclusion of nanocarriers allowed for a better and controlled drug release, ensuring that the active compound was delivered at optimal concentrations within the tumor's environment [36].

2.2. Gold Complexes: Targeting Redox Enzymes and Pathways

Gold (Au) complexes (Figure 4), in particular in the +1 and +3 oxidation states, have attracted focus thanks to their ability to disrupt TrxR, an important enzyme in maintaining redox balance. Cancerous cells secrete excess TrxR to neutralize ROS and sustain their rapid growth. Au(I) complexes, such as auranofin, selectively bind to the selenocysteine part in TrxR, permanently inhibiting its activity and inducing apoptosis (Figure 5) [37,38].

Lu et al. (2022) highlighted the enhanced efficiency of Au complexes in targeting redox-sensitive signaling pathways, leading to synergistic effects with traditional chemotherapies. The researchers demonstrated that Au(I) complexes attached to AuNPs via redox-sensitive linkers significantly improved the targeting of cancer cells. The ROS-rich environment in tumor tissues facilitated the release of the Au(I) complexes, leading to enhanced cytotoxicity and apoptosis induction. The multipotential nature of AuNPs further enhances their therapeutic efficacy. Beyond drug delivery, AuNPs can be synthesized to include imaging agents, allowing for simultaneous diagnostics and treatment, also known as theranostics.

For instance, AuNPs functionalized with fluorescent markers or radiolabels enable real-time monitoring of drug delivery and tumor progression. Such integration of therapy and diagnostics provides a powerful tool for personalized cancer treatment [39].

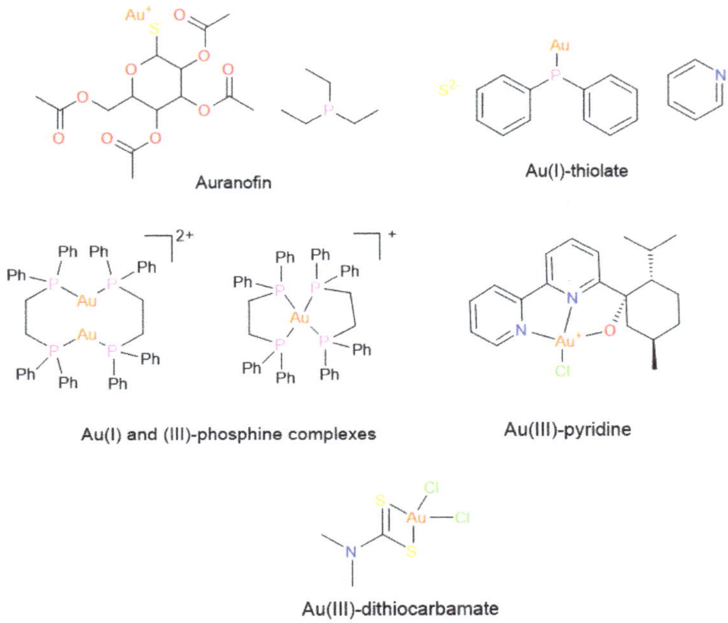

Figure 4. Au(I) and (III) complexes reported in the literature for their effects as possible anticancer drugs.

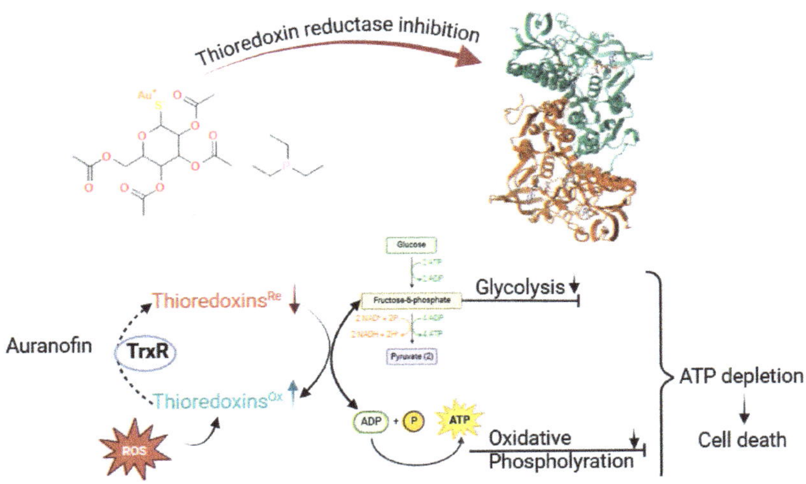

Figure 5. Mechanism of action of auranofin and gold complexes by interfering with TrxR and destabilizing ATP levels, causing cell death.

Au(III) complexes (Table 2) show additional redox properties due to their higher oxidation state, enabling them to interact with biomolecules like DNA, proteins, and ROS-generating enzymes. Thota et al. (2018) demonstrated that Au(III) complexes functionalized with phosphine ligands exhibit strong cytotoxicity in breast and prostate cancer models, primarily through oxidative mitochondrial damage [40].

Table 2. Au-based drugs with high redox potential.

Gold Complex	Mechanism	Key Findings	References
Auranofin	TrxR inhibition	ROS generation, selective redox disruption, apoptosis	[41,42]
Au (III)-phosphine	ROS generation, DNA binding	Mitochondrial dysfunction, DNA binding, apoptosis	[43]
Au(I)-thiolates	ROS generation, mitochondrial damage	ROS-induced mitochondrial damage, apoptosis	[44,45]
Au(I)-NHC complexes	Redox cycling, DNA intercalation	ROS generation, selective cytotoxicity, apoptosis	[46]
Au(III)-pyridine	DNA interaction, oxidative stress	DNA intercalation, ROS-induced apoptosis	[47,48]
Au(III)-dithiocarbamates	Metal ion exchange, ROS generation	ROS-induced DNA damage, apoptosis	[49–51]

Mechanisms targeting cancer cell redox vulnerabilities show that gold complexes have great anticancer potential. The ROS levels are dramatically increased by Au(III) complexes, reaching a maximum of 200% over baseline in sensitive cancer lines after six hours of treatment. Proteins and mitochondrial membranes are immediately damaged by this ROS influx, which leads to depolarization and functional loss in cells. Further disturbance of the redox balance may occur via the suppression of TrxR. At nanomolar doses (IC_{50}: 1–10 nM), the enzyme's activity is reduced by 85–95% when Au(III) complexes bind to selenocysteine and cysteine residues inside its active site. Apoptosis and mitochondrial cytochrome c release are both aided by the quick reduction in decreased thioredoxin brought about by this inhibition [52].

Even in cisplatin-resistant mice, structural investigations reveal that Au(I) complexes, especially those with phosphine or N-heterocyclic carbene ligands, exert strong TrxR inhibition with IC_{50} values in the 100–300 nM range. Their one-of-a-kind ligand architectures make them more selective against cancer cells while reducing collateral damage to healthy organs [53]. According to Tolbatov et al. (2024), these complexes cause a dose-dependent increase in ROS, leading to a two- to threefold rise in colon and breast cancer cells after one day of exposure. In response to this elevated oxidative stress, stress-activated MAPKs, including JNK and p38, phosphorylate and activate factors that promote cell death. The subsequent activation of caspase-3 causes cancer cells to undergo cell death [54]. The effectiveness of gold complexes is generally related to their electrical characteristics, as highlighted in a study by De Franco et al. (2022). This is because nucleophilic amino acids bind well to electron-deficient gold centers, which enhances the mechanistic precision of the complexes [55].

AuNPs have revolutionized the therapeutic opportunities of Au-based anticancer medicine, offering a platform for precise and targeted drug delivery systems. Their nanoscale size, high biocompatibility, and ability to be functionalized with a variety of ligands and biomolecules make them uniquely suited for cancer therapy. Functionalization with redox-sensitive outer layers exploits the oxidative stress and acidic environment commonly found in tumors, ensuring the controlled release of the therapeutic dose directly at the tumor site. One of the key innovations in AuNP-based systems is their ability to respond to the unique redox conditions of cancer cells. The tumor microenvironment typically exhibits elevated levels of ROS compared to normal tissues. Redox-sensitive coatings, such as disulfide linkages or thiol-based ligands, undergo cleavage in this oxidative environment, triggering the release of conjugated anticancer agents. This selective release reduces off-target effects and enhances the therapeutic index [56–58].

Additionally, AuNPs themselves exhibit intrinsic therapeutic properties; they have been shown to disrupt cellular redox homeostasis by interacting with thiol-containing proteins and enzymes, further enhancing oxidative stress within cancerous cells. This dual-action mechanism, combining the therapeutic effects of both the nanoparticle and the conjugated drug, maximizes the anticancer potential [59,60].

2.3. Silver Complexes: Exploiting Oxidative Stress

Silver complexes (Figure 6) containing Ag(I) exhibit unique redox properties that contribute to their anticancer potential; these complexes can interact with biomolecules such as DNA, proteins, and lipids, leading to cellular dysfunction and later apoptosis. The ability of silver to generate ROS is especially crucial in cancer therapy, where high oxidative stress is a hallmark of the disease. By further enhancing this stress, silver compounds selectively induce cell death in cancer cells while sparing normal, healthy cells [61–63].

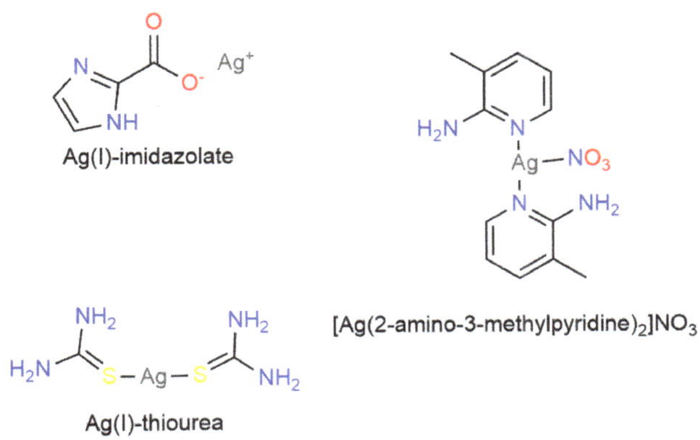

Figure 6. Ag(I) complexes used to induce oxidative stress in tumor cells.

One important mechanism (1) by which silver complexes exert their anticancer activity is the generation of ROS. Silver ions, especially in the +1-oxidation state, have high reactivity and can interact with cellular components to produce ROS, including hydrogen peroxide (H_2O_2), superoxide anions (O_2^-), and hydroxyl radicals (OH•). These ROS initiate oxidative damage to critical cellular structures, including DNA, proteins, and lipids, disrupting cellular functions and metabolism and contributing to apoptosis [64–66].

$$\begin{gathered} Ag^+ \text{ (outside)} \rightarrow Ag^+ \text{ (inside)} \\ Ag^+ + R\text{-}SH \rightarrow R\text{-}S\text{-}Ag + H^+ \\ O_2 + e^- \rightarrow O_2\bullet^- \\ 2\,O_2\bullet^- + 2\,H^+ \rightarrow H_2O_2 + O_2 \\ H_2O_2 + Fe^{2+} \rightarrow \bullet OH + OH^- + Fe^{3+} \\ \text{(Due to ROS)} \rightarrow \text{Cytochrome c} \rightarrow \text{Caspase activation} \\ \text{Caspase-9} \rightarrow \text{Caspase-3} \rightarrow \text{Apoptosis} \end{gathered} \tag{1}$$

Silver complexes, like Ag(I)-imidazolate, have been shown to induce significant cytotoxic effects in vitro by promoting oxidative DNA damage. Ag(I) ions can bind to the phosphate backbone of DNA (DNA-P), causing single- and double-strand breaks. Silver ions can also interact with sulfur-containing amino acids in proteins, disrupting cellular functions, reactions represented in (2). This accumulation of oxidative stress ultimately leads to apoptosis, which is particularly beneficial for targeting cancer cells [67–70].

$$\begin{gathered}
Ag^{+} + \text{DNA-P} \rightarrow \text{DNA-P-Ag} \\
\text{DNA-P-Ag} \rightarrow \text{DNA-P}' + Ag^{+} + \text{free radicals (single-strand break in DNA)} \\
\text{DNA-P}' + \text{DNA-P} \rightarrow \text{DNA-P-P}' + Ag^{+} \text{ (double-strand break in DNA)} \\
Ag^{+} + \text{R-SH} \rightarrow \text{R-S-Ag} + H^{+} \\
\text{Oxidative Stress} \rightarrow \text{Cellular Damage} \\
\text{Oxidative Damage} \rightarrow \text{Caspase Activation} \rightarrow \text{Apoptosis}
\end{gathered} \tag{2}$$

Through pathways including oxidative stress and disturbance of protein homeostasis, silver complexes have strong anticancer action. Research emphasizes that compounds containing Ag(I) suppress the growth of cancer cells with IC_{50} values ranging from 1 to 10 μM. These complexes trigger cell death by interfering with mitochondrial activity and increasing ROS production. The effect of Ag(I) complexes on cellular energy consumption was confirmed when, after 24 h of exposure to dosages as low as 5 μM, mitochondrial membrane potential dropped by 40–50%. Additionally, by taking advantage of the heightened susceptibility of cancers to ROS, silver complexes preferentially harm cancer cells while avoiding healthy ones [71].

Research by Fabbrini et al. (2019) showed that anticancer effectiveness is enhanced by proteasome inhibition in Ag(I) complexes including N-heterocyclic ligands. Ovarian cancer cells showed IC_{50} values ranging from 1 to 4 μM in Ag(I)-NHC complexes, according to in vitro investigations. At these dosages, proton pump inhibitor-like activity in the proteasome was reduced by 70–80% after 12 h of therapy. Further reducing cancer cell viability, these complexes induce endoplasmic reticulum stress and polyubiquitinated protein accumulation. The results from these investigations show that silver complexes work at low micromolar concentrations by increasing oxidative damage and compromising proteostasis [72].

After 6 h of treatment with silver NPs (AgNPs), human glioblastoma cells showed a threefold increase in ROS creation at 20 μg/mL, indicating that the ROS-generating effect is concentration dependent. This triggers caspase-dependent cell death, which in turn causes DNA damage and lipid peroxidation, which in turn activates the ATM/ATR pathway (Figure 7). Additionally, the research demonstrated that AgNPs promote apoptosis by increasing the Bax/Bcl-2 ratio by 50%. The release of mitochondrial cytochrome c enhances the activation of caspase-3. The impact of Ag(I) complexes on cancer cells is supported by the fact that they attach to thiol groups in enzymes, which in turn disturbs cellular signaling pathways and redox equilibrium [73,74].

Birtekocak et al. (2021) further elucidated the role of AgNPs in cancer therapy, particularly in human colon cancer cells. Their study showed that AgNPs, like silver complexes, amplify ROS production, significantly increasing the oxidative stress within the cancerous cells. This ROS overload activates various biochemical pathways that culminate in apoptosis, effectively killing the cancerous cells. The study found that AgNPs were capable of generating ROS at levels sufficient to induce mitochondrial dysfunction, a key event in initiating cell death. One of the special features of silver complexes is their ability to directly interact with DNA, leading to strand breaks and disrupting the integrity of the genetic material. The DNA damage induced by silver ions is thought to be a key driver of their anticancer activity [75].

Another study by Abu-Youssef et al. (2010) showed that silver complexes with 2-amino-3-methylpyridine ligands exhibited strong binding affinity to DNA. This interaction resulted in the formation of silver-DNA complexes, which hindered DNA replication and transcription, contributing to the cytotoxic effects. These DNA-intercalating properties of silver complexes further highlight their potential as anticancer agents, as the prevention of DNA replication and repair can effectively halt the proliferation of cancerous cells [76].

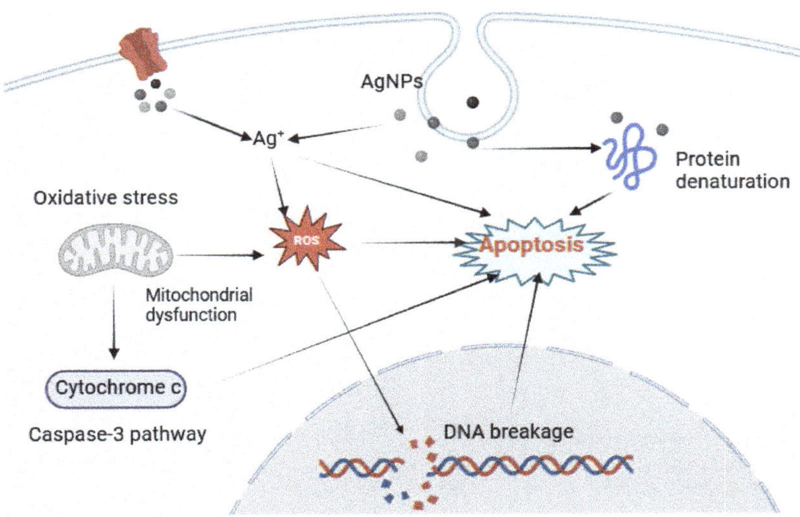

Figure 7. The general mechanism of silver ions in activating the caspase-3 pathway in inducing apoptosis.

Moreover, Ag(I)-based complexes (Table 3) can induce DNA damage through the generation of ROS, which in turn leads to the oxidation of the DNA bases, especially guanine, resulting in mutagenic lesions. This oxidative DNA damage has been implicated in the activation of DNA repair mechanisms, such as base excision repair and nucleotide excision repair, which can overwhelm the cancerous cells' ability to maintain genomic integrity, leading to apoptosis [77,78].

Table 3. Ag-based complexes used as chemotherapy drugs for different types of cancer.

Silver Complex	Primary Mechanism	Applications	References
Ag(I)-thiourea	ROS-mediated DNA damage	Lung, breast	[79–81]
Ag(I)-imidazolate	Oxidative stress, apoptosis	Prostate, colon	[82,83]
AgNPs	Amplified ROS, mitochondrial dysfunction	Colon, cervical, breast	[84,85]
Ag(I)-peptide/antibody conjugates	ROS generation, selective targeting	Various cancers (lung, breast)	[86–88]

Silver complexes present flexibility and adaptability, which allows for the fine-tuning of their redox properties, making them highly compatible for use in cancer treatment. By modifying the ligands attached to silver, researchers can influence the complex's ability to generate ROS, interact with cellular molecules, and overcome resistance mechanisms commonly found in tumor cells.

For example, El-Naggar et al. (2022) synthesized Ag(I) complexes with quinoxaline derivatives; these complexes exhibited increased DNA-binding affinity and redox activity, enabling them to efficiently disrupt DNA and induce cytotoxic effects in cancer cells. Such structural modifications have proven effective in expanding the applicability of silver complexes, particularly in targeting multi-drug-resistant cancer cell lines. This is significant given the growing problem of resistance to conventional chemotherapy agents, which often limits the efficacy of cancer treatment [89].

Silver complexes, with their ability to generate high levels of ROS and selectively interact with biomolecules, represent a promising alternative for overcoming this challenge.

2.4. Palladium Complexes: Catalysts for Redox Disruption

Despite not having been studied as thoroughly as compounds based on platinum, palladium (Pd) complexes have gained more and more interest due to their special qualities in redox-based cancer treatments. A key mechanism for causing cancer cell death is the catalytic activity of Pd(II) in boosting redox reactions, particularly the formation of ROS. Apoptosis, necrosis, and cellular malfunction may result from ROS, which are very reactive chemicals that can oxidatively damage proteins, lipids, and DNA. Rapidly proliferating cancer cells are the ones this process zeroes in on because they are more susceptible to oxidative stress than normal cells (Figure 8) [90,91].

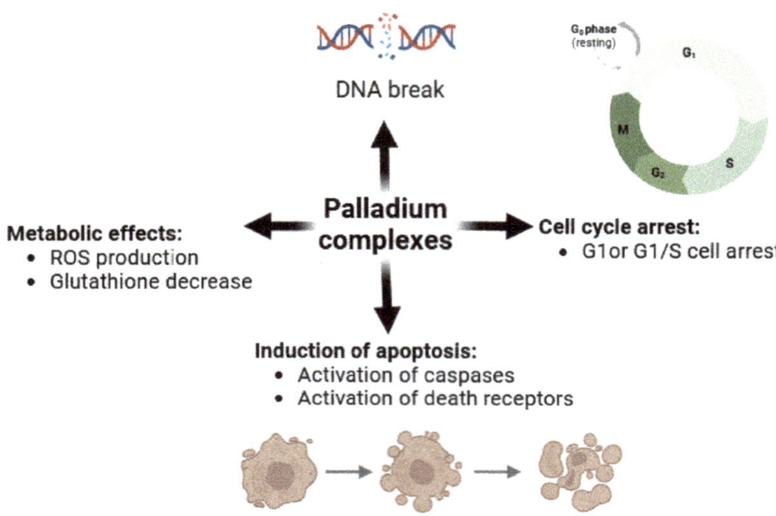

Figure 8. The multitude of mechanisms in which palladium complexes induce apoptosis in cancer cells.

Due to their capacity to promote ROS generation, Pd(II) complexes with ligands like terpyridine have become attractive prospects for the creation of anticancer drugs. The cytotoxic effects of palladium complexes on several cancer cell types provide evidence of their medicinal potential. One important step in the start of apoptosis is mitochondrial malfunction, which Pd(II)-terpyridine complexes have been shown to cause. When Pd(II) and mitochondrial components interact, cytochrome c is released, which starts the caspase cascade and eventually results in programmed cell death. The remarkable selectivity of Pd(II)-terpyridine complexes for cancer cells, which permits less damage to healthy tissues, is one of its main benefits. Pd(II)-terpyridine complexes were shown to preferentially accumulate in breast cancer cells while sparing normal cells in research by Savić et al. (2019) [92]. The complexes' distinctive physicochemical properties, such as their lipophilicity and membrane-crossing capabilities, improve their targeting capabilities and allow for selective absorption. Furthermore, terpyridine ligands' highly flexible structure permits the fine-tuning of their redox characteristics and interactions with biological targets. It is feasible to maximize the therapeutic efficiency of these complexes and customize them to treat various cancer types by altering the palladium coordination environment [93–95].

The creation of Pd(II) complexes (Table 4) that combine targeted delivery methods with increased ROS production is the subject of further investigation. For instance, Pd(II)-terpyridine complexes may be functionalized with biomolecules that bind to cancer cell receptors or conjugated with nanoparticles to enhance their accumulation at tumor locations and lessen systemic toxicity. Because palladium complexes may interact with important biological components, including proteins, lipids, and DNA, while still exhibiting selective cytotoxicity, they hold promise as a platform for next-generation cancer treatments.

Table 4. Pd-based drugs with high redox potential against cancer cells.

Palladium Complex	Mechanism of Action	Target Cells/Cancer Types	Key References
Pd(II)-terpyridine complexes	ROS-induced mitochondrial dysfunction and apoptosis	Breast cancer, colorectal cancer	[92]
Pd(II)-terpyridine complexes	DNA binding, oxidative stress-induced apoptosis	Ovarian cancer, lung cancer	[96]
Pd(II) complexes with phosphine ligands	DNA intercalation, ROS generation	Melanoma, liver cancer	[97]
Pd(II) complexes with heterocyclic ligands	Redox cycling, ROS production	Pancreatic cancer, breast cancer	[98,99]

The promise of palladium-based complexes (Figure 9) as future anticancer medicines is highlighted by their flexibility, especially when terpyridine and other redox-active ligands are included. A number of questions remain, including how well Pd(II)-terpyridine complexes work in vivo, what dosages work best, and whether or not they may be used in combination with other anticancer medications.

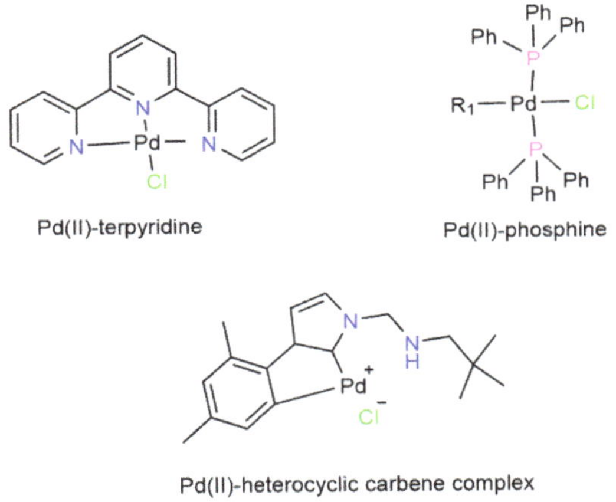

Figure 9. Pd(II) complexes that induce apoptosis in tumor cells.

Romashev et al. (2022) discovered that heteroleptic Pd(II) complexes with redox-active quinonoid ligands attacked cancer cells in a number of different ways. Cytotoxicity was produced by the complexes at IC_{50} values between 1.5 and 3.2 μM, and they showed a significant selectivity for cancer cells. Following 12 h of exposure to 2 μM of Pd(II) complex, detailed ROS quantification experiments revealed a 150% rise in ROS levels compared to the unexposed control group. Increased ROS cause oxidative stress on lipids and proteins, which compromises the integrity of the mitochondrial membrane. Additional evidence of DNA intercalation by these Pd(II) complexes comes from the substantial hypochromic effects and bathochromic shifts seen in UV-visible spectra. Approximately 2.3 times 10^5 M^{-1} was determined for the binding affinity constants, indicating a strong interaction with DNA that impedes the processes of replication and transcription [100].

Conducting further research into arylamide-liganded cyclometallated Pd(II) complexes, Dolengovski et al. (2024) built upon this earlier investigation. Reversible redox peaks indicated electron transfer capacity conducive to oxidative stress, which was validated using cyclic voltammetry, indicating that these complexes displayed redox cycling behavior. There was a clear relationship between redox potential and cytotoxicity, as shown by their IC_{50} values ranging from 2.8 to 6.5 μM in lung and colon cancer models. A maximal 2.5-fold increase at 4 μM after 24 h was achieved in ROS generation, which was dependent on both time and dosage. There was a 60% drop in mitochondrial membrane potential and cellular viability tests connected this oxidative damage to apoptosis in the mitochondrial pathway. Furthermore, apoptosis via the intrinsic route was confirmed by Western blotting, which showed an increase in Bax and cleaved caspase-3 [91].

Other studies focused on Pd(II) complexes with Schiff base ligands and their DNA-binding and enzyme-inhibiting capabilities. Results from fluorescence quenching tests showed that there was substantial intercalation and disturbance of DNA stability, with binding constants reaching up to 3.1×10^6 M^{-1}. The resolution of DNA supercoiling was impaired, as topoisomerase inhibition experiments demonstrated an enzymatic suppression of over 85% at 5 μM [101]. Lastly, in cell-specific lines, Vojtek et al. (2019) showed that Pd(II) complexes were effective against triple-negative breast cancer, with IC_{50} values < 1 μM. Highlighting the multi-faceted anticancer tactics of Pd complexes, cell cycle arrest in the G2/M phase was the result of the synergy between ROS production and DNA-binding capacities [102].

The creation of prodrugs based on Pd(II) that are activated in the tumor microenvironment might also lead to improved outcomes during treatment, fewer side effects, and more precise targeting of cancer cells. By extending the therapeutic window of palladium-based medicines, these initiatives have the potential to solidify their position as an essential class of metallodrugs in cancer treatment.

There are still issues with palladium complexes' stability and bioavailability, despite the fact that they show encouraging redox activity.

Researchers are exploring ligand designs that enhance their pharmacokinetic properties, such as incorporating bulky, electron-donating groups to improve solubility and tumor specificity.

3. Preparation Process—Literature Review

This literature review on noble metal complexes in cancer treatment includes papers that were handpicked after a long and methodical selection process to guarantee high quality, reliability, and coverage of the topic. To begin, we combed through a large number of scientific publications' worth of research papers utilizing databases like PubMed, Scopus, and Google Scholar. To narrow the search, we used keywords like "cancer therapy", "noble metal complexes", "platinum-based drugs", "gold complexes", and "redox mechanisms" to find the best research. With an emphasis on more recent developments in the subject, only publications published in the previous 20 years were taken into account.

Included articles contributed directly to our knowledge of noble metal complexes' function in cancer treatment; in particular, studies that investigated these compounds' biological processes, molecular interactions, and therapeutic promise were given preference. Research on complexes based on palladium, gold, silver, and platinum that shed light on their synthesis, redox behavior, and cytotoxic effects was given precedence. Both experimental investigations, including in vitro and in vivo models, and clinical trial data were included to make sure the review addressed a variety of viewpoints. The trustworthiness of the source was also considered as an additional factor for inclusion. For this evaluation, we strictly regarded peer-reviewed publications in medicinal chemistry, oncology, and biochemistry journals with a high impact factor. This made sure that the data were solid and backed by solid science. Additionally, articles detailing new complicated designs, innovative methodology, and developing treatment approaches (such as targeted drug delivery systems and ways to combat drug resistance) were given preference.

To guarantee a comprehensive overview, articles were selected to reflect both historical perspectives on established treatments, like cisplatin, and newer advancements in the field, including the exploration of AuNPs and novel palladium complexes. Special attention was given to studies that focused on understanding the molecular targets of these metal-based compounds, such as DNA, proteins, and redox-sensitive enzymes, as well as their ability to induce oxidative stress, a key mechanism in cancer cell destruction. Only studies that were relevant to the review's aim—providing a detailed understanding of the potential and challenges of noble metal complexes in cancer therapy—were ultimately selected.

By employing this rigorous selection process, a well-rounded and current body of literature was identified, offering valuable insights into the therapeutic applications and limitations of noble metal complexes in the treatment of cancer.

4. Conclusions

In summary, the distinct redox properties of noble metal complexes (e.g., palladium, platinum, gold, and silver) make them very promising cancer treatments. The literature review concludes that the ability of platinum-based medicines to inhibit DNA replication and generate ROS makes them an important component of cancer chemotherapy. These treatments include cisplatin, oxaliplatin, and the more modern Pt(IV) prodrugs. New platinum complexes with improved pharmacokinetics, selectivity, and reduced side effects have been developed in response to the persistent problems of resistance and toxicity. The therapeutic potential of these platinum-based medicines has been enhanced by innovations such as tailored drug delivery systems, better ligands, and nanocarriers.

Research has shown that gold complexes, particularly those containing the +1 and +3 oxidation states, have a considerable inhibitory effect on TrxR. For cancer cells, this means a redox imbalance and cell death by selective killing. Their anticancer effectiveness is further enhanced when used in conjunction with AuNPs, which decrease systemic toxicity and allow for targeted medication delivery. The dual-action mechanism of AuNPs, which combines therapy and diagnostics, highlights their potential in customized medicine.

The ability of silver complexes to generate ROS and enhance oxidative stress has made them famous for their anticancer effects, and this is especially true in cancer cells that have developed resistance to many medicines. They can interact with proteins and DNA, and they can increase oxidative damage; therefore, they have potential as cancer treatment agents. AgNPs enhance therapeutic efficacy and provide additional advantages by facilitating targeted drug distribution and boosting ROS production.

Palladium complexes' ability to generate ROS and speed up redox reactions gives them great potential, despite the fact that they have received less attention from researchers than platinum-based medicines. In particular, complexes functionalized with redox-active ligands, such as terpyridine, have a specific cytotoxic effect on cancer cells. This is particularly true in colorectal and breast cancer models. Future studies on palladium-based prodrugs and customized delivery methods have the potential to enhance their effectiveness while reducing adverse effects, transforming them into valuable resources for cancer therapy.

Although noble metal complexes have shown encouraging results, they will not be able to fulfill their therapeutic promise unless advances in drug transport, molecular design, and the elimination of issues related to stability, toxicity, and resistance are addressed. Combining these metal-based medications with modern technologies, such as combination therapy and nanomedicine, is the key to improving clinical outcomes in cancer treatment. Future research should focus on optimizing these complexes for specific cancer types, enhancing their bioavailability, and decreasing systemic toxicity to make them more effective in cancer treatment plans as a whole.

Funding: This research received no external funding.

Conflicts of Interest: The authors declare no conflicts of interest.

References

1. Ahles, T.A.; Root, J.C.; Ryan, E.L. Cancer- and Cancer Treatment–Associated Cognitive Change: An Update on the State of the Science. *J. Clin. Oncol.* **2012**, *30*, 3675–3686. [CrossRef]
2. Abdurhman, A.A.M.; Zhang, Y.; Zhang, G.; Wang, S. Hierarchical Nanostructured Noble Metal/Metal Oxide/Graphene-Coated Carbon Fiber: In Situ Electrochemical Synthesis and Use as Microelectrode for Real-Time Molecular Detection of Cancer Cells. *Anal. Bioanal. Chem.* **2015**, *407*, 8129–8136. [CrossRef]
3. Arruebo, M.; Vilaboa, N.; Sáez-Gutierrez, B.; Lambea, J.; Tres, A.; Valladares, M.; González-Fernández, Á. Assessment of the Evolution of Cancer Treatment Therapies. *Cancers* **2011**, *3*, 3279–3330. [CrossRef] [PubMed]
4. Debela, D.T.; Muzazu, S.G.; Heraro, K.D.; Ndalama, M.T.; Mesele, B.W.; Haile, D.C.; Kitui, S.K.; Manyazewal, T. New Approaches and Procedures for Cancer Treatment: Current Perspectives. *SAGE Open Med.* **2021**, *9*, 20503121211034366. [CrossRef] [PubMed]
5. Armstrong, D.K.; Bundy, B.; Wenzel, L.; Huang, H.Q.; Baergen, R.; Lele, S.; Copeland, L.J.; Walker, J.L.; Burger, R.A. Intraperitoneal Cisplatin and Paclitaxel in Ovarian Cancer. *N. Engl. J. Med.* **2006**, *354*, 34–43. [CrossRef]
6. Akhlaghipour, I.; Moghbeli, M. Matrix Metalloproteinases as the Critical Regulators of Cisplatin Response and Tumor Cell Invasion. *Eur. J. Pharmacol.* **2024**, *982*, 176966. [CrossRef] [PubMed]
7. Ali, S.; Sharma, A.S.; Ahmad, W.; Zareef, M.; Hassan, M.M.; Viswadevarayalu, A.; Jiao, T.; Li, H.; Chen, Q. Noble Metals Based Bimetallic and Trimetallic Nanoparticles: Controlled Synthesis, Antimicrobial and Anticancer Applications. *Crit. Rev. Anal. Chem.* **2020**, *51*, 454–481. [CrossRef] [PubMed]
8. Dreaden, E.C.; El-Sayed, M.A. Detecting and Destroying Cancer Cells in More than One Way with Noble Metals and Different Confinement Properties on the Nanoscale. In *Nanomaterials and Neoplasms*; Jenny Stanford Publishing: Singapore, 2021; pp. 1–29.
9. Kuo, C.-C.; Lan, W.-J.; Chen, C.-H. Redox Preparation of Mixed-Valence Cobalt Manganese Oxide Nanostructured Materials: Highly Efficient Noble Metal-Free Electrocatalysts for Sensing Hydrogen Peroxide. *Nanoscale* **2014**, *6*, 334–341. [CrossRef] [PubMed]
10. Bindoli, A.; Rigobello, M.P.; Scutari, G.; Gabbiani, C.; Casini, A.; Messori, L. Thioredoxin Reductase: A Target for Gold Compounds Acting as Potential Anticancer Drugs. *Coord. Chem. Rev.* **2009**, *253*, 1692–1707. [CrossRef]
11. Liu, Y.; Li, Y.; Yu, S.; Zhao, G. Recent Advances in the Development of Thioredoxin Reductase Inhibitors as Anticancer Agents. *Curr. Drug Targets* **2012**, *13*, 1432–1444. [CrossRef]
12. Brown, A.; Kumar, S.; Tchounwou, P.B. Cisplatin-Based Chemotherapy of Human Cancers. *J. Cancer Sci. Ther.* **2019**, *11*, 97. [PubMed]
13. Dasari, S.; Tchounwou, P.B. Cisplatin in Cancer Therapy: Molecular Mechanisms of Action. *Eur. J. Pharmacol.* **2014**, *740*, 364–378. [CrossRef]
14. Galluzzi, L.; Senovilla, L.; Vitale, I.; Michels, J.; Martins, I.; Kepp, O.; Castedo, M.; Kroemer, G. Molecular Mechanisms of Cisplatin Resistance. *Oncogene* **2012**, *31*, 1869–1883. [CrossRef] [PubMed]
15. Cummings, S.D. Platinum Complexes of Terpyridine: Interaction and Reactivity with Biomolecules. *Coord. Chem. Rev.* **2009**, *253*, 1495–1516. [CrossRef]
16. Cai, L.; Yu, C.; Ba, L.; Liu, Q.; Qian, Y.; Yang, B.; Gao, C. Anticancer Platinum-based Complexes with Non-classical Structures. *Appl. Organomet. Chem.* **2018**, *32*, e4228. [CrossRef]
17. Coluccia, M.; Natile, G. Trans-Platinum Complexes in Cancer Therapy. *Anti-Cancer Agents Med. Chem. (Former. Curr. Med. Chem.-Anti-Cancer Agents)* **2007**, *7*, 111–123. [CrossRef] [PubMed]
18. Krause-Heuer, A.M.; Grünert, R.; Kühne, S.; Buczkowska, M.; Wheate, N.J.; Le Pevelen, D.D.; Boag, L.R.; Fisher, D.M.; Kasparkova, J.; Malina, J.; et al. Studies of the Mechanism of Action of Platinum(II) Complexes with Potent Cytotoxicity in Human Cancer Cells. *J. Med. Chem.* **2009**, *52*, 5474–5484. [CrossRef]
19. Callari, M.; Aldrich-Wright, J.R.; de Souza, P.L.; Stenzel, M.H. Polymers with Platinum Drugs and Other Macromolecular Metal Complexes for Cancer Treatment. *Prog. Polym. Sci.* **2014**, *39*, 1614–1643. [CrossRef]
20. Huang, J.-Y.; Chen, M.-H.; Kuo, W.-T.; Sun, Y.-J.; Lin, F.-H. The Characterization and Evaluation of Cisplatin-Loaded Magnetite–Hydroxyapatite Nanoparticles (mHAp/CDDP) as Dual Treatment of Hyperthermia and Chemotherapy for Lung Cancer Therapy. *Ceram. Int.* **2015**, *41*, 2399–2410. [CrossRef]
21. Ghosh, S. Cisplatin: The First Metal Based Anticancer Drug. *Bioorg. Chem.* **2019**, *88*, 102925. [CrossRef] [PubMed]

22. Tang, C.; Livingston, M.J.; Safirstein, R.; Dong, Z. Cisplatin Nephrotoxicity: New Insights and Therapeutic Implications. *Nat. Rev. Nephrol.* **2023**, *19*, 53–72. [CrossRef]
23. Alcindor, T.; Beauger, N. Oxaliplatin: A Review in the Era of Molecularly Targeted Therapy. *Curr. Oncol.* **2011**, *18*, 18–25. [CrossRef] [PubMed]
24. Cersosimo, R.J. Oxaliplatin-Associated Neuropathy: A Review. *Ann. Pharmacother.* **2005**, *39*, 128–135. [CrossRef]
25. Göschl, S.; Schreiber-Brynzak, E.; Pichler, V.; Cseh, K.; Heffeter, P.; Jungwirth, U.; Jakupec, M.A.; Berger, W.; Keppler, B.K. Comparative Studies of Oxaliplatin-Based Platinum (IV) Complexes in Different *in Vitro* and *in Vivo* Tumor Models. *Metallomics* **2017**, *9*, 309–322. [CrossRef] [PubMed]
26. O'Dowd, P.D.; Sutcliffe, D.F.; Griffith, D.M. Oxaliplatin and Its Derivatives–An Overview. *Coord. Chem. Rev.* **2023**, *497*, 215439. [CrossRef]
27. Stein, A.; Arnold, D. Oxaliplatin: A Review of Approved Uses. *Expert Opin. Pharmacother.* **2012**, *13*, 125–137. [CrossRef] [PubMed]
28. Fei Chin, C.; Yuan Qiang Wong, D.; Jothibasu, R.; Han Ang, W. Anticancer Platinum (IV) Prodrugs with Novel Modes of Activity. *Curr. Top. Med. Chem.* **2011**, *11*, 2602–2612. [CrossRef] [PubMed]
29. Gibson, D. Platinum (IV) Anticancer Prodrugs–Hypotheses and Facts. *Dalton Trans.* **2016**, *45*, 12983–12991. [CrossRef] [PubMed]
30. Gao, Z.; Han, Y.; Gao, Z.; Wang, F. Multicomponent Assembled Systems Based on Platinum(II) Terpyridine Complexes. *Acc. Chem. Res.* **2018**, *51*, 2719–2729. [CrossRef]
31. Hoefer, D.; Cseh, K.; Hejl, M.; Roller, A.; Jakupec, M.A.; Galanski, M.S.; Keppler, B.K. Synthesis, Characterization, Cytotoxic Activity, and 19F NMR Spectroscopic Investigations of (OC-6-33)-Diacetato (Ethane-1, 2-Diamine) Bis (3, 3, 3-Trifluoropropanoato) Platinum (IV) and Its Platinum (II) Counterpart. *Inorg. Chim. Acta* **2019**, *490*, 190–199. [CrossRef]
32. Zhang, F.; Bai, Y.; Yang, X.; Li, J.; Peng, J. N-Heterocyclic Carbene Platinum Complexes Functionalized with a Polyether Chain and Silyl Group: Synthesis and Application as a Catalyst for Hydrosilylation. *Phosphorus Sulfur Silicon Relat. Elem.* **2017**, *192*, 1271–1278. [CrossRef]
33. Lo, Y.-C.; Ko, T.-P.; Su, W.-C.; Su, T.-L.; Wang, A.H.-J. Terpyridine–Platinum (II) Complexes Are Effective Inhibitors of Mammalian Topoisomerases and Human Thioredoxin Reductase 1. *J. Inorg. Biochem.* **2009**, *103*, 1082–1092. [CrossRef]
34. Zhao, Y.; Liu, J.; He, M.; Dong, Q.; Zhang, L.; Xu, Z.; Kang, Y.; Xue, P. Platinum–Titania Schottky Junction as Nanosonosensitizer, Glucose Scavenger, and Tumor Microenvironment-Modulator for Promoted Cancer Treatment. *ACS Nano* **2022**, *16*, 12118–12133. [CrossRef] [PubMed]
35. Marchenko, N. Ligand Effects on the Properties of Ultra-Small Platinum-Based Nanoparticles. Ph.D. Thesis, INSA de Toulouse, Toulouse, France, 2023.
36. Johnstone, T.C.; Suntharalingam, K.; Lippard, S.J. The Next Generation of Platinum Drugs: Targeted Pt(II) Agents, Nanoparticle Delivery, and Pt(IV) Prodrugs. *Chem. Rev.* **2016**, *116*, 3436–3486. [CrossRef] [PubMed]
37. Abdalbari, F.H.; Telleria, C.M. The Gold Complex Auranofin: New Perspectives for Cancer Therapy. *Discov. Oncol.* **2021**, *12*, 42. [CrossRef]
38. Galassi, R.; Burini, A.; Ricci, S.; Pellei, M.; Rigobello, M.P.; Citta, A.; Dolmella, A.; Gandin, V.; Marzano, C. Synthesis and Characterization of Azolate Gold (I) Phosphane Complexes as Thioredoxin Reductase Inhibiting Antitumor Agents. *Dalton Trans.* **2012**, *41*, 5307–5318. [CrossRef] [PubMed]
39. Lu, Y.; Ma, X.; Chang, X.; Liang, Z.; Lv, L.; Shan, M.; Lu, Q.; Wen, Z.; Gust, R.; Liu, W. Recent Development of Gold (I) and Gold (III) Complexes as Therapeutic Agents for Cancer Diseases. *Chem. Soc. Rev.* **2022**, *51*, 5518–5556. [CrossRef] [PubMed]
40. Thota, S.; Crans, D.C. *Metal Nanoparticles: Synthesis and Applications in Pharmaceutical Sciences*; John Wiley & Sons: Hoboken, NJ, USA, 2018.
41. Gamberi, T.; Chiappetta, G.; Fiaschi, T.; Modesti, A.; Sorbi, F.; Magherini, F. Upgrade of an Old Drug: Auranofin in Innovative Cancer Therapies to Overcome Drug Resistance and to Increase Drug Effectiveness. *Med. Res. Rev.* **2022**, *42*, 1111–1146. [CrossRef] [PubMed]
42. Capparelli, E.V.; Bricker-Ford, R.; Rogers, M.J.; McKerrow, J.H.; Reed, S.L. Phase I Clinical Trial Results of Auranofin, a Novel Antiparasitic Agent. *Antimicrob. Agents Chemother.* **2017**, *61*, e01947-16. [CrossRef] [PubMed]
43. Carlos Lima, J.; Rodriguez, L. Phosphine-Gold (I) Compounds as Anticancer Agents: General Description and Mechanisms of Action. *Anti-Cancer Agents Med. Chem. (Former. Curr. Med. Chem.-Anti-Cancer Agents)* **2011**, *11*, 921–928. [CrossRef]
44. Fereidoonnezhad, M.; Mirsadeghi, H.A.; Abedanzadeh, S.; Yazdani, A.; Alamdarlou, A.; Babaghasabha, M.; Almansaf, Z.; Faghih, Z.; McConnell, Z.; Shahsavari, H.R. Synthesis and Biological Evaluation of Thiolate Gold (i) Complexes as Thioredoxin Reductase (TrxR) and Glutathione Reductase (GR) Inhibitors. *New J. Chem.* **2019**, *43*, 13173–13182. [CrossRef]
45. Shichibu, Y.; Negishi, Y.; Tsukuda, T.; Teranishi, T. Large-Scale Synthesis of Thiolated Au_{25} Clusters via Ligand Exchange Reactions of Phosphine-Stabilized Au_{11} Clusters. *J. Am. Chem. Soc.* **2005**, *127*, 13464–13465. [CrossRef] [PubMed]
46. Sun, R.W.-Y.; Lok, C.-N.; Fong, T.T.-H.; Li, C.K.-L.; Yang, Z.F.; Zou, T.; Siu, A.F.-M.; Che, C.-M. A Dinuclear Cyclometalated Gold (III)–Phosphine Complex Targeting Thioredoxin Reductase Inhibits Hepatocellular Carcinoma in Vivo. *Chem. Sci.* **2013**, *4*, 1979–1988. [CrossRef]

47. Rubbiani, R.; Zehnder, T.N.; Mari, C.; Blacque, O.; Venkatesan, K.; Gasser, G. Anticancer Profile of a Series of Gold(III) (2-phenyl)Pyridine Complexes. *ChemMedChem* **2014**, *9*, 2781–2790. [CrossRef] [PubMed]
48. Solvi, T.N.; Reiersølmoen, A.C.; Orthaber, A.; Fiksdahl, A. Studies towards Pyridine-Based *N,N,O*-Gold(III) Complexes: Synthesis, Characterization and Application. *Eur. J. Org. Chem.* **2020**, *2020*, 7062–7068. [CrossRef]
49. Marta Nagy, E.; Ronconi, L.; Nardon, C.; Fregona, D. Noble Metal-Dithiocarbamates Precious Allies in the Fight against Cancer. *Mini Rev. Med. Chem.* **2012**, *12*, 1216–1229. [CrossRef]
50. Ronconi, L.; Giovagnini, L.; Marzano, C.; Bettìo, F.; Graziani, R.; Pilloni, G.; Fregona, D. Gold Dithiocarbamate Derivatives as Potential Antineoplastic Agents: Design, Spectroscopic Properties, and in Vitro Antitumor Activity. *Inorg. Chem.* **2005**, *44*, 1867–1881. [CrossRef]
51. Williams, M.R.; Bertrand, B.; Hughes, D.L.; Waller, Z.A.; Schmidt, C.; Ott, I.; O'Connell, M.; Searcey, M.; Bochmann, M. Cyclometallated Au (III) Dithiocarbamate Complexes: Synthesis, Anticancer Evaluation and Mechanistic Studies. *Metallomics* **2018**, *10*, 1655–1666. [CrossRef] [PubMed]
52. Karaaslan, M.G.; Aktaş, A.; Gürses, C.; Gök, Y.; Ateş, B. Chemistry, Structure, and Biological Roles of Au-NHC Complexes as TrxR Inhibitors. *Bioorg. Chem.* **2020**, *95*, 103552. [CrossRef]
53. Schlagintweit, J.F.; Jakob, C.H.G.; Wilke, N.L.; Ahrweiler, M.; Frias, C.; Frias, J.; König, M.; Esslinger, E.-M.H.J.; Marques, F.; Machado, J.F.; et al. Gold(I) Bis(1,2,3-Triazol-5-Ylidene) Complexes as Promising Selective Anticancer Compounds. *J. Med. Chem.* **2021**, *64*, 15747–15757. [CrossRef] [PubMed]
54. Tolbatov, I.; Umari, P.; Marrone, A. Mechanism of Action of Antitumor Au(I) N-Heterocyclic Carbene Complexes: A Computational Insight on the Targeting of TrxR Selenocysteine. *Int. J. Mol. Sci.* **2024**, *25*, 2625. [CrossRef] [PubMed]
55. De Franco, M.; Saab, M.; Porchia, M.; Marzano, C.; Nolan, S.P.; Nahra, F.; Van Hecke, K.; Gandin, V. Unveiling the Potential of Innovative Gold(I) and Silver(I) Selenourea Complexes as Anticancer Agents Targeting TrxR and Cellular Redox Homeostasis. *Chemistry* **2022**, *28*, e202201898. [CrossRef]
56. Dreaden, E.C.; Alkilany, A.M.; Huang, X.; Murphy, C.J.; El-Sayed, M.A. The Golden Age: Gold Nanoparticles for Biomedicine. *Chem. Soc. Rev.* **2012**, *41*, 2740–2779. [CrossRef] [PubMed]
57. Giljohann, D.A.; Seferos, D.S.; Daniel, W.L.; Massich, M.D.; Patel, P.C.; Mirkin, C.A. Gold Nanoparticles for Biology and Medicine. In *Spherical Nucleic Acids*; Jenny Stanford Publishing: Singapore, 2020; pp. 55–90.
58. Jennings, T.; Strouse, G. Past, Present, and Future of Gold Nanoparticles. In *Bio-Applications of Nanoparticles*; Chan, W.C.W., Ed.; Advances in Experimental Medicine and Biology; Springer: New York, NY, USA, 2007; Volume 620, pp. 34–47, ISBN 978-0-387-76712-3.
59. Sztandera, K.; Gorzkiewicz, M.; Klajnert-Maculewicz, B. Gold Nanoparticles in Cancer Treatment. *Mol. Pharm.* **2019**, *16*, 1–23. [CrossRef] [PubMed]
60. Gandubert, V.J.; Lennox, R.B. Assessment of 4-(Dimethylamino)Pyridine as a Capping Agent for Gold Nanoparticles. *Langmuir* **2005**, *21*, 6532–6539. [CrossRef] [PubMed]
61. Hegedhus, C.; Kovács, K.; Polgár, Z.; Regdon, Z.; Szabó, É.; Robaszkiewicz, A.; Forman, H.J.; Martner, A.; Virág, L. Redox Control of Cancer Cell Destruction. *Redox Biol.* **2018**, *16*, 59–74. [CrossRef]
62. Curran, R.J. Silver (I) Complexes as Antimicrobial and Anticancer Drugs. Ph.D. Thesis, National University of Ireland Maynooth, Maynooth, Ireland, 2009.
63. Iacopetta, D.; Mariconda, A.; Saturnino, C.; Caruso, A.; Palma, G.; Ceramella, J.; Muià, N.; Perri, M.; Sinicropi, M.S.; Caroleo, M.C.; et al. Novel Gold and Silver Carbene Complexes Exert Antitumor Effects Triggering the Reactive Oxygen Species Dependent Intrinsic Apoptotic Pathway. *ChemMedChem* **2017**, *12*, 2054–2065. [CrossRef]
64. Li, S.; Zhang, S.; Jin, X.; Tan, X.; Lou, J.; Zhang, X.; Zhao, Y. Singly Protonated Dehydronorcantharidin Silver Coordination Polymer Induces Apoptosis of Lung Cancer Cells via Reactive Oxygen Species-Mediated Mitochondrial Pathway. *Eur. J. Med. Chem.* **2014**, *86*, 1–11. [CrossRef] [PubMed]
65. Batarseh, K.I.; Smith, M.A. Synergistic Activities of a Silver(I) Glutamic Acid Complex and Reactive Oxygen Species (ROS): A Novel Antimicrobial and Chemotherapeutic Agent. *CMC* **2012**, *19*, 3635–3640. [CrossRef] [PubMed]
66. Carlson, C.; Hussain, S.M.; Schrand, A.M.; Braydich-Stolle, L.K.; Hess, K.L.; Jones, R.L.; Schlager, J.J. Unique Cellular Interaction of Silver Nanoparticles: Size-Dependent Generation of Reactive Oxygen Species. *J. Phys. Chem. B* **2008**, *112*, 13608–13619. [CrossRef] [PubMed]
67. Jalil, M.A.; Yamada, T.; Fujinami, S.; Honjo, T.; Nishikawa, H. An Imidazole-Based P–N Bridging Ligand and Its Binuclear Copper (I), Silver (I) and Palladium (I) Complexes: Synthesis, Characterizations and X-Ray Structures. *Polyhedron* **2001**, *20*, 627–633. [CrossRef]
68. Bauman, J.E.; Wang, J.C. Imidazole Complexes of Nickel(II), Copper(II), Zinc(II), and Silver(I). *Inorg. Chem.* **1964**, *3*, 368–373. [CrossRef]
69. Koskinen, L.; Jääskeläinen, S.; Oresmaa, L.; Haukka, M. Argentophilic Interactions in Multinuclear Ag Complexes of Imidazole Containing Schiff Bases. *CrystEngComm* **2012**, *14*, 3509–3514. [CrossRef]

70. Kalinowska-Lis, U.; Felczak, A.; Chęcińska, L.; Szabłowska-Gadomska, I.; Patyna, E.; Małecki, M.; Lisowska, K.; Ochocki, J. Antibacterial Activity and Cytotoxicity of Silver (I) Complexes of Pyridine and (Benz) Imidazole Derivatives. X-Ray Crystal Structure of [Ag (2,6-Di(CH_2OH)Py)$_2$]NO_3. *Molecules* **2016**, *21*, 87. [CrossRef] [PubMed]
71. Raju, S.K.; Karunakaran, A.; Kumar, S.; Sekar, P.; Murugesan, M.; Karthikeyan, M. Silver Complexes as Anticancer Agents: A Perspective Review. *Ger. J. Pharm. Biomater.* **2022**, *1*, 6–28. [CrossRef]
72. Fabbrini, M.G.; Cirri, D.; Pratesi, A.; Ciofi, L.; Marzo, T.; Guerri, A.; Nistri, S.; Dell'Accio, A.; Gamberi, T.; Severi, M.; et al. A Fluorescent Silver(I) Carbene Complex with Anticancer Properties: Synthesis, Characterization, and Biological Studies. *ChemMedChem* **2019**, *14*, 182–188. [CrossRef]
73. Allison, S.J.; Sadiq, M.; Baronou, E.; Cooper, P.A.; Dunnill, C.; Georgopoulos, N.T.; Latif, A.; Shepherd, S.; Shnyder, S.D.; Stratford, I.J. Preclinical Anti-Cancer Activity and Multiple Mechanisms of Action of a Cationic Silver Complex Bearing N-Heterocyclic Carbene Ligands. *Cancer Lett.* **2017**, *403*, 98–107. [CrossRef] [PubMed]
74. Thornton, L.; Dixit, V.; Assad, L.O.N.; Ribeiro, T.P.; Queiroz, D.D.; Kellett, A.; Casey, A.; Colleran, J.; Pereira, M.D.; Rochford, G.; et al. Water-Soluble and Photo-Stable Silver(I) Dicarboxylate Complexes Containing 1,10-Phenanthroline Ligands: Antimicrobial and Anticancer Chemotherapeutic Potential, DNA Interactions and Antioxidant Activity. *J. Inorg. Biochem.* **2016**, *159*, 120–132. [CrossRef]
75. Birtekocak, F.; Demirbolat, G.M.; Cevik, O. TRAIL Conjugated Silver Nanoparticle Synthesis, Characterization and Therapeutic Effects on HT-29 Colon Cancer Cells. *Iran. J. Pharm. Res. IJPR* **2021**, *20*, 45.
76. Abu-Youssef, M.A.M.; Soliman, S.M.; Langer, V.; Gohar, Y.M.; Hasanen, A.A.; Makhyoun, M.A.; Zaky, A.H.; Öhrström, L.R. Synthesis, Crystal Structure, Quantum Chemical Calculations, DNA Interactions, and Antimicrobial Activity of [Ag(2-Amino-3-Methylpyridine)$_2$]NO_3 and [Ag(Pyridine-2-Carboxaldoxime)NO_3]. *Inorg. Chem.* **2010**, *49*, 9788–9797. [CrossRef]
77. Aktepe, N.; Kocyigit, A.; Yukselten, Y.; Taskin, A.; Keskin, C.; Celik, H. Increased DNA Damage and Oxidative Stress Among Silver Jewelry Workers. *Biol. Trace Elem. Res.* **2015**, *164*, 185–191. [CrossRef]
78. Kruszewski, M.; Grądzka, I.; Bartłomiejczyk, T.; Chwastowska, J.; Sommer, S.; Grzelak, A.; Zuberek, M.; Lankoff, A.; Dusinska, M.; Wojewódzka, M. Oxidative DNA Damage Corresponds to the Long Term Survival of Human Cells Treated with Silver Nanoparticles. *Toxicol. Lett.* **2013**, *219*, 151–159. [CrossRef] [PubMed]
79. Segura, D.F.; Netto, A.V.; Frem, R.C.; Mauro, A.E.; da Silva, P.B.; Fernandes, J.A.; Paz, F.A.; Dias, A.L.; Silva, N.C.; de Almeida, E.T. Synthesis and Biological Evaluation of Ternary Silver Compounds Bearing N, N-Chelating Ligands and Thiourea: X-Ray Structure of [{Ag(Bpy)(μ-Tu)}$_2$](NO_3)$_2$ (Bpy = 2, 2′-Bipyridine; Tu = Thiourea). *Polyhedron* **2014**, *79*, 197–206. [CrossRef]
80. Borges, A.P.; Obata, M.M.; Libardi, S.H.; Trevisan, R.O.; Deflon, V.M.; Abram, U.; Ferreira, F.B.; Costa, L.A.S.; Patrocínio, A.O.; da Silva, M.V. Gold (I) and Silver (I) Complexes Containing Hybrid Sulfonamide/Thiourea Ligands as Potential Leishmanicidal Agents. *Pharmaceutics* **2024**, *16*, 452. [CrossRef]
81. Khan, U.A.; Badshah, A.; Tahir, M.N.; Khan, E. Gold (I), Silver (I) and Copper (I) Complexes of 2, 4, 6-Trimethylphenyl-3-Benzoylthiourea; Synthesis and Biological Applications. *Polyhedron* **2020**, *181*, 114485. [CrossRef]
82. Gao, H.; Liu, L.; Luo, Y.; Jia, D. In-Situ Preparation of Epoxy/Silver Nanocomposites by Thermal Decomposition of Silver–Imidazole Complex. *Mater. Lett.* **2011**, *65*, 3529–3532. [CrossRef]
83. Marinelli, M.; Pellei, M.; Cimarelli, C.; Dias, H.R.; Marzano, C.; Tisato, F.; Porchia, M.; Gandin, V.; Santini, C. Novel Multicharged Silver (I)–NHC Complexes Derived from Zwitterionic 1, 3-Symmetrically and 1, 3-Unsymmetrically Substituted Imidazoles and Benzimidazoles: Synthesis and Cytotoxic Properties. *J. Organomet. Chem.* **2016**, *806*, 45–53. [CrossRef]
84. Ávalos, A.; Haza, A.I.; Morales, P. Manufactured Silver Nanoparticles of Different Sizes Induced DNA Strand Breaks and Oxidative DNA Damage in Hepatoma and Leukaemia Cells and in Dermal and Pulmonary Fibroblasts. *Folia Biol.* **2015**, *61*, 33. [CrossRef]
85. Darviot, C.; Gosselin, B.; Martin, F.; Patskovsky, S.; Jabin, I.; Bruylants, G.; Trudel, D.; Meunier, M. Multiplexed Immunolabelling of Cancer Using Bioconjugated Plasmonic Gold–Silver Alloy Nanoparticles. *Nanoscale Adv.* **2024**, *6*, 4385–4393. [CrossRef]
86. Gakiya-Teruya, M.; Palomino-Marcelo, L.; Pierce, S.; Angeles-Boza, A.M.; Krishna, V.; Rodriguez-Reyes, J.C.F. Enhanced Antimicrobial Activity of Silver Nanoparticles Conjugated with Synthetic Peptide by Click Chemistry. *J. Nanopart Res.* **2020**, *22*, 90. [CrossRef]
87. Tobi, A.; Willmore, A.A.; Kilk, K.; Sidorenko, V.; Braun, G.B.; Soomets, U.; Sugahara, K.N.; Ruoslahti, E.; Teesalu, T. Silver Nanocarriers Targeted with a CendR Peptide Potentiate the Cytotoxic Activity of an Anticancer Drug. *Adv. Ther.* **2021**, *4*, 2000097. [CrossRef]
88. Pollok, N.E.; Rabin, C.; Smith, L.; Crooks, R.M. Orientation-Controlled Bioconjugation of Antibodies to Silver Nanoparticles. *Bioconjugate Chem.* **2019**, *30*, 3078–3086. [CrossRef] [PubMed]
89. El-Naggar, M.A.; Abu-Youssef, M.A.; Soliman, S.M.; Haukka, M.; Al-Majid, A.M.; Barakat, A.; Badr, A.M. Synthesis, X-Ray Structure, Hirshfeld, and Antimicrobial Studies of New Ag (I) Complexes Based on Pyridine-Type Ligands. *J. Mol. Struct.* **2022**, *1264*, 133210. [CrossRef]

90. Chong, Y.; Dai, X.; Fang, G.; Wu, R.; Zhao, L.; Ma, X.; Tian, X.; Lee, S.; Zhang, C.; Chen, C. Palladium Concave Nanocrystals with High-Index Facets Accelerate Ascorbate Oxidation in Cancer Treatment. *Nat. Commun.* **2018**, *9*, 4861. [CrossRef] [PubMed]
91. Dolengovski, E.L.; Fayzullin, R.R.; Kholin, K.V.; Voloshina, A.D.; Lyubina, A.P.; Sapunova, A.S.; Shamsutdinova, L.R.; Rizvanov, I.K.; Sinyashin, O.G.; Budnikova, Y.H. Stable Cyclometallated Palladium Complexes of Arylamides: Synthesis, Redox and Anticancer Activities. *Eur. J. Inorg. Chem.* **2024**, *27*, e202400426. [CrossRef]
92. Savić, A.; Marzo, T.; Scaletti, F.; Massai, L.; Bartoli, G.; Hoogenboom, R.; Messori, L.; Van Deun, R.; Van Hecke, K. New Platinum(II) and Palladium(II) Complexes with Substituted Terpyridine Ligands: Synthesis and Characterization, Cytotoxicity and Reactivity towards Biomolecules. *Biometals* **2019**, *32*, 33–47. [CrossRef]
93. Eryazici, I.; Moorefield, C.N.; Newkome, G.R. Square-Planar Pd(II), Pt(II), and Au(III) Terpyridine Complexes: Their Syntheses, Physical Properties, Supramolecular Constructs, and Biomedical Activities. *Chem. Rev.* **2008**, *108*, 1834–1895. [CrossRef]
94. Winter, A.; Schubert, U.S. Metal-Terpyridine Complexes in Catalytic Application—A Spotlight on the Last Decade. *ChemCatChem* **2020**, *12*, 2890–2941. [CrossRef]
95. Kostenko, E.A.; Baykov, S.V.; Novikov, A.S.; Boyarskiy, V.P. Nucleophilic Properties of the Positively Charged Metal Center in the Solid State Structure of Palladium (II)-Terpyridine Complex. *J. Mol. Struct.* **2020**, *1199*, 126957. [CrossRef]
96. Ren, R.; Bi, S.; Wang, L.; Zhao, W.; Wei, D.; Li, T.; Xu, W.; Liu, M.; Wu, Y. Terpyridine-Based Pd (II)/Ni (II) Organometallic Framework Nano-Sheets Supported on Graphene Oxide—Investigating the Fabrication, Tuning of Catalytic Properties and Synergetic Effects. *RSC Adv.* **2020**, *10*, 23080–23090. [CrossRef]
97. Frezza, M.; Dou, Q.P.; Xiao, Y.; Samouei, H.; Rashidi, M.; Samari, F.; Hemmateenejad, B. In Vitro and In Vivo Antitumor Activities and DNA Binding Mode of Five Coordinated Cyclometalated Organoplatinum(II) Complexes Containing Biphosphine Ligands. *J. Med. Chem.* **2011**, *54*, 6166–6176. [CrossRef] [PubMed]
98. Aziz, I.; Sirajuddin, M.; Nadeem, S.; Tirmizi, S.A.; Imad, R.; Sajjad, W. Synthesis, Spectral Characterization and Evaluation of Biological Properties of Pd (II) Tris-4-Cholorophenyl Phosphine Complexes of Thioureas and Heterocyclic Thiones. *J. Chem. Soc. Pak.* **2018**, *40*, 171.
99. Wang, C.-H.; Shih, W.-C.; Chang, H.C.; Kuo, Y.-Y.; Hung, W.-C.; Ong, T.-G.; Li, W.-S. Preparation and Characterization of Amino-Linked Heterocyclic Carbene Palladium, Gold, and Silver Complexes and Their Use as Anticancer Agents That Act by Triggering Apoptotic Cell Death. *J. Med. Chem.* **2011**, *54*, 5245–5249. [CrossRef]
100. Romashev, N.F.; Abramov, P.A.; Bakaev, I.V.; Fomenko, I.S.; Samsonenko, D.G.; Novikov, A.S.; Tong, K.K.H.; Ahn, D.; Dorovatovskii, P.V.; Zubavichus, Y.V.; et al. Heteroleptic Pd(II) and Pt(II) Complexes with Redox-Active Ligands: Synthesis, Structure, and Multimodal Anticancer Mechanism. *Inorg. Chem.* **2022**, *61*, 2105–2118. [CrossRef]
101. Jahromi, E.Z.; Divsalar, A.; Saboury, A.A.; Khaleghizadeh, S.; Mansouri-Torshizi, H.; Kostova, I. Palladium Complexes: New Candidates for Anti-Cancer Drugs. *J. Iran. Chem. Soc.* **2016**, *13*, 967–989. [CrossRef]
102. Vojtek, M.; Marques, M.P.M.; Ferreira, I.M.P.L.V.O.; Mota-Filipe, H.; Diniz, C. Anticancer Activity of Palladium-Based Complexes against Triple-Negative Breast Cancer. *Drug Discov. Today* **2019**, *24*, 1044–1058. [CrossRef] [PubMed]

MDPI AG
Grosspeteranlage 5
4052 Basel
Switzerland
Tel.: +41 61 683 77 34

Inorganics Editorial Office
E-mail: inorganics@mdpi.com
www.mdpi.com/journal/inorganics

www.ingramcontent.com/pod-product-compliance
Lightning Source LLC
Chambersburg PA
CBHW061622210926
51850CB00009B/2410
* 9 7 8 3 7 2 5 8 7 1 2 0 9 *